An Introduction to
Stellar Astrophysics

An Introduction to Stellar Astrophysics

FRANCIS LEBLANC

Université de Moncton, Canada

A John Wiley & Sons, Ltd., Publication

Library of Congress Cataloging-in-Publication Data

LeBlanc, Francis.
   An introduction to stellar astrophysics / Francis LeBlanc.
      p. cm.
   Includes bibliographical references and index.
   ISBN 978-0-470-69957-7 (cloth) – ISBN 978-0-470-69956-0 (pbk.)   1. Stars–Textbooks.
2. Astrophysics--Textbooks.   I. Title.
   QB801.L43 2010
   523.8–dc22

                                                                                    2009052138

A catalogue record for this book is available from the British Library.

ISBN: H/bk 978-0470-699577 P/bk 978-0470-699560

Set in 10/12pt Times by Toppan Best-set Premedia Limited

Cover photo:
Image courtesy of NASA images.org

To Marise

Contents

Preface

This textbook is designed to be used by students following a first course on stellar astrophysics. It is mostly aimed at the advanced undergraduate students in physics or astronomy programs. It may also serve as a basic reference for researchers working in fields other than stellar astrophysics.

This work is not encyclopaedic in nature and therefore does not cover, for example, all type of stars that exist in the universe. This book aspires to give intermediate knowledge on stars in a relatively concise format. It focuses mostly on the explanation of the functioning of stars by using basic physical concepts and observational results. A large number of graphs and figures are included to better explain the concepts covered. Only essential astronomical data are given. The amount of observational results shown is deliberately limited in scope since a too large quantity of observational data can be overwhelming and be counterproductive to newcomers to the field of stellar astrophysics.

This book is written in the scope of the students' needs. Although the students using this book should have seen all the physical concepts needed for exploring stellar astrophysics, brief recalls of the most important ones are given. No prior astronomical knowledge is assumed. This work can therefore be used not only by astronomy students but also by students in a physics program. This book aims to explain stellar astrophysics with clarity and is written in a manner so that it could be read and understood by a physics or astronomy student with little or no outside help. Detailed examples are given throughout the book to help the reader better grasp the most important concepts. A list of exercises is given at the end of each chapter and answers to a selection of these are given. A summary for each chapter is also presented.

Some historical snippets are added to give some perspective on the chronology of various discoveries along with giving merited acknowledgments to the researchers that made these advancements possible. For a complete historical review of stellar astrophysics, the reader is referred to Tassoul, J.-L. and Tassoul, M., *A Concise History of Solar and Stellar Physics*, Princeton University Press, Princeton (2004).

The book is divided in seven chapters: basic concepts, stellar formation, radiative transfer in stars, stellar atmospheres, stellar interiors, nucleosynthesis and stellar evolution and chemically peculiar stars and diffusion. The topics seen in the last chapter are rarely covered in such textbooks and distinguish it from others on stellar astrophysics. This chapter encompasses many concepts seen throughout the book.

The book is divided in core content (approximately 75 %) which is considered crucial for a global understanding of stars and in optional content (about 25 %). Some optional sections also contain more advanced topics. Sections marked [†] are optional, while those marked [††] are optional sections containing advanced topics. These sections may be skipped without interfering in the normal progression of the core topics.

This book is mainly designed to cover the most important aspects of stellar astrophysics inside a one-semester (or half-year) course. The book is, however, somewhat too lengthy to be covered in totality in a single semester. The professor may then choose to skip a certain number of the optional or advanced sections in according to the length of the course given.

Some universities have two one-semester introductory courses (or a full-year course) in stellar astrophysics. They are usually divided into a course on stellar atmospheres, and a second one, pertaining to stellar structure and evolution. This book could be used as the main reference book for two such courses. Chapters 1, 3, 4 along with the first three sections of Chapter 2 could be given as a stellar atmosphere course, while the remainder of Chapter 2 and Chapters 5 and 6 could be given as a stellar interior and evolution course. Chapter 7 could also be seen at the end of either of these courses.

This book could also be used as the main reference for a first course on stellar astrophysics at the graduate level where the professor could choose to give additional selected readings to students to deepen their understanding of certain topics.

Francis LeBlanc
Moncton, Canada
October 2009

Acknowledgments

Since the writing of this book encapsulates many years of study and research on the subject, it is natural that I extend my warmest thanks to the many professors I encountered during my studies, especially, Georges Bader, Guilio Bosi, Claude Carignan, Donald Duplain, Gilles Fontaine, Georges Michaud, Jean-Louis Tassoul, Hubert Reeves, Thomas Richard, François Söler and François Wesemaël, who have guided me and often stoked my interest in physics and astronomy.

In view of the fact that this book is an offshoot of lecture notes that I have prepared for physics and astrophysics courses at Université de Moncton, I thank the many students who have contributed to improving some of the material presented. I wish to underline the contribution of Issouf Kafando, Luc LeBlanc, Marc Richard and Mouhamadou Thiam.

I also thank my colleagues and the staff at our department who were very supportive in this endeavour, especially Francine Maillet. I also thank the many colleagues from all over the world with whom I have collaborated in my research and who graciously shared their passion and wisdom. I am also grateful to the National Sciences and Engineering Research Council of Canada and La Faculté des Études Supérieures et de la Recherche de l'Université de Moncton for funding my research projects.

I also want to express my gratitude to the following people: Georges Alecian, Gibor Basri, Normand Beaudoin, Martin Bolduc, David Branch, Robert Duncan, Robert Hawkes, Gregory Laughlin, Jaymie Matthews, Art McDonald, John R. Percy, Jacques Richer, Ruben Sandapen, John Sichel, Christopher Thompson, Mathieu Vick and Francis Weil, who have given helpful comments on various parts of this book. I especially thank Viktor Khalack who has graciously read most of this book. His many comments led to major improvements in the manuscript. Of course, these individuals are in no way responsible for any errors or omissions that might appear in this book.

I also wish to thank Alexandra Carrick, Richard Davies, Judith Irwin and Sophia Travis for helping me navigate through the publication process.

I also express my gratitude to my family who has supported me in many ways throughout the years. Finally, my warmest thanks go to my dear wife Marise, who has shown great patience and has graciously accepted my relative absence during the writing of this book.

1

Basic Concepts

1.1 Introduction

First, a definition must be given for what constitutes a star. *A star can be defined as a self-gravitating celestial object in which there is, or there once was (in the case of dead stars), sustained thermonuclear fusion of hydrogen in their core.* For example, in the Sun, hydrogen, which is the most abundant element in the Universe, is fused into helium via the nuclear reaction $4^1\text{H} \rightarrow {}^4\text{He} + \text{energy}$. Fusion is only present in the central regions of stars, because there exists a minimum threshold temperature at which this exothermic reaction can be ignited (which is of the order of ten million degrees for this particular reaction). For hydrogen nuclei (protons) to be fused, they must have a close approach on the order of distance at which the strong nuclear force comes into play.[1] The strong nuclear force is responsible for binding the nucleons (protons and neutrons) in the nucleus and contrary to gravity, for instance, its field of action is limited to a distance on the order of 10^{-15} m. At the high temperatures found in the centres of stars, the kinetic energy of the protons is sufficient to vanquish the repulsive Coulomb force between them and bring the protons within the distance where the attractive strong nuclear force becomes dominant. Protons can then fuse together while emitting energy.

The energy emitted by thermonuclear reactions is given by Einstein's famous $E = \Delta m c^2$ formula, where Δm is the difference in mass between the species on the left-hand and right-hand sides of the arrow found in the nuclear reaction given above and c is the speed of light in vacuum. However, the hydrogen burning reaction given above can be a bit misleading, since it suggests that four protons meet to form a helium nucleus. In reality, a series of nuclear reactions is needed to give this global reaction. On another note, even though only a small fraction of a star's mass will be transformed to energy during its lifetime, it will suffice to compensate for the energy irradiated at its surface.

---

[1]Here, a simple phenomenological explanation of nuclear fusion is given. In reality, quantum tunnelling intervenes. This will be discussed in more detail in Chapter 6.

An Introduction to Stellar Astrophysics Francis LeBlanc
© 2010 John Wiley & Sons, Ltd

Details concerning various nuclear reactions of importance in stars will be discussed in Chapter 6.

Stars are formed following the gravitational collapse of cold molecular clouds found in the Universe. As the cloud or portions of it collapses, it can be shown (see Chapter 2) that approximately half of the gravitational energy gained is used to increase the internal temperature of the cloud and the remaining energy is irradiated as electromagnetic radiation in space. If the mass of the collapsed cloud is sufficient (i.e. more than approximately 8 % of the mass of the Sun), the central temperatures will attain a value superior to the threshold temperature for sustained hydrogen fusion, which would by definition, lead to star birth. The solar mass is $M_\odot = 1.989 \times 10^{33}$ g, where the symbol $\odot$ represents the Sun.[2] The physical properties of stars are often given in units of the corresponding value for the Sun. The gravitational collapse will continue until equilibrium is reached, where the nuclear energy generated per unit time (or its power) at the centre of the star equals the power output at its surface due to radiation emission. A star at this stage of its life is commonly called a main-sequence star. Since gravity has radial symmetry, a star will have a spherical shape (unless it has a high rotational speed). More details concerning stellar formation will be given in Chapter 2.

A star shines (or emits radiation) because of its high surface temperature. For example, the surface temperature of the Sun is approximately 5800 K, while its central temperature is approximately 16 million K. The decrease of the temperature as a function of distance from the centre is a natural occurrence that causes energy transport from the central regions to the surface of the Sun. Since the gas composing a star is characterized by an opacity to radiation, an observer looking at a star can only see its exterior regions, which is commonly called the photosphere or stellar atmosphere, having a geometrical depth of up to a few per cent of the stellar radius. This is similar to looking in a cloud of fog, being able to see only a certain distance before light signals are attenuated. The radiative field exiting a star depends on the temperature of these outer layers and is associated to their blackbody spectra. The physical properties of blackbodies will be discussed in Section 1.3 and will lead to an explanation why stars have different colours.

There are three modes of transportation of energy in stars. The most important is radiation. For this mode, the energy is transported when electromagnetic radiation diffuses from the central regions of stars towards its exterior. In regions where the radiative opacity becomes large, convection can dominate energy transport. Convection is the transport of energy by the vertical movements of cells of matter in the stars. Conduction is the third mode of transportation of energy in stars. However, this mode is rarely important. More details concerning energy transport will be discussed in Chapters 3 and 5.

As mentioned above, a star begins its life by transforming hydrogen to helium in its core. As time passes, the abundance of hydrogen gradually decreases in the star's core, and eventually, the fuel for this particular nuclear process, namely hydrogen, will all be spent. As hydrogen is transformed into helium, the structure of the star readjusts. The core contracts causing an increase of the central temperatures until possibly, depending on the initial mass of the star, helium fuses to produce carbon via the well-known triple-α reaction: $3^4\text{He} \rightarrow {}^{12}\text{C} + \text{energy}$. Meanwhile, the outer regions of the star expand. The star then becomes what is called a red giant. The final destiny of a star depends almost solely on

---

[2]Other physical properties of the Sun are given in Appendix C.

its initial mass; it will either become a white dwarf, a neutron star or a black hole. More details concerning stellar evolution will be given in Chapter 6.

For massive stars, a succession of nuclear reactions will occur during their different stages of evolution. The thermonuclear reactions in these stars are responsible for the synthesis of various elements, such as carbon, oxygen, silicon, etc. up to iron. This process is called nucleosynthesis. As known from the Big-Bang theory, at the beginning of the Universe, only hydrogen, helium and trace amounts of lithium were created. The formation of the other elements takes place in stars. Stars can therefore be seen as the Universe's production factories, generating all atoms heavier than helium, except for some lithium. In astronomy, elements heavier than helium are called metals and the fraction of the mass composed of metals is called the metallicity (Z). The metallicity of outer layers of the Sun is approximately $Z = 0.0169$. Meanwhile, the mass fraction of hydrogen (X) and helium (Y) at the surface of the Sun are, respectively, $X = 0.7346$ and $Y = 0.2485$ (and therefore $X + Y + Z = 1$). All of the atoms of these heavy elements found on Earth were created in stars, which then exploded in the form of supernovae ejecting this enriched matter into space. Some of this enriched matter was later found in the primordial cloud from which the Sun and the Earth were created. Life itself would be impossible without the creation of the elements in stars.

This is why stars are fundamental for our existence and can be considered as the main building blocks of the Universe. It is then crucial to understand them via the study of stellar astrophysics. This field of study is fascinating since it incorporates all major fields of physics (see Figure 1.1): nuclear, atomic, molecular and quantum physics, electromagnetism, relativity, thermodynamics, hydrodynamics, etc. This book aims to give the reader an introduction to this fundamental subject by emphasising the physical concepts involved and their specific importance in stars.

1.2 The Electromagnetic Spectrum

As is known from quantum mechanics, electromagnetic radiation has two personalities. It sometimes behaves like waves and at other times like particles. These particles are called photons. These two aspects of radiation are known as the wave–particle duality. For most radiative processes in stars, like an atomic absorption of a photon for example, radiation will act like a photon, rather than a wave. The wave–particle duality also applies to matter.

The energy (E) of photons is related to the frequency (v) and wavelength (λ) of the associated electromagnetic wave via the following expression

$$E = hv = \frac{hc}{\lambda} \qquad (1.1)$$

where h is the Planck constant and c is the speed of light in vacuum.

Even though a photon of wavelength λ has no mass, it possesses momentum p equal to

$$p = \frac{E}{c} = \frac{h}{\lambda} \qquad (1.2)$$

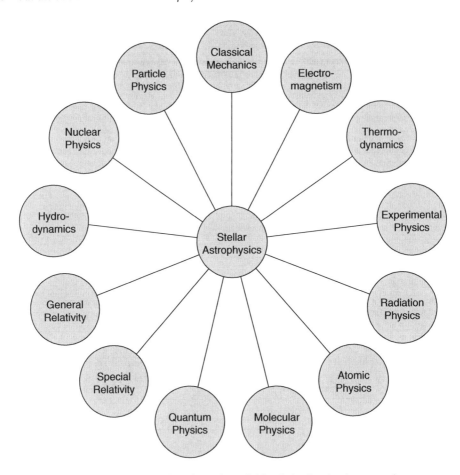

Figure 1.1   Figure illustrating the various fields of physics that intervene in stars.

As will be shown later, this physical quantity is of great importance in stars. Momentum transfer occurs from the radiation field to the stellar plasma following atomic absorption of photons, and this causes what is called radiation pressure.

The electromagnetic spectrum can be divided into a number of regions (see Table 1.1). It should be noted that the boundaries of these regions can vary from one source to another. For example, in astronomy the radio region often includes microwaves ($0.1\,cm < \lambda < 100\,cm$). The visible part of the electromagnetic spectrum is in the range $4000\,\text{Å} < \lambda < 7000\,\text{Å}$ where Å represents a unit of length called the angstrom and is equal to $10^{-8}\,cm$. Within the visible part of the spectrum, several colours (blue, yellow, etc.) can be observed that are defined by wavelength. The approximate (or representative) wavelengths of these colours are given in Table 1.2. The most energetic photons in the visible spectrum are violet; whereas the least energetic are red.

Earth's atmosphere is opaque to most wavelengths except those in the visible part of the spectrum and in some parts of the radio. This is why Earth-based observatories detect either visible or radio waves, while ultraviolet or X-ray observatories are placed in orbit around the Earth. Since the vast majority of the information gathered from the Universe

Table 1.1   The electromagnetic spectrum.

Region	Wavelength range
Radio	>0.1 cm
Infrared	7000 Å to 0.1 cm
Visible	4000 to 7000 Å
Ultraviolet	100 to 4000 Å
X-ray	0.1 to 100 Å
Gamma-ray	<0.1 Å

Table 1.2   Approximate wavelength of colours.

Colour	Wavelength (Å)
Violet	4200
Blue	4700
Green	5300
Yellow	5800
Orange	6100
Red	6600

comes in the form of electromagnetic radiation, it is imperative to properly understand the interaction between radiation and matter.

1.3   Blackbody Radiation

In everyday life, when observing an object, what is detected is the light that it is reflecting. For instance, if when looking at a red object, the reason why it is red is that the object in question is absorbing most colours except red, which is being reflected. In sunlight or light emitted by most household bulbs, there exist all of the colours of visible part of the electromagnetic spectrum. That is why it is preferable to wear light clothing (optimally white) in hot weather, since it will reflect most of the light that falls upon it. Meanwhile, black objects absorb most of the visible light they receive.

A body will also emit radiation whose spectra will depend on its temperature. By definition, a blackbody is a physical entity that absorbs all radiation that falls upon it. Radiation emanating from a blackbody is due uniquely to its thermal energy.

The German physicist Max Planck (1858–1947) showed that a blackbody with temperature T emits a continuous spectrum of radiation characterized by a function $B_\nu(T)$, commonly called the Planck function. The units of this function are[3] erg/s/Hz/cm^2/sr and are those of the physical quantity called specific intensity (I_ν, see Section 3.3 for more details). In the field of astrophysics the cgs (standing for centimetre-gram-second) unit system is

---

[3]The unit erg is the unit of energy in the cgs system while sr is the unit of solid angle (see Chapter 3 for more details). One erg equals 10^{-7} J (see Appendix B).

the norm. The main physical constants in cgs used throughout this book can be found in Appendix A, while both cgs and S.I. (or the international system) units and conversion factors are given in Appendix B.

The monochromatic flux (F_v) is defined as the quantity of energy in the spectral range between v and $v + dv$ emitted per unit surface, per unit time in units of erg/s/Hz/cm^2. In Chapter 3, it will be shown that for a blackbody, this quantity is given by the simple relation $F_v = \pi B_v$. It should be noted that in some physics textbooks, the Planck function given is the flux instead of the specific intensity and a factor π will then appear there.

The Planck function depends only on T and v and is given by the following expression

$$B_v(T) = \frac{2hv^3}{c^2} \frac{1}{e^{\frac{hv}{kT}} - 1} \tag{1.3}$$

where k is the Boltzmann constant. This function is isotropic and thus independent of the direction.

The Planck distribution can also be written per unit wavelength (B_λ). Since, for a given blackbody, the integration over the entire spectra of B_v and B_λ must be equal

$$B_v dv = -B_\lambda d\lambda \tag{1.4}$$

and

$$B_\lambda = -B_v \frac{dv}{d\lambda} = \frac{c}{\lambda^2} B_v = \frac{2hc^2}{\lambda^5} \frac{1}{e^{\frac{hc}{\lambda kT}} - 1} \tag{1.5}$$

The cgs units of B_λ are erg/s/cm/cm^2/sr. Sometimes, units per unit wavelength in Å, erg/s/Å/cm^2/sr are used instead. Figure 1.2 illustrates Planck functions for several temperatures.

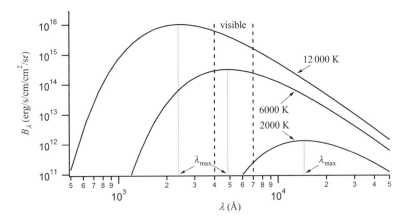

Figure 1.2 Planck distributions (B_λ) as a function of wavelength for $T = 2000$, 6000 and 12 000 K. The λ_{max} associated to each function and the visible part of the spectrum are also identified in this figure.

The energy distribution emitted by a blackbody leads to two laws. The first, the Stefan–Boltzmann law, gives the total power output per unit area F (or integrated flux in units of erg/s/cm^2) of a blackbody with temperature T is

$$F = \int_0^\infty F_\nu d\nu = \int_0^\infty \pi B_\nu d\nu = \sigma T^4 \qquad (1.6)$$

where σ is the Stefan–Boltzmann constant. To obtain this result, an integration of the monochromatic flux over the entire electromagnetic spectrum has been carried out (see Exercise 1.1). It shows that the energy output of a blackbody increases very rapidly with temperature. It should be noted that a blackbody with a higher temperature emits more energy at *all* wavelengths than a cooler one (see Figure 1.2). Since a star can be approximated by a blackbody (see Figure 1.8 in Section 1.6), a massive star having a high surface temperature, will emit much more power than a low-mass star that possesses a lower surface temperature. Massive stars will then have a shorter lifespan than smaller ones, since they burn their hydrogen at a much faster rate to compensate for their high brightness (this higher rate of nuclear burning is actually due to higher central temperatures). This topic will be discussed in more detail in Chapter 6.

A second law can also be derived from B_λ. It can be shown (see Exercise 1.2), that the wavelength λ_{max}, at which the function B_λ is at its maximum, varies inversely with temperature (see Figure 1.2)

$$\lambda_{max} = \frac{0.290 \, \text{K cm}}{T} \qquad (1.7)$$

This equation is called Wien's law. It explains why hotter blackbodies (or stars) are blue and cooler ones are red. For example, when a blacksmith puts a piece of iron in the fire, it first starts glowing red. Then, as it gets hotter, it becomes white and even blue, hence the term *white hot*. When the piece of iron is at room temperature, it emits almost no visible light since the maximum of its energy distribution is found in the infrared. For that reason, when a person is lost in the forest, a search can be undertaken using infrared detectors. The body of a human being has a temperature of about 310 K (or 37 °C) and is hotter than the surrounding nature with a temperature of about 293 K (or 20 °C) depending on the season. A human body emits much more infrared radiation than these surroundings.

Figure 1.2 shows that a blackbody with a temperature of 2000 K has its λ_{max} in the infrared part of the electromagnetic spectrum, a 6000-K blackbody has its maximum emission in the visible region of the spectrum, while a 12 000-K blackbody has its λ_{max} in the ultraviolet. Since the human eye is more sensitive to photons with wavelengths in the blue part of the electromagnetic spectrum than those in the violet portion, the hottest stars in the sky seem blue, even though the maximum of the energy distribution of these stars is in the violet or even in the ultraviolet. They seem blue, because they emit more blue light than the other less energetic colours, due to the slope of the Planck distribution. The Sun is yellow, because its λ_{max} lies in the visible part of the electromagnetic spectrum (see Example 1.1).

Example 1.1: Calculate λ_{max} for the Sun.

Answer:

The surface temperature of the Sun is approximately 5800 K. If the radiation field of the Sun is approximated by that of a blackbody

$$\lambda_{max} = \frac{0.290 \, \text{K cm}}{T} = \frac{0.290 \, \text{K cm}}{5800 \, \text{K}} = 5 \times 10^{-5} \, \text{cm} = 5000 \, \text{Å} \qquad (1.8)$$

This wavelength lies in the green part of the visible region of the spectrum. But since the Sun also emits a lot of blue, yellow and red light, the human eye, which is not equally sensitive to all wavelengths, incorporates all of these colours and sees the Sun as yellow.

Special Topic – The Greenhouse Effect

The average temperature on the Earth's surface is regulated by the amount of energy it receives from the Sun and the amount irradiated to space. The Earth's atmosphere is transparent to the visible part of the electromagnetic spectrum. Since the temperature at the Sun's surface is approximately 5800 K, its spectrum maximum is in the visible region and thus a lot of energy crosses the atmosphere and reaches the Earth's surface. Meanwhile the Earth's surface has an approximate temperature of 290 K and emits mostly infrared radiation. However, molecules such as H_2O and CO_2 can absorb infrared radiation and thus keep some heat in the terrestrial system. If it wasn't for the atmosphere, the temperature at our planet's surface would be more than 30 degrees cooler than it is now.

Unfortunately, human activity, such as the burning of fossil fuels, has increased the amount of pollutants (mostly CO_2) in our atmosphere. The increase of the abundances of these gases, called greenhouse gases, amplifies the opacity of the atmosphere to infrared radiation, which decreases the amount of energy lost to space. This process leads to a slight increase of the Earth's temperature and is called the greenhouse effect. Even the relatively small temperature increases expected are predicted to have important negative ecological impacts.

1.4   Luminosity, Effective Temperature, Flux and Magnitudes

The luminosity of a star is defined as the radiative power output emanating from its surface and is given in units of erg/s. The luminosity is an intrinsic value of a star and is not related to its distance from the observer. To obtain the luminosity, one must integrate the radiation field emitted over the entire electromagnetic spectrum and over the entire surface of the star. In the cases treated here, the flux will be assumed to be constant over the entire stellar surface. The luminosity is then obtained by simply multiplying the integrated flux (F) by the value of the star's surface area.

The effective temperature T_{eff} of a given star is defined as being the temperature needed for a blackbody with the same radius R_* as this star, to have the same luminosity L_* as this star. Since the integrated flux at the surface of this hypothetical blackbody is σT_{eff}^4, its luminosity is

$$L_* = 4\pi R_*^2 \sigma T_{eff}^4 \tag{1.9}$$

and the effective temperature of a star is

$$T_{eff} = \left(\frac{L_*}{4\pi R_*^2 \sigma} \right)^{1/4} \tag{1.10}$$

The integrated radiative flux at the surface of a star, in units of erg/s/cm^2, can also be written as a function of luminosity

$$F = \frac{L_*}{4\pi R_*^2} = \sigma T_{eff}^4 \tag{1.11}$$

At a distance r larger than R_* from the centre of the star, the integrated flux is

$$F(r) = \sigma T_{eff}^4 \left(\frac{R_*}{r} \right)^2 \tag{1.12}$$

Contrarily to the luminosity, the flux depends on the distance of the observer from the star. This equation shows the effect of the geometrical dilution of the flux as a function of distance from a star. This results from the fact that the luminosity is being distributed over a spherical surface of value $4\pi r^2$.

The human eye has a nonlinear response to light intensity. For example, a star that has an observed flux 10 times greater than a neighbouring star will not seem ten times brighter to the human eye. Thus, for practical and technological reasons, ancient astronomers divided the visible stars into a number of magnitude classes that better measures brightness with respect to the human eye than does flux. Unfortunately, these astronomers chose an unconventional scale such that the brighter stars have a lower magnitude. Magnitude is a relative scale that measures the logarithmic value of the radiative flux. A modern definition of magnitude is given by the formula

$$m_1 - m_2 = 2.5 \log \left(\frac{F_2}{F_1} \right) \tag{1.13}$$

which gives the difference of magnitudes of two stars as a function of their observed flux. This formula was chosen so that two stars with flux ratio of 100 will have a magnitude difference of 5 and, again for historical reasons, so that magnitude decreases when flux increases. Since the magnitude depends on the flux, it also depends on the distance separating the observer from the star. The magnitude m observed from Earth is called the apparent magnitude. An absolute magnitude M is then defined as the magnitude at a distance of 10 parsecs ($1\,\text{pc} = 3.26$ light years[4]). Since the formula above is given on a relative scale, its usefulness is limited unless it is calibrated by fixing a magnitude for a given flux. Historically, the star Vega was chosen to have a magnitude of zero, so any object brighter than this standard star will have a negative magnitude.

It can be easily demonstrated (see Example 1.2) that the difference between the apparent and the absolute magnitude of a star is related to its distance d (in parsecs) to the observer via the equation

$$m - M = 5\log\left(\frac{d}{10}\right) \tag{1.14}$$

The value m–M is often called the distance modulus.

Example 1.2: Demonstrate the distance modulus equation given above.

Answer:

The definition of the magnitude is

$$m_1 - m_2 = 2.5\log\left(\frac{F_2}{F_1}\right) \tag{1.15}$$

For a given star with an apparent magnitude of m and an absolute magnitude of M, the magnitudes in the equation above may be defined as $m_1 = m$ and $m_2 = M$. Also, the flux at distance d from the star of luminosity L is $F_1 = L/(4\pi d^2)$. Finally, the flux at a distance $d_{10} = 10\,\text{pc}$, $F_2 = L/(4\pi d_{10}^2)$. Therefore

$$m - M = 2.5\log\left(\frac{d}{d_{10}}\right)^2 \tag{1.16}$$

and if d is expressed in parsecs, this equation becomes

$$m - M = 5\log\left(\frac{d}{10}\right) \tag{1.17}$$

However, since it is impossible to observe the entire spectrum of a star, it is useful to define a magnitude for a given portion of the electromagnetic spectrum. The study of radiation inside a certain range of wavelength, commonly called a photometric band, is

----

[4]The parsec is a unit of distance defined in Section 6.9.5, while the light year is the distance travelled by light in vacuum during a one-year period.

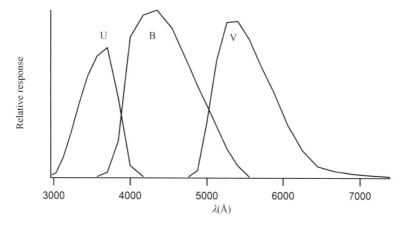

Figure 1.3   Response of U, B and V photometric indices (data from Arp, H.C.,*The Astrophysical Journal*, 133, 874 (1961)).

Table 1.3   Visual magnitudes of various astronomical objects.

Object name	m_V
Sun	−26.73
Full Moon	−12.7
Venus[#]	−4.5
Jupiter[#]	−2.5
Sirius	−1.44
Rigel	0.12
Saturn[#]	0.7
Deneb	1.23
Polaris	1.97

[#]At maximum brightness.

called photometry. To obtain the flux inside a given photometric band, a filter that is transparent to the radiation found inside this band and opaque to the photons outside of it, is placed in front of a photon detector.

Since radiation at different energies reacts with materials in different ways, telescopes and detectors must be adapted to the energy range of interest. Naturally, in the visible region of the spectrum, an optical telescope is used to accumulate the light on the detector. Figure 1.3 illustrates the transparency of such filters in the visible (V), blue (B) and ultraviolet (U) portions of the visible spectrum. These transparency functions must be taken into account when comparing observed magnitudes to theoretical values.

The brightest star in the sky is Sirius, while the faintest stars that are visible by the human eye have an apparent visual magnitude of approximately 6. Table 1.3 shows the apparent visual magnitudes of several well-known astronomical objects.

Example 1.3: Knowing that the apparent visual magnitude of the Sun is −26.73, calculate its absolute magnitude.

Answer:

The Sun is by definition at a distance of one astronomical unit (AU) from the Earth. Since $1\,\mathrm{AU} = 1.496 \times 10^{13}\,\mathrm{cm} = 4.848 \times 10^{-6}\,\mathrm{pc}$, the distance modulus equation

$$m_\mathrm{V} - M_\mathrm{V} = 5\log\left(\frac{d}{10}\right) \tag{1.18}$$

may be used to find the solution.
Replacing the known values in the equation above

$$-26.73 - M_\mathrm{V} = 5\log\left(\frac{4.848 \times 10^{-6}\,\mathrm{pc}}{10\,\mathrm{pc}}\right) \tag{1.19}$$

leads to $M_\mathrm{V} = 4.84$.

Later, it will be shown that the absolute magnitude of a star can be determined by spectroscopy. Spectroscopy is defined as the study of radiation with respect to wavelength. Since the apparent magnitude can be obtained by photometric observations, the distance to stars can then be determined with the distance modulus equation (Eq. 1.14).

The definition of magnitude given above (Eq. 1.13) can also be applied to magnitudes of two photometric bands of a single star. If one obtains photometric measurements of two photometric bands for a star, the flux ratio of these bands can be used to obtain its effective temperature. To better illustrate this, an example is shown in Figure 1.4, where the flux of

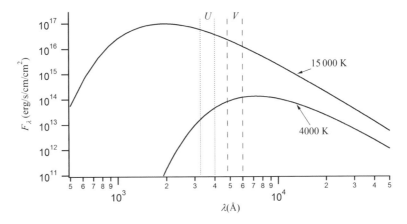

Figure 1.4   Monochromatic flux (F_λ) as a function of wavelength for two stars with $T_\mathrm{eff} = 4000$ and $15\,000\,\mathrm{K}$ approximated by blackbody radiation. The approximate positions of two photometric filters (U and V) are also shown.

a star is approximated by that of a blackbody with temperature T_{eff}. Two photometric bands for two blackbodies of different temperatures are shown. From this illustration, it is found that the ratio F_U/F_V (and thus $m_V - m_U$) increases with temperature. Since the blackbody flux is a well-known quantity, a value F_U/F_V is associated to each temperature. Assuming that the theoretical fluxes of stars with various effective temperatures can be calculated via the study of stellar atmospheres (see Chapter 4), the observed values of two apparent photometric magnitudes can be used to obtain T_{eff}. If nothing obstructs the light coming from the stars (interstellar clouds for example), $m_V - m_U$ is independent of distance to the observer. Typically, however, the presence of interstellar absorption or scattering necessitates certain corrections to be brought to the observed photometric magnitudes.

1.5   Boltzmann and Saha Equations

A star is composed of gaseous plasma containing both neutral and ionised atoms as well as free electrons. These free electrons come from ionisation. Ionisation is a process by which an atom loses one or more of its bound electrons. The atoms of a given element in various states of ionisation are called ions. In spectroscopy, ions are represented by the elemental nomenclature followed by a roman number. For example, CI is neutral carbon, CII is singly ionised carbon, and CVII is carbon ionised six times (i.e. a bare nucleus). Each ion of an element has its specific atomic energy levels. For reasons that will become clearer in later chapters, it is important to know the relative population of the various states of ionisation for each element present as a function of stellar depth, as well as the population among the various atomic energy levels for each of these ions. These quantities are critical for calculating the radiative opacity, which is the capacity of matter to absorb electromagnetic radiation. Opacity affects how radiation is transported from the inner to the outer portions of a star (see Chapter 3 for more details).

The field of statistical physics shows that the atomic energy levels of a given ion are populated inversely exponentially as a function of their energy: lower energy levels are naturally more populated than higher-lying energy levels. This being said, a bound electron can be excited to a higher energy level by two processes. Firstly, the energy needed for the bound electron to change levels can be obtained during a collision of the atom with another particle, for instance, a free electron. In this case, the kinetic energy of the free electron is used to excite the bound electron. The second process that can cause an excitation of an ion, is the absorption of a photon with energy equal to that of the electron transition (i.e. of energy equal to the difference between the two levels under consideration). These are called bound–bound transitions, since an electron goes from one bound state to another; whereas ionisation is a bound–free transition since the electron goes form a bound to a free state (see Figure 1.5). When collisions are the dominant processes that influence the energy-level populations (which is often the case in stars), the ratio of the population of two energy levels of a given ion in a gas at temperature T is given by the Boltzmann equation

$$\frac{n_i}{n_j} = \frac{g_i}{g_j} e^{-\frac{(E_i - E_j)}{kT}}$$
(1.20)

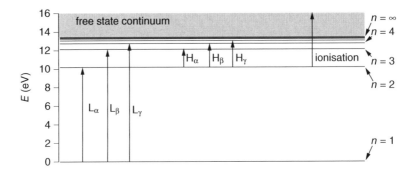

Figure 1.5   Energy levels of hydrogen in eV. Various bound-bound transitions are also shown, as well as a bound–free transition from level $n = 2$ (see Section 1.6 for more details).

where k is the Boltzmann constant, n_i is the number of atoms per unit volume (or population) in energy level i of the ion under consideration and g_i is the degeneracy of this level. The reader is reminded that the degeneracy of an energy level is the number of quantum states with the same energy. The quantity E_i is the energy of level i relative to the fundamental level, *which is set to zero*.

However, this form of the Boltzmann equation is not often useful. Instead, the ratio of the population of a given energy level to the total population of the ion under consideration is more useful. This quantity, which is useful for radiative opacity calculations (see Chapter 3), can be written (see Example 1.4)

$$\frac{n_i}{n_{\text{ion}}} = \frac{g_i}{U_{\text{ion}}} e^{-\frac{E_i}{kT}} \tag{1.21}$$

with

$$U_{\text{ion}} = \sum_{n=1}^{\infty} g_n e^{-\frac{E_n}{kT}} \tag{1.22}$$

where U_{ion} is called the partition function of the ion under consideration, and n_{ion} is its total population. This form of the Boltzmann equation shows that the fraction of ions in a given energy level is equal to the portion of the partition function related to this level.

Example 1.4: Demonstrate the equation

$$\frac{n_i}{n_{\text{ion}}} = \frac{g_i}{U_{\text{ion}}} e^{-\frac{E_i}{kT}} \tag{1.23}$$

Answer:

From the equation

$$\frac{n_i}{n_j} = \frac{g_i}{g_j} e^{-\frac{(E_i - E_j)}{kT}} \tag{1.24}$$

and since $E_1 = 0$, n_i with respect to the population of the fundamental level n_1 is written

$$\frac{n_i}{n_1} = \frac{g_i}{g_1} e^{-\frac{E_i}{kT}} \tag{1.25}$$

Meanwhile, the total population of the ion under consideration is

$$n_{\text{ion}} = \sum_{m=1}^{\infty} n_m = \frac{n_1}{g_1} \sum_{m=1}^{\infty} g_m e^{-\frac{E_m}{kT}} = \frac{n_1}{g_1} U_{\text{ion}} \tag{1.26}$$

The two equations above can be used to show that

$$\frac{n_i}{n_{\text{ion}}} = \frac{g_i}{U_{\text{ion}}} e^{-\frac{E_i}{kT}} \tag{1.27}$$

To better understand these concepts, it is instructive to apply them to hydrogen, which has well-known energy levels that can be calculated analytically via Bohr's atomic model. In units of electronvolts (eV),[5] E_n for the hydrogen atom is

$$E_n = 13.6\left[1 - \frac{1}{n^2}\right] \tag{1.28}$$

where n is the principal quantum number of the atomic energy level under consideration. Figure 1.5 shows the energy levels of hydrogen, and some transitions that can take place among them (see next section for more details). The degeneracy of a given level n is equal to $g_n = 2n^2$ for hydrogen.

To calculate the partition function, an infinite number of terms, related to the energy levels, must be summed. Unfortunately, for large values of n, the degeneracy (g_n) increases rapidly while the exponential found in the partition function equation ($e^{-\frac{E_n}{kT}}$) tends towards a constant value. The sum will then diverge for any temperature. Luckily, some simple physical considerations can alleviate this problem.

To better illustrate this problem, the case of hydrogen will be discussed. According to the Bohr model of the atom, the radius of the hydrogen atom in level n is $r = a_0 n^2$, where $a_0 = 0.529$ Å is the radius of the fundamental level of hydrogen (called the Bohr radius). The infinite sum needed to calculate the partition function is not physical, since for high-lying levels, the electron will eventually be closer to another nucleus than its own. An infinite sum for the partition function makes sense only if the atom in question is alone in

---

[5] 1 eV = 1.6×10^{-12} erg.

the Universe, which is obviously not the case! It should also be noted that in the analytical development leading to the Bohr radius equation, it is usually supposed that the only force on the electron is the attractive Coulomb force between the nucleus and the electron. So here again, the Universe is approximated to be composed only of the atom under consideration. A cut-off level of quantum number n_{max}, where the levels superior to this energy level are no longer bound to the nucleus, can be defined and used to approximate the value of the partition function. This can also be interpreted as a lowering of the continuum shown in Figure 1.5. It can be shown that for a pure hydrogen gas, $n_{max} = (2a_0)^{-1/2}(N)^{-1/6}$ where N is the number density of hydrogen atoms (see Example 1.5). The partition function can then be approximated by a finite sum

$$U = \sum_{n=1}^{n_{max}} g_n e^{-\frac{E_n}{kT}} \tag{1.29}$$

Example 1.5: Show that for a pure hydrogen gas the cut-off value of the energy levels can be approximated by $n_{max} = (2a_0)^{-1/2}(N)^{-1/6}$ when calculating the partition function and where N is the number density of hydrogen atoms in the gas.

Answer:

By supposing that the average distance between two hydrogen atoms in the gas is $2d$, the number density is thus one atom per $(2d)^3$ volume

$$N = \frac{1}{(2d)^3} \tag{1.30}$$

The maximum value of n where the electron is still closer to the initial nucleus than a neighbouring one is $r_n \le d$ where $r_n = a_0 n^2$. The variable n_{max} may be defined by the following

$$r_{max} = a_0 n_{max}^2 = d = \frac{1}{2N^{1/3}} \tag{1.31}$$

and thus

$$n_{max} = \frac{1}{\sqrt{2a_0}\, N^{1/6}} \tag{1.32}$$

Since ionised hydrogen has no atomic energy levels because it has lost its only electron, its partition function equals unity (i.e. it may be assumed that this ion has a single state of energy equal to $0\,eV$). This partition function is necessary to solve the equations describing ionisation of hydrogen shown below. At low temperatures, the partition function of neutral hydrogen can be approximated by the statistical weight of the fundamental energy level $g_1 = 2$ since the other terms in the sum (see Eq. 1.29) become small.

Example 1.6: Find the temperature at which the number density of hydrogen atoms in the fundamental state is equal to that of its second excited state ($n = 3$).

Answer:

From the Boltzmann equation

$$\frac{n_1}{n_3} = \frac{g_1}{g_3} e^{-\frac{(E_1 - E_3)}{kT}} = 1 \tag{1.33}$$

and since $g_1 = 2$, $g_3 = 18$, $E_1 = 0\,\text{eV}$ and $E_3 = 12.09\,\text{eV}$,

$$\frac{2}{18} e^{\frac{12.09\,\text{eV}}{kT}} = 1 \tag{1.34}$$

This becomes

$$\frac{12.09\,\text{eV}}{kT} = \ln(9) \tag{1.35}$$

and by using the value $k = 8.617 \times 10^{-5}\,\text{eV/K}$, the temperature is thus $T = 63\,900\,\text{K}$.

In stars, the local temperature increases as a function of depth. Moreover, deeper inside the stars, more energetic collisions will take place. This is due to the fact that according to statistical physics, the average thermal velocity of the particles in the stellar plasma is proportional to $T^{1/2}$. These collisions will cause excitations of atoms to higher energy levels (as described by the Boltzmann equation) and can also lead to ionisation of these atoms. Another process that can lead to an atom losing an electron is the absorption of a sufficiently energetic photon (see Figure 1.5). This process is called photoionisation. The freed electrons will contribute to the total gas pressure P. The reader is reminded that for an ideal gas, the equation of state is $P = n_{\text{tot}}kT$, where n_{tot} is the total number density of particles in the gas. This number density includes both the free electrons and the ions that are present in the plasma. A new physical quantity μ called the mean molecular weight of the particles in the gas may be defined by writing $n_{\text{tot}} = \dfrac{\rho}{\mu m_{\text{H}}}$, where ρ is the gas mass density (often simply called the density) and m_{H} is the mass of the hydrogen atom. Therefore, since density is given by the following equation

$$\rho = \sum_i n_i m_i \tag{1.36}$$

the mean molecular weight is

$$\mu = \frac{1}{m_{\text{H}} n_{\text{tot}}} \sum_i n_i m_i \tag{1.37}$$

where the sum over i runs over all types of particles present in the plasma including free electrons. The mean molecular weight gives the average mass of the particles in units of m_H. For instance, in a completely ionised hydrogen gas, $\mu = \dfrac{m_p + m_e}{2m_H} \approx \dfrac{1}{2}$, where m_p and m_e are respectively the proton and electron masses. The mean molecular weight is a useful concept that is used in stellar astrophysics and will be employed on several occasions in this book.

When collision processes dominate (which is often the case inside stars), the equation that regulates ionisation is called the Saha equation. It can be written

$$\frac{n_{i+1}}{n_i} = \frac{1}{n_e} \left(\frac{2\pi m_e kT}{h^2} \right)^{\frac{3}{2}} \frac{2U_{i+1}}{U_i} e^{-\frac{E_{ion}}{kT}} \tag{1.38}$$

where n_i and n_{i+1} are the populations of neighbouring ions of a given element, n_e is the number density of free electrons in the gas (often called the electronic density), T the local temperature, U_i and U_{i+1} are the corresponding partition functions and E_{ion} is the ionising energy of ion i *from its fundamental energy level*. Here, ion $i + 1$ is the more highly ionised ion.

From this equation, it may be deduced that ionisation increases with temperature. This is related to the fact that more energetic collisions are possible in hotter plasma. Also, for a given temperature, ionisation decreases with increasing electronic density. An increase in n_e fills the phase space of free electrons and increases recombination of free electrons with ions (i.e. deionisation).

The equation shown above gives the relative populations of two neighbouring ionisation states. However, this quantity is not often useful in astrophysical applications. As will be discussed in Chapter 3, to calculate the radiative opacity for a given elemental species, the population of each energy level needs to be known, which necessitates the knowledge of the population of each ionisation state. A quantity that is critical for such calculations is the ionisation fraction. The ionisation fraction is the portion of atoms in a given ionisation state of the element under consideration. The ionisation fraction f_i of ionisation state i can be written

$$f_i = \frac{n_i}{n_1 + n_2 + n_3 + n_4 + \ldots} \tag{1.39}$$

and by dividing both the numerator and the denominator by the neutral state's population n_1

$$f_i = \frac{\left(\dfrac{n_i}{n_1} \right)}{1 + \left(\dfrac{n_2}{n_1} \right) + \left(\dfrac{n_3}{n_1} \right) + \left(\dfrac{n_4}{n_1} \right) + \ldots} = \frac{\left(\dfrac{n_i}{n_{i-1}} \right) \left(\dfrac{n_{i-1}}{n_{i-2}} \right) \cdots \left(\dfrac{n_2}{n_1} \right)}{1 + \left(\dfrac{n_2}{n_1} \right) + \left(\dfrac{n_3}{n_2} \right) \left(\dfrac{n_2}{n_1} \right) + \left(\dfrac{n_4}{n_3} \right) \left(\dfrac{n_3}{n_2} \right) \left(\dfrac{n_2}{n_1} \right) + \ldots} \tag{1.40}$$

A series of multiplications of Saha equations (Eq. 1.38) is thus obtained, that once calculated, will give the value of the ionisation fraction (assuming n_e and T are known).

Special Topic – Ionisation Energies

Ionisation energies for the first five ionisation stages for a large number of elements are given in Appendix D. Figure 1.6 shows the ionisation energy for neutral atoms as a function of atomic number. It is shown that there exists a local maximum of the ionisation energy for noble gases (He, Ne, Ar, etc ...). These maxima are shifted to other elements for higher stages of ionisation. For example, for the singly ionised ion, maxima are found for LiII, NaII and KII (see Appendix D). These ions possess electronic configurations having respectively 2, 10 and 20 electrons and have filled electronic shells. They are also called noble gas electronic configurations.

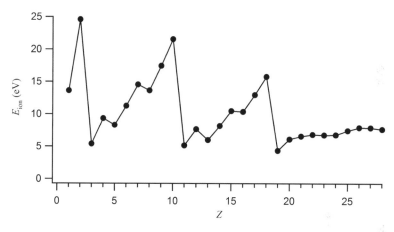

Figure 1.6   Ionisation energy (from the fundamental atomic energy state) as a function of atomic number for neutral atoms.

Of course, the ionisation fraction will vary with depth in stars, with more highly ionised ions appearing in deeper stellar layers. Figure 1.7 shows the ionisation fractions of the first 13 calcium ions in a star with $T_{eff} = 7600\,K$. In this figure, there exists a large plateau for CaIII and CaXI ionisation fractions. These ions have large ionisation energies since they are in noble-gas configurations (respectively, those of Ar and Ne). These noble-gas configurations stay populated for a large domain of temperatures compared to other electronic configurations because of their large ionisation energy. Since the atomic energy levels (and therefore the absorption transitions) are different for each ionisation state, the radiative opacity of a given element will also vary with depth.

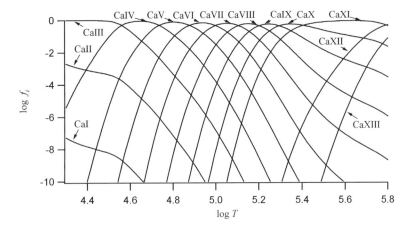

Figure 1.7 Ionisation fractions (f_i) of Ca ions as a function of temperature (or depth) in the interior of a star with $T_{\text{eff}} = 7600\,\text{K}$. The surface of the star is found at the left side of the horizontal axis.

Example 1.7: For a given star, calculate the fraction of neutral atoms in a gas composed of pure hydrogen at a depth where $T = 12\,000\,\text{K}$ and $n_e = 2.0 \times 10^{15}\,\text{cm}^{-3}$ (assume that the partition function of neutral hydrogen $U_I = 2$).

Answer:

In a pure hydrogen gas, the free electrons come exclusively from hydrogen ionisation and therefore $n_e = n_{II}$ where n_{II} represents the population of HII ions.
From the Saha equation

$$\frac{n_{II}}{n_I} = \frac{1}{n_e}\left(\frac{2\pi m_e kT}{h^2}\right)^{\frac{3}{2}} \frac{2U_{II}}{U_I} e^{-\frac{E_{ion}}{kT}} \tag{1.41}$$

where $E_{ion} = 13.6\,\text{eV}$ and $U_{II} = 1$. By inserting the appropriate values into this equation, n_I is obtained

$$n_I = \frac{n_e^2}{6.156 \times 10^{15}\,\text{cm}^{-3}} = 6.498 \times 10^{14}\,\text{cm}^{-3} \tag{1.42}$$

The ionisation fraction of neutral hydrogen is then

$$f_I = \frac{n_I}{n_I + n_{II}} = 0.245 \text{ or } 24.5\% \tag{1.43}$$

Example 1.8: Calculate the electronic density (n_e) in a gas at $T = 14\,000\,\text{K}$ composed of pure hydrogen where 70 % of the atoms are ionised (assume $U_I = 2$).

Answer:

Since

$$\frac{n_{II}}{n_I + n_{II}} = 0.7 \tag{1.44}$$

therefore, $n_I = 0.428\,n_{II}$. Also, since the gas under consideration is made of pure hydrogen $n_{II} = n_e$.

From the Saha equation

$$\frac{n_{II} n_e}{n_I} = \left(\frac{2\pi m_e kT}{h^2}\right)^{\frac{3}{2}} \frac{2 U_{II}}{U_I} e^{-\frac{E_{ion}}{kT}} \tag{1.45}$$

where $E_{ion} = 13.6\,\text{eV}$ and $U_{II} = 1$. By inserting the appropriate values into this equation

$$\frac{n_{II} n_e}{n_I} = \frac{n_e^2}{0.428 n_e} = 2.33 n_e = 5.08 \times 10^{16}\,\text{cm}^{-3} \tag{1.46}$$

and $n_e = 2.18 \times 10^{16}\,\text{cm}^{-3}$.

It will be shown in Chapter 4 that the application of the Saha equation in real stars is more complex than the relatively simple examples shown above. In stellar models, since a large number of elements are present a large series of Saha equations has to be solved simultaneously. Atomic data included in the calculation of the partition functions and the Saha equations must then be known for all elements present. Such calculations therefore necessitate considerable computing resources.

Finally, it should be mentioned that the Boltzmann and Saha equations, respectively, give, statistically speaking, the portion of atoms in a given atomic level and in the various ionisation states. However, a single atom's state (atomic or ionisation) will constantly change as a function of time due to interactions with other particles. Generally, these interactions are induced by collisions, but radiative excitations and ionisations can sometimes be important. This will be discussed further in Chapter 3.

1.6   Spectral Classification of Stars

In astronomy, many objects, be it meteorites, galaxies or stars are classified. These classifications aim at a better understanding of the group of objects under consideration. In this section, one such classification will be discussed, namely the spectral classification of stars.

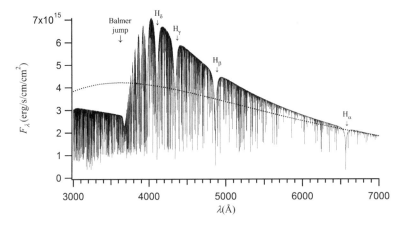

Figure 1.8   Theoretical monochromatic flux emerging form an A type star with $T_{\text{eff}} = 8000\,\text{K}$. The first four Balmer absorption lines, as well as the Balmer jump, are identified in this figure. Thousands of other absorption atomic lines can also be seen. This theoretical flux was obtained with the Phoenix stellar atmosphere code (Hauschildt, P.H., Allard, F. and Baron, E., *The Astrophysical Journal*, 512, 377 (1999)) while using the elemental abundances found in the Sun. The flux at the surface of a blackbody with $T = 8000\,\text{K}$ (dotted curve) is also shown.

As photons diffuse towards the surface of a star, they can interact with the atoms present in the stellar plasma. A photon can, for example, be absorbed when its energy is used to excite an electron from a lower to an upper bound state of an atom. The absorption features seen in the spectrum from these transitions are called atomic lines (see Figure 1.8). If the atomic energy levels were precisely defined, only photons with a single value of λ could be absorbed by the transition under consideration. The value of λ is related to the energy difference between the upper and lower levels associated to the transition. The photon wavelength necessary for an electronic excitation from level n to level m is

$$\lambda_{n \to m} = \frac{hc}{E_m - E_n} \tag{1.47}$$

However, because of the uncertainty principle of the quantum theory, the energy levels cannot be precisely defined, thus giving an absorption profile with a certain width. Additionally, since the atoms in the star have a velocity distribution associated to the local temperature, called the Maxwell distribution, the Doppler effect as well as broadening by pressure (or collisions) will also play a role in the widening of the atomic lines (see Chapter 4 for more details).

For a given absorption line of an ion to be present in the spectra, the lower (or initial) level must be populated (i.e. Boltzmann equation) and of course, the ion must also be present (i.e. the Saha equation). Since a star's spectrum emerges from its photosphere, its effective temperature will play a pivotal role in determining which atomic lines are present in the spectrum.

Table 1.4 Lyman and Balmer series.

	Lyman Series			Balmer Series	
Name	Transition $(n \rightarrow m)$	Wavelength	Name	Transition $(n \rightarrow m)$	Wavelength
L_α	$1 \rightarrow 2$	1216 Å	H_α	$2 \rightarrow 3$	6563 Å
L_β	$1 \rightarrow 3$	1025 Å	H_β	$2 \rightarrow 4$	4861 Å
L_γ	$1 \rightarrow 4$	972 Å	H_γ	$2 \rightarrow 5$	4341 Å
.			.		
.			.		
.			.		
Lyman jump	$1 \rightarrow \infty$	911 Å	Balmer jump	$2 \rightarrow \infty$	3646 Å

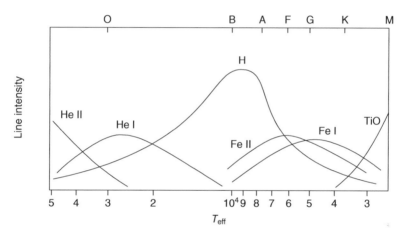

Figure 1.9 Approximate line intensity as a function of T_{eff} for several ions. The spectral types (these are positioned at the coolest temperature for each class) and the intensity of the TiO molecular bands are also shown.

Let's first discuss the behaviour of hydrogen lines in stellar spectra. Figure 1.5 shows the energy levels of hydrogen and some of the transitions that can occur. These transitions can be grouped as per their initial level. The lines emanating from the $n = 1$ level are called the Lyman lines (L_α, L_β, L_γ, etc.) and are found in the ultraviolet part of the spectrum. The Balmer series (H_α, H_β, H_γ, etc.) emanate from $n = 2$ and are in the visible part of the spectrum, while the Paschen lines (from $n = 3$) are found in the infrared. More details concerning the Lyman and Balmer series are given in Table 1.4. At the surface of cool stars, almost all of the hydrogen atoms are in the fundamental level and the Balmer lines (found in the visible spectrum) are very weak. The Lyman lines are also weak since relatively few ultraviolet photons exist in the spectrum of such a cool star. For hotter stars (say $T_{\text{eff}} = 8000\,\text{K}$ or so), the hydrogen atoms found in the $n = 2$ level begin to be significantly populated and the Balmer lines are then quite intense (see Figure 1.9). For even hotter stars, the intensity of the Balmer lines decreases, owing to the fact that the quantity of neutral hydrogen atoms contributing to the presence of the Balmer lines diminishes due to ionisation.

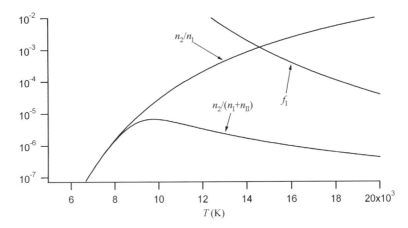

Figure 1.10   Illustration showing the portion of neutral hydrogen atoms found in the $n = 2$ level (n_2/n_1), the neutral ionisation fraction ($f_1 = \dfrac{n_1}{n_1 + n_{II}}$) and the product of these two factors that give the portion of all hydrogen atoms found in the $n = 2$ level (i.e. $\dfrac{n_2}{n_1 + n_{II}}$).

Figure 1.10 illustrates the two contributing factor explaining why hydrogen Balmer lines are at their strongest for stars with surface temperatures around 10 000 K. The portion of neutral hydrogen atoms found in the $n = 2$ level increases with temperature, while the neutral ionisation fraction decreases. The line strength depends on the product of these two factors which has a maximum at $T \approx 10\,000$ K.

Similar tendencies are observed for the atomic lines of the other elements (see Figure 1.9). For example, FeI lines are strong in cool stars. But for hotter stars, FeII, FeIII, etc., eventually dominate. The position, with respect to T_{eff}, of maximum strength of the atomic lines of various ions is related to their ionisation energy. For example, the ionisation energy of FeI is 7.9 eV, while it is 24.6 eV for HeI, the FeI atomic transitions are thus at lower energies than those of HeI. This explains why FeI lines are more prominent in cooler stars than those of HeI. The relative strength of atomic lines of different ions (either of the same or of a different element) can be used to estimate the surface temperature of stars. Such studies fall in the field of research called stellar spectroscopy.

Photons can also be absorbed during photoionisation. For hydrogen, the ionisation energy from its fundamental level is 13.6 eV, whereas it is 3.4 eV from its first excited state. The synthetic spectrum of Figure 1.8 shows a large flux decrease near $\lambda = 3646\,\text{Å}$, due to the ionisation of hydrogen from level $n = 2$. This spectral feature is called the Balmer jump. As mentioned previously, the minimum energy of photons that can ionise hydrogen from this level is 3.4 eV. When more energetic photons are absorbed by this bound–free transition, the excess of energy is transformed to kinetic energy transferred to the ejected electron.

Stars are generally divided into seven spectral classes or types: O, B, A, F, G, K and M going from hotter (bluer) to cooler (redder) effective temperatures. This classical categorization of stellar spectra, based mainly on the strength of hydrogen Balmer lines, is called the Harvard classification. The A-type stars fall where the strongest (or deepest)

Table 1.5 Spectral classes.

Spectral class	T_{eff}	Spectral characteristics	Colour	Example
O	>30 000 K	HeII strong, H faint, multiply-ionised metals strong	blue	λ Ori
B	10 000–30 000 K	HeI strong, H moderate	blue-white	Rigel
A	7500–10 000 K	H lines at their maximum	white	Vega
F	6000–7500 K	Singly ionised metals strong, H moderate	white-yellow	Procyon
G	5000–6000 K	Singly ionised metals strong, H faint	yellow	Sun
K	3500–5000 K	Strong neutral and Singly ionised metals, H faint	orange	Arcturus
M	<3500 K	Strong molecule bands (i.e. TiO), strong neutral metals, H very faint	red	Betelgeuse

hydrogen lines are observed. As discussed above, two processes, excitation and ionisation, conspire to give the largest portion of hydrogen atoms in the $n = 2$ level in A-type stars (see Figure 1.10). A useful mnemonic to remember the order of the spectral classes is *'Oh Be A Fine Girl (or Guy, depending on the reader's preference), Kiss Me'*. The spectral features and T_{eff} of the different spectral classes are given in Table 1.5.

Simple molecules (TiO, CH, H_2O, etc.) can also exist in cooler stars and may absorb radiation not only through electronic transitions but also via rotational or vibrational transitions. These transitions are called bands instead of lines and are found in the infrared region of the spectrum. In hotter stars, the molecules are destroyed by photodissociation due to energetic photons, or by energetic collisions; hence, no molecular bands are observed in the spectra of such stars.

Hot stars are often called early-type stars, while cooler stars are called late-type stars. These terms came about when astronomers erroneously thought that stars began their lives as hot stars and cooled down during their lifespan.

The spectral classes can also be subdivided into 10 partitions. These subdivisions are identified by a single Arabic digit increasing from the hotter end to the cooler end of the spectral class (i.e. F0 stars are hotter than F9 stars). The spectral class of the Sun is G2.

All spectral types are not equally populated. There are fewer high-mass stars (i.e. type O and B) than less massive ones (i.e. type K and M). This is associated to the process of stellar formation that does not uniformly create stars with respect to their mass. This will be discussed in Chapter 2.

Several types of stars do not fit into the classical spectral classification given above. For instance, ApBp stars (p standing for peculiar) are A and B type stars with strong magnetic fields and large observed abundance anomalies. Abundance anomalies, are defined as when the abundances of some elements are very different from those expected (either those found in the Sun, or in the vicinity of the star under consideration). These abundance anomalies or peculiarities strongly modify their spectra which differentiate them from normal A-type stars. For example, in the case of an overabundance for a given element,

Table 1.6   Solar abundances of the most abundant elements.

Element	N_{elem}/N_{tot}
H	9.097×10^{-1}
He	8.890×10^{-2}
O	7.742×10^{-4}
C	3.303×10^{-4}
Ne	1.119×10^{-4}
N	1.021×10^{-4}
Mg	3.458×10^{-5}
Si	3.228×10^{-5}
Fe	3.154×10^{-5}
S	1.475×10^{-5}

its lines are much stronger. Abundances are often given relative to those of the Sun. Table 1.6 shows the abundances of the most abundant elements found in the Sun. A more complete set of solar-abundance data is given in Appendix E.

Abundance anomalies are believed to be caused by diffusion of the elements within the star, caused partly by the radiative force transferred to ions. The radiative force is due to momentum transfer from photons to atoms during line absorption for instance. The diffusion process can cause an accumulation or depreciation of certain species at different depths (see Chapter 7 for more details). Abundances observed at the surface of a star are not always indicative of the average abundances of the elements within the whole star.

Among other types of stars with peculiar spectra are Am (m standing for metallic) and HgMn stars (where Hg and Mn are generally overabundant by several orders of magnitudes at their surface as compared to their solar abundance). Another example of stars that can't be classified in the types shown in Table 1.5 are Be stars (e standing for emission). These stars are surrounded by gas, and emission lines are observed in their spectra. Emission lines are spectral features that resemble inverted absorption lines or spikes in the flux. Many other peculiar spectral types not mentioned here also exist.

Abundances found in stars are also used to define their population. There are three types of stellar populations. Population I stars are young stars with relatively large metallicity, while population II stars are older stars with a smaller value of metallicity. Population III stars are the oldest stars that, hypothetically, have zero metallicity. However, the stars of this population have never been directly observed. The Sun is a population I star. The relation between the age of a star and its metallicity can be explained by results from the Big-Bang theory and stellar evolution. As mentioned previously, at the beginning of the Universe, only hydrogen and helium were present, with the exception of a trace of lithium. Therefore, the first generation of stars (population III) did not contain any metals except for this trace element. As this generation of stars evolved, some become supernovae thereby enriching the interstellar medium with the newly synthesized heavy elements. Following generations of stars were then composed of this enriched matter, which translated into increasing metallicities. This process will be explained in more detail in Chapter 6.

1.7   The Hertzsprung–Russell Diagram

As discussed in Section 1.4, the luminosity of a star depends on both its radius and effective temperature. A famous diagram, called the Hertzsprung–Russell (hereafter H–R) diagram, shows the relation between the luminosity and the effective temperature of stars. In such diagrams, the direction of the abscissa (T_{eff}) is reversed (see Figures 1.11 and 1.12). This tool for studying stars was developed by the Danish astronomer Ejnar Hertzsprung

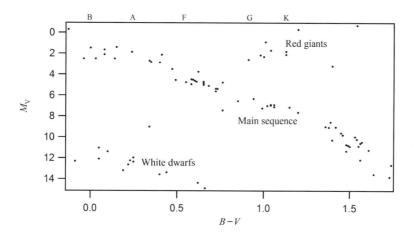

Figure 1.11   A sample taken among the 1000 nearest stars on a color-magnitude H–R diagram. The spectral types are also shown (these are positioned at the coolest temperature for each class).

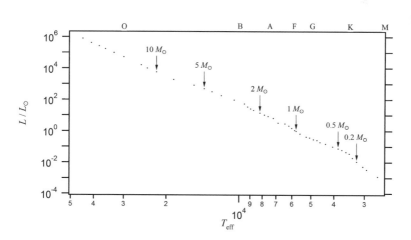

Figure 1.12   The main sequence on an H–R diagram. Several values of the mass are given. The spectral types are also shown (these are positioned at the coolest temperature for each class).

(1873–1967) and the American astronomer Henry Norris Russell (1877–1957) at the beginning of the twentieth century. The H–R diagram is extremely useful when studying the evolution of stars, since there are well-determined paths along which stars should travel as they evolve. These paths depend mostly on stellar mass (see Figure 6.10). During evolution, both the T_{eff} and the radius of a star change. Its spectral type will also be time dependent. Observational astronomers often use an absolute magnitude scale instead of luminosity, and $m_B - m_V$ instead of effective temperature. These are called colour-magnitude diagrams (see Figure 1.11). The colour index $m_B - m_V$ is usually written as $B - V$.

Figure 1.11 represents an observational H–R diagram containing a sample taken among the 1000 nearest stars, obtained from the Gleise star catalogue. A large portion of these stars are concentrated on a branch called the main sequence. This is where stars begin their lives and stay while burning hydrogen in their core. In this figure, the stars found above the main sequence are red giants; whereas those below are white dwarfs. These regions of the H–R diagram are often called branches. During its evolution, a star eventually leaves the main sequence, its radius increases and its T_{eff} at first decreases, giving a red giant star. It can then become a supergiant and possibly a white dwarf, depending on the value its initial mass (see Chapter 6 for more details).

For many reasons, a certain scatter is observed along each branch. For example, as time evolves, stars move in the H–R diagram. Even stars on the main sequence branch move slightly during their hydrogen-burning phase, their structure changes as more helium is produced in their core. Another factor that causes scatter is the varying metallicity among the stars. This leads to structural changes that modify their position on the H–R diagram. Observational errors can also add to the observed scatter.

When moving from the upper left to the lower right along the main sequence, the stars found there have lower masses and T_{eff} (see Figure 1.12). High-mass stars are more luminous because their central temperatures are higher and therefore they fuse hydrogen and produce nuclear energy at a higher rate. Their central temperatures are higher due to the large amount of gravitational energy that can be released during their formation (see Chapter 2). Figure 1.12 shows main-sequence stars of various masses within an H–R diagram. The range of masses for stars is approximately $0.08 \, M_\odot \leq M \leq 120 \, M_\odot$. The upper limit is related to the fact that high radiation pressure present at the surface of such massive stars pushes out any additional mass that would otherwise be gravitationally attracted to the star during its formation. However, the value of this upper limit is quite uncertain. The lower limit of this range exists because the central temperature of astronomical objects with $M \leq 0.08 \, M_\odot$ does not attain the value needed for substantive and sustained hydrogen fusion. Objects with masses just below this limit are called brown dwarfs. These astronomical objects will be described in Chapter 6. Meanwhile, the range of effective temperature of main-sequence stars is approximately $2000 \, K \leq T_{eff} \leq 60\,000 \, K$.

For main-sequence stars, the relation between the mass and radius is nearly linear (see Figure 1.13); whereas the luminosity increases much faster than mass (see Figure 1.12). This stems from the dependence of luminosity on R_* and T_{eff}, $L_* = 4\pi R_*^2 \sigma T_{eff}^4$. As mentioned previously, more fundamentally, the luminosity of a star is determined by the nuclear power generated in its core, which itself depends on the central temperature. The relation between the luminosity and stellar mass is critical for estimating the lifespan of main-sequence stars (see Chapter 6).

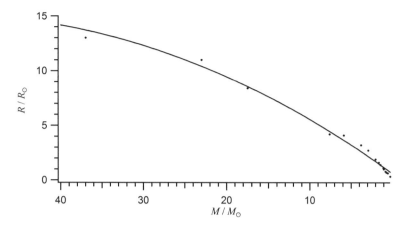

Figure 1.13 Relation between mass and radius for main-sequence stars (dots). Also shown is a curve fitted to the data.

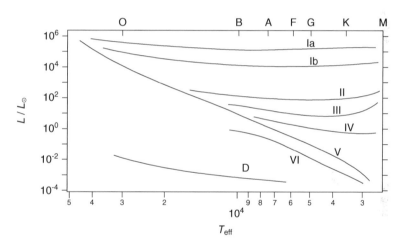

Figure 1.14 Luminosity classes of the H–R diagram. These are identified in Table 1.7. The spectral types are also shown (these are positioned at the coolest temperature for each class).

For a given $T_{\rm eff}$, stars can have very different luminosities due to differing radii. A star of a given $T_{\rm eff}$ can, for instance, be a white dwarf, a main-sequence, or a supergiant star. A supergiant can have a radius up to the order of $1000\,R_\odot$ (where $R_\odot = 6.955 \times 10^{10}\,\rm cm$), while white dwarfs typically have $R \approx 0.01\,R_\odot$. This explains their position in the H–R diagram vis-à-vis the luminosity axis. The spectral class of a star is thus not sufficient to correctly specify its evolutionary status, since its spectral type depends solely on the physical properties of its photosphere. To solve this problem, luminosity classes (see Figure 1.14 and Table 1.7) are defined as a second parameter to the spectral classification of stars. These luminosity classes are related to differing evolutionary stages. For example, the Sun has a spectral type G2V, V being the luminosity class of a main-sequence star.

Table 1.7 Luminosity classes.

Ia	Bright supergiants
Ib	Supergiants
II	Bright giants
III	Giants
IV	Subgiants
V	Main-sequence stars (or dwarfs)
VI (or sd)	Subdwarfs
D (or VII)	White dwarfs

It should be noted that the T_{eff} range given in Table 1.5 for the spectral classes and shown in the figures found in this chapter are those of main-sequence stars. These T_{eff} ranges are slightly shifted for other luminosity classes (see Exercise 1.14).

Main-sequence stars are also called dwarfs. As shown in Figure 1.14, there exists a class of stars called subdwarfs found just below the main sequence. Subdwarf stars have low metallicities. This leads to a smaller radius and higher T_{eff} than a main-sequence star with the same mass. This larger T_{eff} can be explained by the fact that the outer layers are closer to the stellar core. In other words, the smaller radius leads to a higher flux, thus a larger T_{eff}.

In conclusion, the global properties of a star can be defined by three fundamental parameters: mass, radius and luminosity. With the luminosity and the radius, the effective temperature is defined by Eq. 1.10. A star found at a given point in the H–R diagram (i.e. with known luminosity and effective temperature) isn't completely defined since stars with different masses can pass at a same point in the H–R diagram during their lifetime. Its mass is needed to define it completely. Secondary parameters such as the abundances of the elements present in the star, the presence of magnetic fields, stellar rotation, etc. can also come into play. The fundamental parameters for main-sequence stars are given in Appendix G.

1.8 Summary

Modes of energy transport in stars: radiation, convection and conduction

$$\text{Planck distribution: } B_\lambda = \frac{2hc^2}{\lambda^5} \frac{1}{e^{\frac{hc}{\lambda kT}} - 1} \tag{1.48}$$

$$\text{Stefan–Boltzmann law: } F = \sigma T^4 \tag{1.49}$$

$$\text{Wien's law: } \lambda_{max} = \frac{0.290\,\text{K cm}}{T} \tag{1.50}$$

$$\text{Luminosity: } L_* = 4\pi R_*^2 \sigma T_{eff}^4 \tag{1.51}$$

$$\text{Integrated flux } \left(\text{for } r \geq R_*\right): F(r) = \sigma T_{eff}^4 \left(\frac{R_*}{r}\right)^2 \tag{1.52}$$

Magnitude: $m_1 - m_2 = 2.5 \log\left(\dfrac{F_2}{F_1}\right)$ (1.53)

Distance modulus: $m - M = 5 \log\left(\dfrac{d}{10}\right)$ (1.54)

Boltzmann equation: $\dfrac{n_i}{n_{\text{ion}}} = \dfrac{g_i}{U_{\text{ion}}} e^{-\frac{E_i}{kT}}$ (1.55)

Saha equation: $\dfrac{n_{i+1}}{n_i} = \dfrac{1}{n_e}\left(\dfrac{2\pi m_e kT}{h^2}\right)^{\frac{3}{2}} \dfrac{2U_{i+1}}{U_i} e^{-\frac{E_{\text{ion}}}{kT}}$ (1.56)

Density: $\rho = \sum_i n_i m_i$ (1.57)

Mean molecular weight: $\mu = \dfrac{\rho}{m_H n_{\text{tot}}}$ (1.58)

Ionisation fraction: $f_i = \dfrac{\left(\dfrac{n_i}{n_{i-1}}\right)\left(\dfrac{n_{i-1}}{n_{i-2}}\right)\cdots\left(\dfrac{n_2}{n_1}\right)}{1 + \left(\dfrac{n_2}{n_1}\right) + \left(\dfrac{n_3}{n_2}\right)\left(\dfrac{n_2}{n_1}\right) + \left(\dfrac{n_4}{n_3}\right)\left(\dfrac{n_3}{n_2}\right)\left(\dfrac{n_2}{n_1}\right) + \cdots}$ (1.59)

Spectral types (in order of decreasing T_{eff}): O, B, A, F, G, K and M

Three fundamental parameters of stars: mass, radius and luminosity

1.9   Exercises

1.1   Demonstrate the Stefan–Boltzmann law.

1.2   Demonstrate Wien's law (numerical problem).

1.3   A binary star system is observed, and since the separation between the two stars is much smaller that the distance of the system from the observer, it can be supposed that both stars are found at the same distance from Earth. The absolute magnitude in a given photometric band of the first star is determined to be −0.5, while its apparent magnitude is 3.5. If the apparent magnitude of the second star is 4.5, what is its absolute magnitude? At what distance (in light-years) is the binary system from the observer?

1.4   What is the numerical difference between the absolute magnitudes of two stars having the same T_{eff}, where one of these stars is in the giant phase and has a radius 15 times larger than the other star, which finds itself on the main sequence?

1.5   At what distance would the Sun have to be to have the same apparent magnitude as a 100-W light bulb found 100 m away? Express your answer in ly.

1.6　Assuming a flux equal that of a blackbody, calculate the percentage of the flux for stars with $T_{\text{eff}} = 5000$, $10\,000$ and $20\,000\,\text{K}$, capable of ionising hydrogen from level $n = 2$ (numerical problem)?

1.7　Calculate the temperature at which the number density of hydrogen atoms in the first excited state is ten times less than the number density of those in the fundamental level.

1.8　A hypothetical ion of an element has a degeneracy equal to $4n^2$, where n is the principal quantum number. At $T = 40\,000\,\text{K}$, the ratio of the number density in level $n = 3$ to that of the fundamental ($n = 1$) is 0.25. Find the energy of level $n = 3$, assuming $E_1 = 0$.

1.9　What is the ionisation fraction of HI at a depth where $T = 9000\,\text{K}$ and $P = 140$ dyn/cm^2 in a star composed of pure hydrogen (assume $U_1 = 2$)?

1.10　Calculate the total number density (n_{tot}) and the density (ρ) at a depth in a star composed of pure hydrogen where $T = 9500\,\text{K}$ and 35 % of the atoms are ionised (assume $U_1 = 2$). What percentage of hydrogen atoms are in the energy level $n = 2$?

1.11　Calculate the pressure in a pure hydrogen gas at $T = 12\,000\,\text{K}$ that has 20 % of its atoms in the ionisation state HII (assume $U_1 = 2$).

1.12　At a certain depth in a star, three ions of a given element have the following ionisation fractions: $f_1 = 0.10$, $f_2 = 0.85$ and $f_3 = 0.05$. Their partition functions are: $U_1 = 1$, $U_2 = 2$ and $U_3 = 8$. The ionisation energy from the fundamental level for ion 1 is 30 eV and it is 55 eV for ion 2. Calculate n_e and T at this depth.

1.13　Figure 1.15 shows a portion of the spectra for two stars named A and B. The two curves shown in this figure are vertically shifted for visual effect. Using the relative inten-

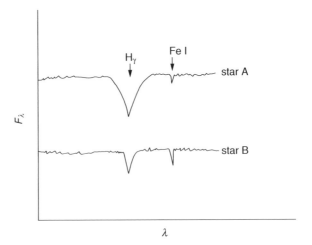

Figure 1.15　Illustration of the spectra of two stars showing the line H$_\gamma$ and an atomic line from the ion FeI. These spectra are vertically shifted for visual effect (see Exercise 1.13).

sities of the hydrogen (H_γ) line and an atomic line from the ion FeI, which of these two stars is hotter? Why?

1.14   The effective temperature of a main-sequence star with spectral type B2 is approximately 22 000 K. Whereas, the effective temperature for a luminosity III class star of the same spectral type (i.e. with the same relative intensities of the various lines) possesses an effective temperature almost 2000 K lower than this value. Using the theoretical concepts seen in this chapter, explain the reason for the discrepancy.

2

Stellar Formation

2.1 Introduction

It is natural to begin the discussion of the various fields of stellar astrophysics by first studying stellar formation. Stars are formed when interstellar gas clouds of sufficient mass collapse under their gravitational pull. This collapse transforms gravitational energy into thermal energy, thus heating the gas. As will be seen in more detail later in the chapter, during the stellar formation process, approximately half of the gravitational energy liberated is transformed into thermal energy and the rest of the energy is emitted in the form of radiation to interstellar space. The reader is reminded that when a mass collapses, the gravitational binding energy is increased. This process liberates energy within the star. In other words the system becomes more stable. This is equivalent to the well-known result of classical mechanics of a mass falling in a gravitational field and where kinetic energy is gained while the potential energy of the mass is decreased (and thus the falling mass is more strongly bound to the body responsible for the gravitational field).

If the mass of the collapsing body is more than approximately 0.08 $M_\odot$, its gravitational energy is sufficient so that the central temperatures will reach the critical value necessary for sustained hydrogen fusion and a star is born. A star in the process of formation is commonly called a protostar.

If a hypothetical homogeneous cloud contracts due to gravitation, a pressure gradient inside the contracted body is created. Pressure within a star or protostar is caused by the weight of the mass above the depth under consideration. Of course, temperature and density gradients will also develop during star formation. Eventually, equilibrium between gravity and the pressure gradient will be attained and the star may then stop contracting. For protostars arriving on the main sequence, another state of equilibrium must also occur, namely energy equilibrium. It is achieved when the rate of nuclear energy produced in their central regions is equal to the luminosity of the star (or in other words the rate at which it loses energy to interstellar space).

An Introduction to Stellar Astrophysics Francis LeBlanc
© 2010 John Wiley & Sons, Ltd

The hydrostatic equilibrium equation will be the first topic discussed in this chapter. This equation describes the pressure gradient found in stars at equilibrium. It will then be used to find another equation that is fundamental for understanding stellar formation called the virial theorem. From this equation, a criterion for the mass of clouds that is necessary for gravitational collapse will be found. A section pertaining to the collapse time of such clouds will also be presented. This chapter will conclude with an optional section discussing the pre-main-sequence evolution of stars where several types of young stars will be discussed.

2.2   Hydrostatic Equilibrium

Before discussing the theory of stellar formation, a fundamental equation of stellar structure called the hydrostatic equilibrium equation must first be considered. This equation will lead to the virial theorem (see Section 2.3) that is crucial for understanding stellar formation. In a star, the gravitational force acting on the plasma causes a pressure-stratification profile. Since the pressure at a given point is due to the weight of the gas above it, pressure increases as a function of depth. At equilibrium, the pressure gradient present in stars counterbalances the gravitational force.

Figure 2.1 shows an element of matter inside a star. Supposing that the star under consideration has spherical symmetry, the local gas pressure, density, temperature, etc. depend solely on the variable r that defines the distance from the star's centre. By assuming that the mass element shown in Figure 2.1 is at equilibrium (i.e. that the total force acting on it is nil), the following equation may be written for the vertical component of the total force

$$P(r)dA - [P(r) + dP]dA - \rho(r)dAdrg(r) = 0 \qquad (2.1)$$

where $P(r)$, $\rho(r)$ and $g(r)$ are, respectively, local gas pressure,[1] gas density and gravitational acceleration. The mass of the element is equal to $\rho(r)dAdr$ and by definition, pressure is

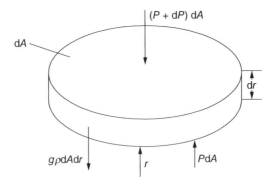

Figure 2.1   A mass element of thickness dr and area dA at a distance r from the centre of the star. The three forces acting on the mass element are shown.

---

[1] Here only gas pressure is considered. However, under certain conditions, radiation pressure can come into play. This topic will be discussed in optional Section 3.12.

the amount of force exerted per unit area thus explaining the various terms found in the equation shown above. The first two terms in this equation, respectively, represent the upward and the downward force on the mass element due to pressure. The third term is simply the weight of the mass element under consideration. Since the pressure on either side of the mass element is equal, the horizontal component of the total force is also nil, but this is of no interest here. This equation can be simplified to obtain the so-called hydrostatic equilibrium equation

$$\frac{dP(r)}{dr} = -\rho(r)g(r) \tag{2.2}$$

Since both ρ and g are positive, $dP/dr < 0$, and as expected P increases as r decreases (or as the centre of the star is approached). Not surprisingly, this equation shows that the pressure as a function of depth (with respect to the stellar surface) will increase more rapidly for larger ρ or g. If ρ and g are known throughout the star, this differential equation can be solved to give the pressure at all depths. This equation is fundamental and is crucial in both stellar-structure and stellar-atmosphere calculations.

Classical mechanics shows that for a spherically symmetric mass distribution the gravitational acceleration at a distance r from the centre of the mass is

$$g(r) = \frac{GM(r)}{r^2} \tag{2.3}$$

where G is the universal constant of gravitation and $M(r)$ is the mass found inside the radius r. This expression will be useful in certain applications seen below and throughout this book.

As shown in the example below, the hydrostatic equilibrium can also be used to calculate the pressure in the atmosphere of a planet. To obtain an analytical solution, this example assumes that the atmosphere is isothermal and that the gravitational acceleration within it is constant.

Example 2.1: Show that the pressure stratification in the atmosphere of a planet, assuming that it is isothermal (with temperature T) and has a constant gravitational acceleration (g), obeys the following expression

$$P(r) = P(0)e^{-\frac{\mu m_H g}{kT}r} \tag{2.4}$$

where r is the distance from the surface, μ is the mean molecular weight of the particles in the atmospheric gas (considered constant throughout the atmosphere), m_H is the mass of the hydrogen atom and $P(0)$ is the pressure at the surface of the planet.

Answer:

The road to the solution begins with the hydrostatic equilibrium equation

$$\frac{dP(r)}{dr} = -\rho(r)g \tag{2.5}$$

Assuming an ideal gas

$$\rho(r) = \frac{\mu m_{\mathrm{H}} P(r)}{kT} \tag{2.6}$$

where m_{H} is the mass of the hydrogen atom, the pressure gradient then becomes

$$\frac{dP(r)}{dr} = -\frac{\mu m_{\mathrm{H}} g}{kT} P(r) \tag{2.7}$$

which may be written as

$$\frac{dP}{P} = -\frac{\mu m_{\mathrm{H}} g}{kT} dr \tag{2.8}$$

If this equation is integrated from the surface ($r = 0$) to an altitude r

$$\int_{P(0)}^{P(r)} \frac{dP}{P} = -\int_{0}^{r} \frac{\mu m_{\mathrm{H}} g}{kT} dr \tag{2.9}$$

it gives

$$\ln(P(r)) - \ln(P(0)) = -\frac{\mu m_{\mathrm{H}} g}{kT} r \tag{2.10}$$

which leads to the solution

$$P(r) = P(0)e^{-\frac{\mu m_{\mathrm{H}} g}{kT} r} \tag{2.11}$$

This equation may be written as follows

$$P(r) = P(0)e^{-\frac{r}{H}}, \text{ where } H = \frac{kT}{\mu m_{\mathrm{H}} g} \tag{2.12}$$

The quantity H is called the pressure scale height and it defines the distance over which the pressure changes by a factor e. For this solution where both the temperature and the mean molecular weight are constant, the density profile has the same form as the pressure profile

$$\rho(r) = \rho(0)e^{-\frac{r}{H}} \tag{2.13}$$

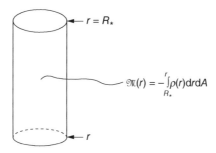

Figure 2.2   Illustration of an imaginary cylinder of area dA extending from the surface of a star (with radius R_*) to a depth located at a distance r from the centre of the star. The variable $\mathfrak{M}(r)$ represents the mass found inside this cylinder.

It might seem strange to the reader to be discussing the atmospheric characteristics of a planet in a book on stars. There are major differences between the example seen above and stars, since in stars the gravity, temperature and mean molecular weight vary with depth. However, such a simple example is very instructive to help understand the pressure (and density) stratification in stars.

The concept of column mass m, which is the mass per unit area above a certain point in a star, can be useful in stellar astrophysics. Its physical units are g/cm^2. Let's first define the mass $\mathfrak{M}(r)$ found inside an imaginary cylinder of area dA and extending from the surface of a star to radius r (see Figure 2.2). The mass inside the cylinder is

$$\mathfrak{M}(r) = -\int_{R_*}^{r} \rho(r)\,dr\,dA \tag{2.14}$$

The column mass $m(r)$ is therefore given by the equation

$$m(r) = \frac{\mathfrak{M}(r)}{dA} = -\int_{R_*}^{r} \rho(r)\,dr = \int dm \tag{2.15}$$

where

$$dm = -\rho(r)\,dr \tag{2.16}$$

A negative sign appears here because the column mass increases with geometrical depth (with respect to the surface). The hydrostatic equilibrium equation can then be written as a function of the column mass

$$\frac{dP}{dm} = g(r) \tag{2.17}$$

This equation predicts that the pressure increases as the column mass increases. In the exterior regions of stars the gravity can be considered approximately constant: $g(r) = g$ for all values of r, and the above equation can be integrated to give $P(m) = gm$, assuming the pressure at the surface is nil. In this case, the pressure is then simply the product of the column mass times the gravitational acceleration. This is due to the fact that pressure at a given point is caused by the weight of the column mass above this point.

When radiation attempting to exit the star is absorbed by the stellar plasma, the momentum of the photons is transferred to matter and it causes radiation pressure. This leads to a radiative acceleration (g_{rad}) of matter and its effect must be taken into account in the hydrostatic equilibrium equation

$$\frac{dP(r)}{dr} = -\rho(r)[g(r) - g_{rad}(r)] \tag{2.18}$$

This equation tells us that if conditions occur where radiative acceleration becomes larger than that of gravity, matter will be pushed out of the star. This equation imposes an upper limit for the mass of stars, since the radiation field in the most massive stars can become large enough to expel their external layers. More information regarding radiative acceleration and radiative pressure is given in the optional Section 3.12.

2.3   The Virial Theorem

A fundamental equation necessary for proper understanding of stellar formation is the virial theorem. The virial theorem gives the relation between the potential and kinetic energies of a system of particles at equilibrium.

The virial theorem for a star (or an interstellar cloud) in hydrostatic equilibrium can be easily derived. For this case, the virial theorem gives the relation between its gravitational energy (Ω) and its thermal energy (U). By inserting the value of $g(r)$ for a spherically symmetric mass distribution in the hydrostatic equilibrium equation gives the following expression

$$\frac{dP}{dr} = -\frac{\rho(r)GM(r)}{r^2} \tag{2.19}$$

When each side of this equation is multiplied by the volume inside radius r, namely $V(r) = 4\pi r^3/3$, the equation above becomes

$$\frac{4}{3}\pi r^3 dP = -\frac{4}{3}\pi r\rho(r)GM(r)dr \tag{2.20}$$

Since the mass inside a spherical shell between r and $r + dr$ is[2]

$$dM = 4\pi r^2\rho(r)dr \tag{2.21}$$

where $4\pi r^2 dr$ is the volume of the shell. Equation (2.20) may then be written as

---

[2] In Section 5.2.2 it will be shown that for spherically symmetrical stars, the mass inside a shell between r and $r + dr$ is $dM(r) = 4\pi r^2 \rho(r)dr$ (see Figure 5.1).

$$VdP = -\frac{GM(r)}{3r}dM \tag{2.22}$$

This equation can then be integrated over the entire volume of the star. First, the left-hand side of the equation can be integrated by parts

$$\int_{P_{\text{centre}}}^{0} VdP = PV\Big|_{\text{centre}}^{\text{surface}} - \int_{0}^{V_*} PdV \tag{2.23}$$

where V_* is the volume of the star. Since the pressure at the surface and the volume at the centre are nil, the first term on the right-hand side of this equation is zero. The two equations above lead to

$$\int_{0}^{V_*} PdV = \frac{1}{3}\int_{0}^{M_*} \frac{GM(r)}{r}dM \tag{2.24}$$

where M_* is the mass of the star. By definition, the potential gravitational energy of the star, which has a negative value, is equal to (more details will be given below)

$$\Omega = -\int_{0}^{M_*} \frac{GM(r)}{r}dM \tag{2.25}$$

and thus

$$3\int_{0}^{V_*} PdV = -\Omega \tag{2.26}$$

According to the equipartition theorem of statistical physics, the average kinetic energy of the particles in a gas at temperature T is equal to $3kT/2$. If N is the total number of particles in a given volume V inside the star small enough so that the temperature is constant within it and equal to T, the thermal energy density ε in units of erg/cm^3 is

$$\varepsilon = \frac{N}{V}\frac{3}{2}kT \tag{2.27}$$

For an ideal gas

$$P = \frac{NkT}{V} = \frac{2}{3}\varepsilon \tag{2.28}$$

which leads to the following relation

$$3\int_0^{V_*} P \mathrm{d}V = 2\int_0^{V_*} \varepsilon \mathrm{d}V = -\Omega \tag{2.29}$$

Since the total thermal energy of the star U is

$$U = \int_0^{V_*} \varepsilon \mathrm{d}V \tag{2.30}$$

the relation between the thermal energy and the gravitational energy or the so-called virial theorem is finally found

$$U = -\frac{\Omega}{2} \quad \text{or} \quad 2U + \Omega = 0 \tag{2.31}$$

During a contraction $-\Omega$ increases, and consequently so does U. The internal temperatures of the body in contraction will then also increase. The virial theorem shows that half of the gravitational energy is used to heat the gas (i.e. $\Delta U = -\Delta\Omega/2$). The other half is radiated in the form of electromagnetic radiation into the surrounding space. However, the virial theorem only applies to a system of particles in equilibrium and therefore cannot apply to collapsing interstellar clouds. Keeping this in mind, the virial theorem can be used to state that *approximately* half of the gravitational energy is transformed to thermal energy during the collapse of a cloud. During such a collapse, the internal temperatures increase, and if the mass of the cloud is sufficient, or in other words if it has enough gravitational energy, the central temperatures attain the critical temperature needed for sustained hydrogen fusion. When this occurs, a star is born. The contraction of the star continues until the power generated by the nuclear reactions equals the luminosity of the star. At that point, the star is in a stable state and finds itself on the main sequence of the H–R diagram.

Earlier, an equation for the gravitational energy of a star was given without further explanation. This equation can be easily understood with well-known results from classical mechanics: the gravitational energy of a two-mass system separated by a distance r is equal to $-Gm_1m_2/r$. This result is found by calculating the work necessary to bring one of these masses from infinity to a distance r from the other mass. Since the work needed to bring a spherical shell of mass $\mathrm{d}M$ to a spherical mass M of radius r is $-GM\mathrm{d}M/r$, the potential energy of a star of mass M_* can be obtained with the following integral

$$\Omega = -\int_0^{M_*} \frac{GM(r)}{r}\mathrm{d}M \tag{2.32}$$

where $\mathrm{d}M$ is the mass of a shell between r and $r + \mathrm{d}r$ and $M(r)$ is the mass within the radius r. This integral measures the potential energy when progressively assembling the mass M_* shell by shell. For stars, the quantity $M(r)$ (or indirectly $\rho(r)$) is not readily known, but Ω can be approximated by making certain assumptions (see Example 2.2). As stated earlier, for spherically symmetrical stars, the mass inside a shell between r and $r + \mathrm{d}r$ is $\mathrm{d}M(r) = 4\pi r^2 \rho(r)\mathrm{d}r$ (see Section 5.2.2).

Example 2.2: Calculate the potential gravitational energy of a star of mass M_* and radius R_* assuming it possesses a constant density.

Answer:

As seen above, the gravitational energy is given by the following expression

$$\Omega = -\int_0^{M_*} \frac{GM(r)}{r} dM \tag{2.33}$$

The average density inside the radius r can be expressed as

$$\bar{\rho}(r) = \frac{M(r)}{\frac{4}{3}\pi r^3} \tag{2.34}$$

Since the density is assumed constant for the star under consideration $\bar{\rho}(r) = \rho$ and r can then be written

$$r = \left(\frac{M(r)}{\frac{4}{3}\pi\rho} \right)^{1/3} \tag{2.35}$$

The quantity Ω can be expressed in terms of $M(r)$

$$\Omega = -G\left(\frac{4}{3}\pi\rho\right)^{1/3} \int_0^{M_*} M^{2/3} dM \tag{2.36}$$

and once the integration is evaluated, this equation becomes

$$\Omega = -G\left(\frac{4}{3}\pi\rho\right)^{1/3} \frac{3}{5} M_*^{5/3} \tag{2.37}$$

Since the density can also be written

$$\rho = \frac{M_*}{\frac{4}{3}\pi R_*^3} \tag{2.38}$$

Ω becomes

$$\Omega = -\frac{3}{5}\frac{GM_*^2}{R_*} \tag{2.39}$$

The gravitational energy of a star with constant density was found in the preceding example. Of course, the density profile (or $\rho(r)$) inside a real star can only be calculated through a complex numerical stellar model. However, the value of the gravitational energy of a typical star can be estimated to be on the order of $-\dfrac{GM_*^2}{R_*}$. This approximation is useful to roughly approximate the gravitational energy of stars. Meanwhile, the exact value of Ω depends on the intricacies of the density profile of the star.

Finally, this section will end by showing that some global characteristics of stars can be estimated with appropriate approximations. The two following examples will use, respectively, the virial theorem and the hydrostatic equilibrium equation to estimate the average temperature and the central pressure of a star.

Example 2.3: With the help of the virial theorem, estimate the average temperature inside a star with mass M_* and radius R_*.

Answer:

With the equipartition theorem, an average temperature $\overline{T}$ can be defined such that the total thermal energy of the star of mass M_* containing N particles is

$$U = \frac{3}{2}k\overline{T}N \tag{2.40}$$

Since stars are made mostly of ionised hydrogen due to their high internal temperatures, the mean molecular weight may be approximated to be $\mu \approx 1/2$, so $N \approx 2M_*/m_H$. The thermal energy can then be written

$$U \approx \frac{3k\overline{T}M_*}{m_H} \tag{2.41}$$

As discussed above, the gravitational energy of a star may be approximated by

$$\Omega \approx -\frac{GM_*^2}{R_*} \tag{2.42}$$

where R_* is the radius of the star.

These last two results can be inserted into the viral theorem

$$2U + \Omega = 0 \tag{2.43}$$

and then the average temperature may then be isolated to give

$$\overline{T} = \frac{1}{6}\frac{Gm_H M_*}{kR_*} \tag{2.44}$$

For the Sun, the average temperature found is $\bar{T} = 4 \times 10^6$ K. This value seems reasonable since its surface temperature is 5800 K; while according to theoretical models its central temperature is approximately equal to 16×10^6 K.

Example 2.4: With the help of the hydrostatic equilibrium, estimate the central pressure in a star with mass M_* and radius R_*.

Answer:

Since the mass inside a spherical shell found between r and $r+dr$ is $dM(r) = 4\pi r^2 \rho(r)dr$ and $g(r) = \dfrac{GM(r)}{r^2}$ the hydrostatic equilibrium equation can be expressed as

$$dP = -\frac{GM}{4\pi r^4}dM \qquad (2.45)$$

This equation may be integrated from the surface to the centre of the star

$$\int_0^{P_c} dP = -\int_{M_*}^0 \frac{GM}{4\pi r^4}dM \qquad (2.46)$$

where P_c is the central pressure and where it is assumed that the pressure at the surface of the star is nil.

The above equation may be rewritten as follows

$$P_c = \frac{GM_*}{4\pi}\left[\frac{1}{M_*}\int_0^{M_*}\frac{M}{r^4}dM\right] \qquad (2.47)$$

The expression in brackets above gives the average value of $M(r)/r^4$ in the star. Its value can be crudely approximated by M_*/R_*^4 as to obtain

$$P_c \approx \frac{GM_*^2}{4\pi R_*^4} \qquad (2.48)$$

For the Sun, this equation gives a central pressure on the order of 10^{15} dyn/cm². In reality, because of the density stratification in stars, the average value of $M(r)/r^4$ is much larger than M_*/R_*^4 since the layers at small radii will contribute more to the integral above. Detailed modelling of the solar structure predicts that its central pressure is approximately 2×10^{17} dyn/cm² (or 10^{11} times the atmospheric pressure at the Earth's surface).

Some of the crude approximations made in the last two examples might seem imprudent. However, in astrophysics, it is sometimes useful to make such assumptions in order to evaluate the order of magnitude of certain physical quantities not readily known. A more precise evaluation often necessitates complex numerical calculations that are outside the scope of this book.

2.4   The Jeans Criterion

In this section, a criterion will be found that can be used to predict under what physical conditions an interstellar cloud can collapse. This criterion is a product of the British scientist Sir James Hopwood Jeans (1877–1946).

As discussed previously, an interstellar cloud in hydrostatic equilibrium obeys the virial theorem. If the gravitational energy dominates the thermal energy such as $-\Omega > 2U$, the cloud will be unstable and collapse due to gravitational pull. The quantity Ω is difficult to precisely calculate since the knowledge of density profile is needed for its assessment. To overcome this obstacle, the gravitational energy can be approximated by the corresponding value for a homogeneous cloud of mass M and radius R (see Example 2.2)

$$\Omega = -\frac{3}{5}\frac{GM^2}{R} \tag{2.49}$$

If the interstellar cloud is approximated as being isothermal (with temperature T) and having a constant density ρ, the thermal energy may be written

$$U = \frac{3}{2}NkT \tag{2.50}$$

where N is the total number of particles contained in the cloud. The number of particles in the cloud may then be written as a function of the mean molecular weight μ of the particles

$$N = \frac{M}{\mu m_{\mathrm{H}}} \tag{2.51}$$

and the cloud can collapse when $-\Omega > 2U$ or

$$\frac{3}{5}\frac{GM^2}{R} > \frac{3MkT}{\mu m_{\mathrm{H}}} \tag{2.52}$$

But since

$$R = \left(\frac{M}{\frac{4}{3}\pi\rho}\right)^{1/3} \tag{2.53}$$

where ρ is the density of the cloud, it can easily be shown that for a cloud to be able to collapse, its mass must respect the following inequality

$$M > \left(\frac{5kT}{\mu m_{\mathrm{H}}G}\right)^{3/2}\left(\frac{3}{4\pi\rho}\right)^{1/2} = M_{\mathrm{J}} \tag{2.54}$$

where Jeans' mass M_J is defined. So, in order for a cloud to collapse, its mass must be larger than Jeans' mass: $M > M_J$. This establishes a quantitative criterion by which the stability of a cloud can be evaluated.

It should be noted that M_J is smaller for clouds with larger densities since a higher gravitational energy is stored there. Also, clouds with higher temperatures have larger M_J because internal pressure counters gravitational collapse. Similarly, an expression for Jeans' density ρ_J can be found. Collapse can take place when the density of the cloud ρ under consideration surpasses Jeans' density

$$\rho > \rho_J = \left(\frac{5kT}{\mu m_H G}\right)^3 \left(\frac{3}{4\pi M^2}\right) \tag{2.55}$$

Jeans' length (or radius) may also be defined by isolating the radius of the cloud in the above theoretical development. It can be shown (see Exercise 2.7) that this quantity is

$$R_J = \left(\frac{15kT}{4\pi\rho\mu m_H G}\right)^{1/2} \tag{2.56}$$

When a given cloud with density ρ has a radius larger than this value, this criterion predicts that it should collapse due to its gravitational pull.

Example 2.5: Calculate Jeans' mass for an average molecular cloud. Typically, molecular clouds have masses on the order of 1000 $M_\odot$ or more, temperatures on the order of 10 K and number densities of approximately 1000 H_2 molecules per cm^3. Discuss the results vis-à-vis star formation.

Answer:

Since the clouds under consideration are composed mainly of H_2 molecules, the mean molecular weight is approximately equal to 2 and the density in these clouds is therefore

$$\rho = 2m_H \times 1000 \text{ cm}^{-3} \approx 3 \times 10^{-21} \text{ g cm}^{-3} \tag{2.57}$$

and Jeans' mass for this type of cloud is

$$M_J = \left(\frac{5kT}{\mu m_H G}\right)^{3/2} \left(\frac{3}{4\pi\rho}\right)^{1/2} \approx 5 \times 10^{-34} \text{ g} \approx 20 \, M_\odot \tag{2.58}$$

Jeans' mass for molecular clouds is smaller than their actual mass. Jeans' criterion predicts that they can collapse and lead to stellar formation.

Example 2.6: Calculate Jeans' density for a diffuse hydrogen (or HI) cloud. Typically, diffuse hydrogen clouds have masses of less than 100 $M_\odot$, temperatures on the order of 100 K and number densities of less than 1000 H atoms per cm^3. Discuss the results vis-à-vis star formation.

Answer:

Since the temperature inside diffuse hydrogen clouds is low, most of the hydrogen atoms are in their neutral ionisation state. The mean molecular weight is therefore approximately equal to 1. The density of these clouds is then

$$\rho = m_H \times 1000 \text{ cm}^{-3} \approx 2 \times 10^{-21} \text{ g cm}^{-3} \tag{2.59}$$

Jeans' density for this kind of interstellar cloud is

$$\rho_J = \left(\frac{5kT}{\mu m_H G} \right)^3 \left(\frac{3}{4\pi M^2} \right) \approx 1 \times 10^{-18} \text{ g cm}^{-3} \tag{2.60}$$

Since Jeans' density for diffuse hydrogen clouds is larger than their typical densities, they will be gravitationally stable and no star formation can take place there.

It was just shown that, according to Jeans' criterion, stellar formation can take place in interstellar molecular clouds. These cold and dense clouds can collapse, thus increasing their density, until smaller portions can also become unstable and collapse. This fragmentation process makes it possible for a large number of stars to form from a single molecular cloud. An example of a molecular cloud, namely M16[3] or better known as the Eagle Nebula, is shown in Figure 2.3. It should be noted that in these so-called stellar nurseries, other molecules besides H_2 can also be present along with dust particles. During the star-formation process, more low-mass stars are generally formed than high-mass ones. The distribution of the stars relative to their mass is called the initial mass function and is discussed below.

---

[3]This object is part of the Messier catalogue that originally contained 103 astronomical objects (mostly nebulae, stellar clusters and galaxies). The French astronomer Charles Messier (1730–1817) created this list of dim objects in order not to confuse them with comets. Seven other astronomical objects were later added to the Messier catalogue that now goes from M1 to M110.

Figure 2.3   Molecular clouds in the M16 nebula where star formation is present (NASA/courtesy of nasaimages.org). (see colour plate.)

Special Topic – The Initial Mass Function

Even when using crude statistical methods with a limited sampling of stars to study stellar-mass distribution, it is easy to find that there exist many more low-mass stars than high-mass ones in the Universe. In an attempt to determine the exact distribution of stars relative to their mass, a distribution function called the initial mass function (IMF) is used. It should, however, be clear that a single function should not be able to reproduce observations of all astronomical bodies (i.e. various types of stellar clusters, galaxies, etc.) since many variables (such as the presence of a magnetic field, rotation or turbulence for instance) can affect the star formation process. An example of the IMF for the stars in the solar neighbourhood is shown below.

The determination of the IMF is complex since it is luminosity (within a certain wavelength range) that is observed rather than stellar masses. This luminosity must than be transformed into stellar masses. Several complications occur. For example, the distance to the various stars must be evaluated with precision. Also, absorption or scattering from the interstellar medium can also come into play.

The Austrian born astronomer Edwin Ernest Salpeter (1924–2008) found that the distribution of stellar masses varies as a power function and has the general shape of $M_*^{-2.35}$. Salpeter's power law is compared to a more detailed determination of the IMF for the solar neighbourhood in Figure 2.4. The variable that defines the IMF is ξ and gives the normalized number of stars formed per solar unit mass. Salpeters's power law well reproduces the results shown in Figure 2.4 for the high-mass tail end of the distribution. Figure 2.4 clearly shows that the star-formation process preferably produces stars of low mass. This confirms the earlier statements about the non-uniform mass distribution of stars.

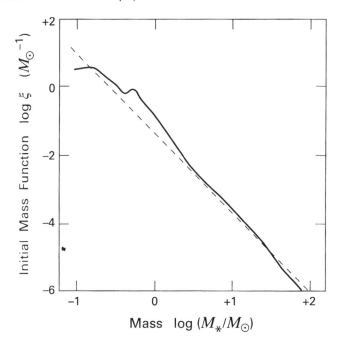

Figure 2.4   The initial mass function of the solar neighbourhood (solid curve) as compared to Salpeter's power law (dashed curve). Figure reproduced with permission of Stahler, S.W. and Palla, F., *The Formation of Stars*, Wiley-VCH Verlag GmbH & Co. KGaA, p.125 (2004).

Contrarily to the results from the basic theoretical framework discussed above, observations show that most interstellar molecular clouds are not collapsing. The framework shown above is therefore incomplete. The reader is reminded that physical phenomena such as rotation, turbulence or the presence of magnetic fields were neglected in our analysis and probably come into play. A triggering mechanism seems necessary to induce gravitational collapse in molecular clouds. Several mechanisms are possible. For example, a neighbouring supernova can cause a shock wave in molecular clouds and trigger collapse and thereafter star formation. A collision between two interstellar clouds could also lead to star formation.

Star formation is a process that is much more complex than that described in this section. Star formation does not simply consist of a spherically symmetrical collapse of an interstellar cloud. For example, circumstellar matter can form an accretion disk around protostars. Jets of matter emanating from these young astronomical objects are also observed. Moreover, debris (planets, asteroids, etc.) is left orbiting the newly formed stars.

Following the fragmentation process during star formation, newly created O- and B-type stars can heat the part of the cloud left over. This hot gas is commonly called an HII region, since hydrogen is ionised there due to the energetic photons emitted by these hot stars. The typical temperature of HII regions is about 8000 K and these astronomical structures are stable against gravitational collapse (see Exercise 2.8).

Finally, it should be mentioned that the framework described here for star formation is quite rudimentary. Detailed study of star formation requires complex hydrodynamical computer simulations. Such advanced computations are outside the scope of this book.

Special Topic – Another Derivation of Jeans' Mass

Earlier in this section, Jeans' mass was found by comparing the gravitational and thermal energy in a spherical gas cloud. Another way of finding Eq. (2.54) is by using Newton's second law. When a cloud is not in hydrostatic equilibrium, a force equation may be written for an element of mass with density ρ that takes into account gravity and the expansive force caused by the pressure gradient

$$\rho \frac{d^2 r}{dt^2} = -\rho g - \frac{dP}{dr} \qquad (2.61)$$

The cloud contracts when $\dfrac{d^2 r}{dt^2} < 0$, or in other words, when

$$\rho g > -\frac{dP}{dr} \qquad (2.62)$$

To simplify the problem, a constant gravitational acceleration equal to the value at the surface

$$g = \frac{GM}{R^2} \qquad (2.63)$$

is assumed throughout the cloud. Since

$$P = \frac{\rho kT}{\mu m_H} \qquad (2.64)$$

and assuming an isothermal gas of temperature T with a constant mean molecular weight μ, the pressure gradient is equal to

$$\frac{dP}{dr} = \frac{kT}{\mu m_H} \frac{d\rho}{dr} \qquad (2.65)$$

Therefore,

$$\rho \frac{GM}{R^2} > -\frac{kT}{\mu m_H} \frac{d\rho}{dr} \qquad (2.66)$$

By crudely estimating the density gradient as

$$\frac{d\rho}{dr} \approx -\frac{\rho}{R} \qquad (2.67)$$

where R is the radius of the spherical cloud and ρ is its average density and since

$$R = \left(\frac{M}{\frac{4}{3}\pi\rho} \right)^{1/3} \tag{2.68}$$

it is trivial to show that

$$M > \left(\frac{kT}{\mu m_{\mathrm{H}} G} \right)^{3/2} \left(\frac{3}{4\pi\rho} \right)^{1/2} \tag{2.69}$$

Except for the constant $5^{3/2}$, this is the same expression previously found for Jeans' mass (Eq. 2.54). This difference is not surprising since several approximations were made in the two methods used to estimate M_{J}.

2.5   Free-Fall Times[†]

In the previous section, the physical conditions under which interstellar clouds can collapse were considered. However, the timescale on which molecular clouds collapse was not discussed. This is the aim of the present section.

The precise evaluation of the collapse time of an interstellar cloud requires detailed calculations that take both the gravitational force as well as the counteracting pressure in the cloud into account. Other factors such as the presence of a magnetic field can also affect the duration of gravitational collapse. Nonetheless, a simpler physical framework may be employed to estimate the duration of such a collapse. An expression for the commonly called free-fall time will be found. The free-fall time is the time it would take for a cloud to collapse from its original shape to a single point due to gravity while neglecting gas pressure that will counteract this force.

Figure 2.5 shows a small mass m at the surface of a spherical cloud of radius R and mass M. The mass m is initially at rest on the surface of this cloud and then follows the surface of the cloud that collapses due to gravity. Under the free-fall assumption, when the surface is contracted to radius r the kinetic energy of this mass is equal to the difference of its gravitational potential energy between the final and initial positions

$$K = \frac{1}{2}m\left(\frac{\mathrm{d}r}{\mathrm{d}t} \right)^2 = \frac{GMm}{r} - \frac{GMm}{R} \tag{2.70}$$

The velocity of the particle m can then be expressed as

$$\frac{\mathrm{d}r}{\mathrm{d}t} = -\left(\frac{2GM}{r} - \frac{2GM}{R} \right)^{1/2} \tag{2.71}$$

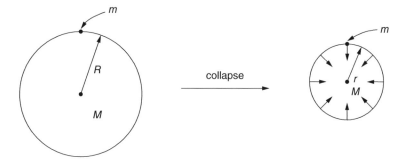

Figure 2.5   A small mass m initially at rest (left-hand side of the figure) at the surface of a spherical cloud of mass M that collapses from a radius R to r. The small inward pointing arrows on the right-hand side of the figure represents the velocity of the surface of the cloud.

The time variable can then be isolated and integrated to obtain the free-fall time t_{ff} (assuming $t = 0$ at the onset of collapse)

$$t_{ff} = -\int_{R}^{0}\left(\frac{2GM}{r} - \frac{2GM}{R}\right)^{-1/2} dr \qquad (2.72)$$

The equation above measures the time for the cloud to collapse from its initial radius R to a radius that tends towards zero (or in other words a geometrical point). The final stages of such a theoretical collapse cannot be treated with classical physics since, in the presence of large gravitational fields, Newton's law of gravity becomes invalid and the theory of general relativity comes into play. If such a collapse were to take place, the astronomical body would become a black hole (see Chapter 6 for more details). In reality, the cloud will stop its collapse before attaining a physical state where general relativity is required due to the pressure inside the protostar. However, the integral shown above gives an estimate for the order of magnitude for the collapse time of a cloud of mass M and initial radius R.

The above integral may be calculated by performing the change of variable $x = r/R$

$$t_{ff} = \left(\frac{R^3}{2GM}\right)^{1/2} \int_{0}^{1}\left(\frac{x}{1-x}\right)^{1/2} dx \qquad (2.73)$$

and since the definite integral above equals $\pi/2$, the free-fall time is

$$t_{ff} = \frac{\pi}{2}\left(\frac{R^3}{2GM}\right)^{1/2} = \left(\frac{3\pi}{32G\rho}\right)^{1/2} \qquad (2.74)$$

where it is assumed that

$$M = \frac{4}{3}\pi R^3 \rho \qquad (2.75)$$

Example 2.7: Calculate the free-fall time for a one-solar-mass molecular cloud.

Answer:

We suppose that the collapse begins when the cloud achieves Jeans' density. For a one-solar-mass cloud

$$\rho_J = \left(\frac{5kT}{\mu m_H G}\right)^3 \left(\frac{3}{4\pi M^2}\right) \approx 2\times10^{-18} \text{ g cm}^{-3} \tag{2.76}$$

The free-fall time is then

$$t_{ff} = \left(\frac{3\pi}{32G\rho_J}\right)^{1/2} \approx 50\,000 \text{ yr} \tag{2.77}$$

The timescale on which stars are formed is much shorter than that of stellar evolution. For instance, in Chapter 6, it will be shown that a one-solar-mass star stays approximately 10^{10} years on the main sequence before evolving towards the red giant phase. Since the star-formation timescale is short, it is difficult to obtain a large quantity of observational data as compared to main-sequence stars, for example.

2.6   Pre-Main-Sequence Evolution[†]

In Chapter 6, the evolution of stars from the main sequence to later stages will be discussed. During evolution, stars change position in the H–R diagram because both their T_{eff} and luminosity vary with time. In this section, the evolutionary tracks (in the H–R diagram) followed by stars before they reach the main sequence will be shown. Two types of pre-main-sequence stars will also be introduced.

Figure 2.6 shows the evolutionary tracks of stars in formation of various masses up to their arrival on the main sequence. The vertical portions of the pathways shown in this H–R diagram are called Hayashi tracks named after the Japanese astrophysicist Chushiro Hayashi (b. 1920) who first discovered these evolutionary tracks. On the vertical part of the curves (or on the so-called Hayashi tracks) shown in Figure 2.6, stars are fully convective, or in other words convection dominates energy transport throughout the star. During the bottom and more horizontal part of the evolutionary tracks, radiative transport appears in these stars. At the endpoints of these curves (at the open circles), the stars are burning hydrogen in their core and are thus on the main sequence. However, before this stage, other secondary nuclear reactions occur. For example, deuterium (^2H) is fused into ^3He (via the reaction $^2H + ^1H \rightarrow ^3He + \gamma$) while Li is fused with hydrogen (via the reaction $^7Li + ^1H \rightarrow ^4He + ^4He$). These nuclear reactions diminish the quantity of deuterium and

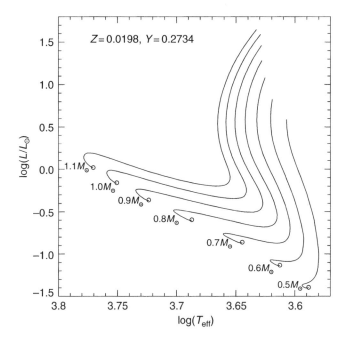

Figure 2.6   Pre-main-sequence evolutionary tracks of stars with various masses. These tracks end at the onset of hydrogen burning (open circles). The metallicity Z and helium abundance Y (in mass fractions) used to model these evolutionary tracks are given. Figure courtesy of Salaris and Cassisi, *Evolution of Stars and Stellar Populations*, Wiley, Chichester (2005).

lithium in these stars, and the abundance of these elements is an important diagnostic tool to study pre-main-sequence stars. However, the energy generated by these nuclear reactions is small compared to the hydrogen fusion energy production rate in the centre of stars and to the radiative energy emitted at their surface. More details concerning nuclear reactions will be given in Chapter 6.

It should be noted that the evolutionary timescales of pre-main-sequence stars are smaller for the more massive stars. More massive protostars possess more gravitational energy and can thus collapse faster. They can also dissipate the gravitational energy at a relatively faster rate during contraction because of their larger luminosity (via radiation emitted to space from their surface). For example, the time it takes for a star with a mass of $1M_\odot$ to make it to the main sequence is approximately 100 times larger than the corresponding time for a $4M_\odot$ star. This along with the fact that the formation process produces many fewer O- and B-type stars than less-massive ones, leads to a selective effect in the observation of pre-main-sequence stars in which predominantly less-massive stars are observed. Two such types of pre-main-sequence stars are T Tauri and Herbig Ae/Be stars.

T Tauri stars are pre-main-sequence stars with masses of approximately $M \leq 2M_\odot$ found mainly in or near molecular clouds. Their name comes from the name of the

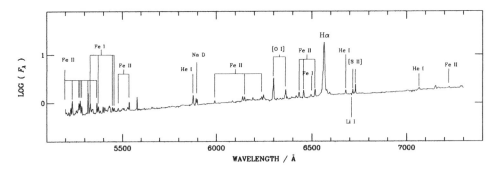

Figure 2.7   Spectra of the T Tauri star DG Tau. Figure reproduced with permission from Hessman, F.V. and Guenther, E.W., *Astronomy & Astrophysics*, 321, 497 (1997).

first of such stars discovered (or the prototype) that is found in the Taurus constellation. T Tauri stars are surrounded by an accretion disk and exhibit emission lines. Some emission lines found in the spectrum of a T Tauri star are shown in Figure 2.7. The lines for which the ion is placed in brackets are called forbidden lines. Forbidden lines come from excited states that have a long life or in other words the excited electron has a low probability of decaying to a lower level. In dense plasma, these forbidden transitions are not visible since collisional de-excitations are more important than radiative decays. However, in low-density plasma such as circumstellar matter, such stable excited levels can stay intact long enough to eventually decay and produce a line called a forbidden line. Both infalling and outflowing matter are detected in these stars, and not surprisingly, their spectra vary irregularly with time. T Tauri stars are therefore classified as variable stars. Their spectra also exhibit an excess of infrared radiation that most likely comes from circumstellar matter surrounding these stars.

Another type of pre-main-sequence stars are Herbig Ae/Be stars with masses of approximately $2\,M_\odot \le M \le 10 M_\odot$ and that possess similar properties to those of T Tauri stars. Their effective temperature ranges within approximately $8000\,\mathrm{K} \le T_{\mathrm{eff}} \le 20000\,\mathrm{K}$. These stars should not be confused with regular Be stars that are main sequence or more evolved peculiar stars. Even though both Herbig Ae/Be- and Be-type stars possess circumstellar matter, the matter surrounding Herbig Ae/Be stars contains dust typically found in molecular clouds. This type of star is named after the American astronomer George Herbig (b. 1920) who first discovered these stars. Some Herbig Ae/Be stars exhibit stellar pulsations that are detected via temporal photometric observations. Figure 2.8 shows the pulsation of the Herbig Ae star V351 Ori. The best-fit curve in this figure is composed by the superposition of five periodic signals, the one with the largest amplitude having a period of 0.0841 days. More details concerning stellar pulsations and the observational determination of their period will be given in Chapter 5.

The concepts seen in this chapter were given at an introductory level. For a more advanced and complete treatise of star formation theory and observations, the reader is referred to Stahler, S.W. and Palla, F., *The Formation of Stars*, Wiley-VCH, Weinheim (2004).

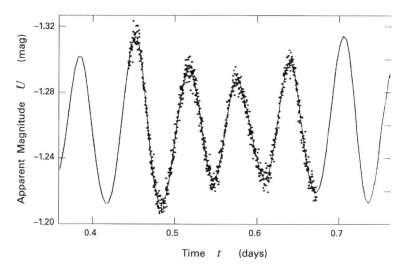

Figure 2.8 Photometric variability of the magnitude in the U band of the Herbig Ae star V351 Ori. The solid curve is a best-fit combination of five pulsation frequencies. The frequency of the signal with the largest amplitude has a period of 0.0841 days (or approximately 2 h). Figure reproduced with permission of Stahler, S.W. and Palla, F., *The Formation of Stars*, Wiley-VCH Verlag GmbH & Co. KGaA, p.702 (2004).

2.7 Summary

Hydrostatic equilibrium: $\dfrac{\mathrm{d}P(r)}{\mathrm{d}r} = -\rho(r)g(r)$ (2.78)

Virial theorem: $2U + \Omega = 0$ (2.79)

Gravitational energy: $\Omega = -\displaystyle\int_0^{M_*} \dfrac{GM(r)}{r}\,\mathrm{d}M$ (2.80)

Jeans' criterion: $M > \left(\dfrac{5kT}{\mu m_\mathrm{H} G}\right)^{3/2}\left(\dfrac{3}{4\pi\rho}\right)^{1/2} = M_\mathrm{J}$ (2.81)

Free-fall time: $t_\mathrm{ff} = \left(\dfrac{3\pi}{32G\rho}\right)^{1/2}$ (2.82)

2.8 Exercises

2.1   The atmosphere of a fictitious planet has a density profile $\rho(r) = \rho_0(1 - r/R_\mathrm{f})$, where and r is the distance from the surface, R_f is the distance from the surface where the density

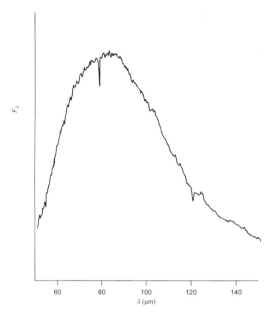

Figure 2.9   Spectra of an interstellar molecular cloud (see Exercise 2.5).

falls to zero and ρ_0 being the density at the surface. If the gravitational acceleration g is constant throughout the atmosphere, calculate the pressure profile $P(r)$.

2.2   Calculate $P(r)$ inside a sphere of radius R_* with a constant density ρ.

2.3   Calculate the gravitational potential energy of a fictitious star with mass M_* and radius R_* that has a density profile $\rho(r) = \rho_0(1 - r/R_*)$. Give your answer in terms of M_* and R_*.

2.4   Show that for a star with mass M_* and radius R_* the result found in Example 2.3 can be written

$$\overline{T} = 4\times 10^6 (M_*/M_\odot)(R_*/R_\odot)^{-1} \text{ K} \tag{2.83}$$

2.5   Estimate the temperature of the interstellar molecular cloud for which the spectra is shown in Figure 2.9.

2.6   Calculate Jeans' mass for diffuse hydrogen clouds (see Example 2.6).

2.7   Show that for a homogeneous cloud with temperature T and density ρ, Jeans' length is

$$R_J = \left(\frac{15kT}{4\pi\rho\mu m_H G} \right)^{1/2} \tag{2.84}$$

2.8 Assume an HII region with a radius of 10 ly, $T \approx 8000\,\text{K}$ and number density $n_{\text{HII}} \approx 10^3\,\text{cm}^{-3}$. Prove that according to Jeans' criterion this interstellar matter is stable against collapse.

2.9 Knowing that the energy density for a magnetic field of value B is equal to $\dfrac{B^2}{8\pi}$, calculate the magnetic field at which the magnetic energy density is comparable to the gravitational energy density for a molecular cloud (see Example 2.5 for the physical data for such a cloud).

3

Radiative Transfer in Stars

3.1 Introduction

Since most of the information that arrives to Earth from stars (and in fact from any astronomical body) is in the form of electromagnetic radiation, it is imperative to properly understand how radiation emanates from the surface of stars. The frequency distribution of the surface monochromatic stellar flux depends on the physical properties of the external regions of stars and on how radiation crosses, and reacts with, stellar plasma. In most instances, the energy created by nuclear fusion in the central parts of the stars is transported to the outer regions by radiation. During this type of energy transport, photons making up the radiation field can be absorbed or scattered by stellar plasma. This modifies its frequency distribution. Since matter also emits radiation due to its thermal energy, this process must also be taken into account when studying radiative transfer in stars. This interplay between radiation and matter is fundamental in stellar astrophysics. The radiation field in the central stellar regions is composed mostly of high-energy photons, while when the radiative energy emerges at the surface of stars, its spectrum depends on the temperature of the outer stellar regions and is comprised of lower-energy photons. Radiative transfer through the various layers in a star therefore strongly modifies the spectrum of the radiation field.

The aim of this chapter is to present the theory of radiative transfer, as it pertains to stars. The equation of radiative transfer will enable, among other things, the determination of the radiative flux emanating from the surface of stars. Its general solution, providing sufficient information about the star's structure is known, will also permit the calculation of the radiative flux at any depth. Precise knowledge of the surface radiative flux is critical for gaining physical insight regarding the exterior regions of stars by comparing, for example, the observed and theoretical stellar fluxes. The radiative-transfer equation is one of the most important equations in the field of stellar astrophysics and merits the close attention that will be brought to it in this chapter. It should be noted that this equation is

An Introduction to Stellar Astrophysics Francis LeBlanc
© 2010 John Wiley & Sons, Ltd

not only important in astrophysics, but is also of interest in several other fields of physics such as photonics. The physical principles presented in this chapter therefore apply to other physical entities besides stars.

3.2 Radiative Opacities

3.2.1 Matter–Radiation Interactions

As previously mentioned, the radiative field in stars interacts with the stellar plasma and this affects its transfer through the star. Radiative opacity is a physical quantity that is a function of frequency and it measures the capability of matter to absorb or scatter photons. The scattering of photons is defined as a deviation of these particles from their trajectory following an interaction with matter. A scattering process does not absorb photons, but still takes away photons from the beams of radiation described by a physical quantity called the specific intensity (its definition will be given in Section 3.3) and must then be taken into account when studying radiative transfer.

When a photon is absorbed, let's say by an atom following the photoexcitation of a bound electron, the energy gained by the atom is eventually redistributed to the radiation field or to the thermal energy bath (i.e. in the form of kinetic energy of the particles in the

Figure 3.1 A *nonexhaustive* list of interactions between radiation and matter is illustrated. (a) A photon is absorbed by an atom and excites it to a higher energy level that is followed by a de-excitation of the bound electron to the same initial level that emits a photon. This process is considered as scattering. In this series of interactions the radiative energy is transformed into atomic energy and then back into radiative energy. (b) A photon excites an atom that is followed by a de-excitation due to a collision with a free electron. The energy of the photon is thus transformed into kinetic energy (please note that the velocity vector of the free electron increases following the collision). (c) A free electron excites an atom via a collision that is followed by the emission of a photon. In this process, kinetic energy is first transformed into atomic energy (please note that the velocity vector of the free electron decreases following the collision) which is thereafter transformed into radiative energy. (d) A first free electron excites an atom. This is followed by a de-excitation due to a collision of the atom with a second incoming free electron (that appears in the middle figure). This whole process transfers the kinetic energy from the first to the second free electron shown in the figure. (e) A photon excites an atom to a high-lying energy level. This is followed by two successive radiative de-excitations. The energy of the original photon is then split among the two emitted photons. (f) A photon ionises an atom. This is followed by a recombination (or electron capture of an incoming free electron) which emits a photon whose energy depends on the velocity of the captured electron. (g) An atom is ionised by a collision of the atom with a free electron having a sufficiently large kinetic energy (this free electron is represented by the one with the largest velocity vector in the figure on the left). This is followed by a recombination (or free electron capture of the second free electron shown on the figure on the left) to the initial energy level. A photon is emitted following the recombination.

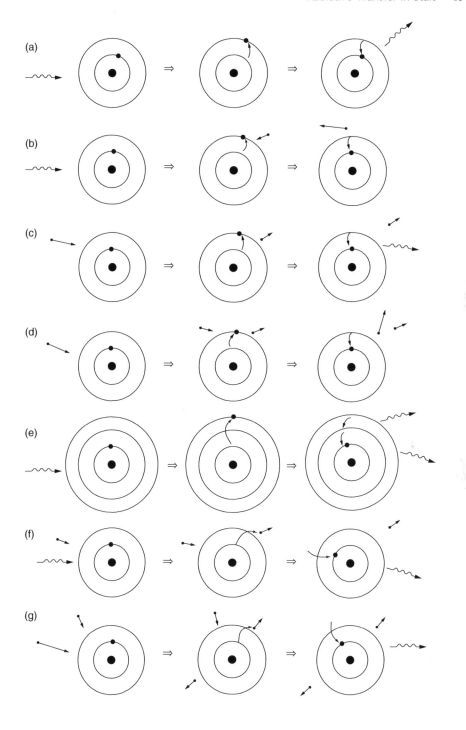

plasma). In the first case just mentioned, the excited electron may, for example, shortly after, fall to a lower level while emitting a photon. When this lower level is the same as the initial level before excitation, the energy of the absorbed photon is returned to the radiation field and this is considered a scattering process since the energy of the absorbed and emitted photon is the same (see Figure 3.1(a)). This type of scattering is sometimes called resonance scattering.

The process by which an excited electron emits a photon is called spontaneous emission. This is due to the fact that an excited atomic state has a finite lifetime and the electron eventually cascades to a lower level. This process will be discussed in more detail in Chapter 4. In the second case mentioned above, where the energy of the absorbed photon is transferred to the thermal bath of the medium, the excited atom may collide with another particle, a free electron for instance, and the excited bound electron can cascade to a lower level, giving energy to the free electron that gains kinetic energy (see Figure 3.1(b)). The energy of the absorbed photon is thus transferred to the thermal bath of the plasma.

The energy of an absorbed photon can also be fractioned by the combination of the emission of several photons of lower energy (see Figure 3.1(e) for instance). In each step of this process, the bound electron that was initially excited is de-excited to a lower energy level.

The kinetic energy of the free particles found in the plasma can also be transformed to radiative energy (see Figure 3.1(c)). Collisions can excite atoms, which in turn can emit photons. In this instance, a portion of the kinetic energy of the colliding particle is transformed into radiation. Under most physical conditions found in stars, the collisional processes dominate radiative processes (excitations and photoionisations) and therefore, the atomic energy and the ionisation state populations can be obtained, respectively, with the Boltzmann and Saha equations.[1]

Other examples of the interplay between the radiation field and surrounding plasma are illustrated in Figure 3.1. This figure shows examples of how atomic, kinetic and radiative energy can be transformed from one form to another. Precise knowledge of the interactions between radiation and matter is therefore critical for proper understanding of radiative transfer in stars.

The free electrons found in stellar plasma generally dominate the excitations and photoionisations caused by collisions for the following reason. The equipartition theorem predicts that the various types of particles in an ideal gas possess the same average kinetic energy equal to $3kT/2$. Since the mass of the electron is much smaller than that of an atom, its average speed will be much larger. There will thus be a much larger number of electron–atom collisions than atom–atom collisions. Therefore, in stars the populations of the atomic energy levels and the ions are regulated by electron–atom collisions rather than by atom–atom collisions.

3.2.2   Types of Radiative Opacities

Radiative opacity measures the capability of matter to absorb or scatter radiation (as a function of frequency) and it depends on both the physical condition of the stellar plasma

---

[1]The reader is reminded that these equations are valid when collisions dominate radiative processes.

and on the type of particles (atoms, molecules and electrons) present there. Radiative opacities can be divided into four types: 1) bound–bound transitions, 2) bound–free transitions, 3) free–free transitions and 4) scattering of photons. These various types of opacities, some of which were already described in Figure 3.1, are briefly discussed below.

1) Bound–bound transitions: Atoms can absorb photons, their energy being used to excite an electron from a lower to an upper bound state. The energy of the absorbed photon must be equal to the difference between the initial and final energy states of the excited electron (see Figure 1.5 for examples of such transitions for hydrogen and Figure 3.1(b) for an illustration of this type of opacity). At the surface of cooler stars, where molecules can exist, these entities can also absorb photons via electronic, rotational or vibrational transitions.

2) Bound–free transitions: Atoms can absorb photons during photoionisation. In such a transition, an electron initially in a bound state is ejected from the atom and becomes a free electron (see Figure 3.1(f) for an illustration of such a transition). The energy of the photon must be at least equal to the difference between the continuum and the initial atomic energy state of the bound electron (see Figure 1.5). If the photon possesses excess energy, it will be transformed into kinetic energy carried away by the ejected electron. Molecules can also be disassociated (i.e. destroyed) by photons, where one or several atoms are released from the initial molecule. In this case, the energy of the absorbed photon is used to dissociate the molecules.

3) Free–free transitions: This process takes place when a photon is absorbed and excites a free electron on a hyperbolic (or unbound) orbit around an atom A to an orbit of higher energy. This is the inverse process of bremsstrahlung radiation, a well-known phenomenon of modern physics. This absorption interaction may be represented by

$$A + e^-(V_i) + \gamma \rightarrow A + e^-(V_f) \tag{3.1}$$

where V_i and V_f, respectively, represent the initial and final velocities of the free electron e^-. This reaction respects energy conservation: $\frac{1}{2}m_e V_f^2 = h v_\gamma + \frac{1}{2}m_e V_i^2$ where v_γ is the frequency of the absorbed photon. Here, the kinetic energy of the atom A was neglected. The reader is reminded that an isolated free electron cannot absorb both the energy and the momentum of a photon. It needs the presence of a nearby nucleus that can absorb the excess momentum.

4) Scattering: Photons can be scattered by free electrons (Thomson scattering) or atoms and molecules (Rayleigh scattering). Rayleigh scattering diffuses photons having energies lower than the transition energies of the atom or molecule. An atom can also momentarily absorb a photon that is followed by the emission of a photon with approximately the same energy, the excited electron returning soon after to its initial state (see Figure 3.1(a)). Since the emitted photon is not necessarily in the same direction as the absorbed photon, the photon is taken away from the initial beam of energy (which is measured by a physical quantity called the specific intensity, which will be defined in Section 3.3). This type of interaction is thus considered a scattering process, sometimes called line scattering. The emitted photon does not necessarily possess the exact same

energy as the absorbed photon since the bound energy levels of atoms are not precisely defined because of the uncertainty principle of quantum mechanics (see Section 4.3 for more details).

The physical quantity that defines radiative opacity k_v is given in units of cm²/g, or in other words, in units of cross section per unit mass. The opacity times the density is related to the following physical quantities

$$k_v \rho = (\text{number density of absorbers}) \times (\text{cross section per absorber}) = \chi_v \qquad (3.2)$$

The quantity χ_v is the opacity in units of cm^{-1} that is sometimes used in certain textbooks. It represents the inverse of the local mean free path of photons of frequency v in the stellar plasma. However, in this book the quantity k_v will be used exclusively. The opacity can be divided in two parts, one for the absorption processes (κ_v) and the other for scattering phenomena (σ_v). The total opacity then becomes

$$k_v = \kappa_v + \sigma_v \qquad (3.3)$$

The cross section of various opacity sources will be discussed throughout this book. For example, since free electrons are often present in stellar plasma, the opacity due to electron scattering is of importance in stars. The scattering cross section due to free electrons (σ_T, T standing for Thomson scattering) is independent of frequency for the relatively low photon energies of importance in stellar astrophysics and is equal to $\sigma_T = 6.6524 \times 10^{-25}$ cm². The scattering opacity (σ_e) due to Thomson scattering is thus

$$\sigma_e = \frac{n_e \sigma_T}{\rho} \qquad (3.4)$$

Beware, the variable σ is used for both cross section and scattering opacity, *be careful not to confuse these two meanings.*

Special Topic – Rayleigh Scattering

Atoms and molecules are able to scatter radiation. This process is called Rayleigh scattering. The cross section of this interaction is proportional to λ^{-4}. Therefore, blue light will be more scattered than red light. This has important implications not only for astronomy but also in everyday life. For instance, when light from stars traverses interstellar space it sometimes crosses interstellar clouds of gas. This interaction with interstellar matter reddens the light signal since, relatively speaking, more photons in the blue part of the spectrum than those in the red part will be scattered or taken away from the signal arriving at a telescope observing the signal. This interstellar reddening must be taken into account during observations.

Also, the sky is blue because more photons from the blue part of the spectrum coming from the Sun are scattered in the Earth's atmosphere than those in the red portion of the spectrum. After many scattering processes, some of the scattered photons in the blue part of the spectrum eventually arrive on our planet's surface from all over the sky, giving it its colour.

The Sun is also sometimes reddened near sunset because sunlight must then travel a longer distance in the atmosphere than at noon time for example. In other words, at sunset, sunlight grazes the surface of the Earth and has to travel a distance larger than the vertical thickness of the Earth's atmosphere to arrive at the observer. Moreover, since the atmosphere is also denser near the Earth's surface, the photons arriving at an observer from the Sun at sunset will encounter more molecules than at noon. The amount of blue light that is scattered is therefore larger at sunset than at noon. The light arriving directly from the solar disc will then look redder than at noon since a larger quantity of photons in the blue part of the spectrum is subtracted from the observed light beam at sunset. During hot and humid days, water molecules and other pollutants are more prevalent in the atmosphere and this increases the efficiency of Rayleigh scattering and thus causes the Sun to be significantly reddened at sunset.

The opacity due to atomic lines (i.e. bound–bound transitions) is of great importance for the study of stellar spectrum. Figure 3.2 shows schematically the cross section of an atomic line. The manifestation of the absorption due to line opacities can be seen in the theoretical spectrum shown in Figure 1.8. Since each ion of the various elements has its specific atomic energy levels, each of these ions will also have its specific atomic lines. The lines of each ion present in the plasma have to be included in any realistic opacity

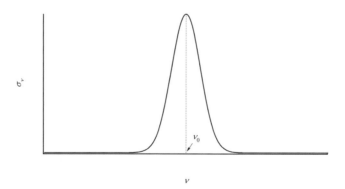

Figure 3.2 Schematic frequency dependence of the cross section per absorber (i.e. per atom in the initial energy state of the transition considered) due to an atomic line (bound–bound transition). The natural frequency of the transition is represented by v_0 (see Eq. (1.47) in Section 1.6). The width of the atomic lines is due to three broadening mechanisms (natural, Doppler and pressure broadening) that will be discussed in Section 4.3.

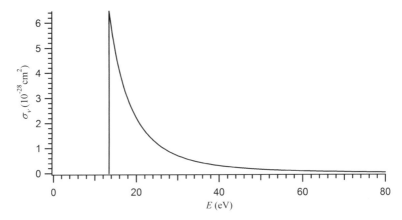

Figure 3.3   Photoionisation cross section as a function of energy of photons for the fundamental hydrogen energy state per absorber (i.e. per atom in the fundamental energy state). This cross section is taken from the Opacity Project database (Cunto, W., *et al.*, *Astronomy & Astrophysics*, 275, L5 (1993)).

calculation. In a typical stellar atmosphere calculation, tens of millions of atomic lines are normally included. The opacity due to atomic lines as well as line-broadening mechanisms that determine the exact shape of the line profiles will be discussed in more detail in Chapter 4.

At the surface of cool stars, molecular absorption must also be taken into account. In hot stars, molecules cannot exist since they are destroyed by energetic collisions or photons from the more intense radiation field.

Another important opacity source is the one due to the photoionisation of atoms. Figure 3.3 shows the photoionisation cross section from the fundamental energy state of hydrogen. For this specific case, one may notice that there is an energy threshold of 13.6 eV under which no photoionisation is possible from this atomic energy state. In other words, photons with energies below this threshold do not possess sufficient energy to free the bound electron (in the fundamental state) from the nucleus (see discussion surrounding Figure 1.5 in Section 1.5). The frequency dependence of the cross sections for the bound–free transitions of hydrogen is $1/\nu^3$. Therefore, high-energy photons possess a smaller probability of ionising an atom than those with energies just above the ionisation threshold of the atomic level under consideration. The effect of bound–free opacity from the $n = 2$ level of hydrogen (called the Balmer jump) on the radiation field can be seen in Figure 1.8. In this case, the radiative flux falls abruptly at $\lambda < 3646\,\text{Å}$ that defines the threshold energy from the $n = 2$ level.

The total bound–free opacity for hydrogen is proportional to the sum of each of its energy level's photoionisation cross section times the population of the respective level. A similar sum must be undertaken for each ion present of every element found in the plasma under consideration.

Special Topic – Negative Hydrogen Ion Opacity

Hydrogen can capture a second electron to form a negatively charge ion called H^-. This ion possesses only one bound state from which the ionisation energy is 0.754 eV. This ion can thus only exist near the surface of cooler stars (G and later spectral types). The threshold of the bound–free transition being at $\lambda \approx 16\,500\,\text{Å}$ implies that its bound–free cross section is important in the part of the spectrum where the flux of these cool stars is relatively large. This ion can then significantly contribute to the opacity not only through its bound–free transition, but also via free–free interactions. Since hydrogen is mostly neutral at the surface of these cool stars, the free electrons that intervene in H^- opacity come mostly from metals that have a lower ionisation energy than hydrogen (see Appendix D for typical values of ionisation energies). Therefore, H^- opacity is more important in stars with larger metallicities. It should be noted that a Saha equation, similar to those seen in Chapter 1, can be written for the H^- ion.

The precise calculation of the opacity spectrum k_ν depends not only on the thermodynamic state of the stellar plasma (i.e. the local values of T and ρ), but also on the abundances of the species present. Such calculations also require the knowledge of atomic and molecular data for each of these species. The local physical conditions in the plasma are crucial since they influence the population of the various atomic energy levels as well as the ionisation fractions of the ions present. Saha and Boltzmann equations must then be solved. In this chapter, the opacity (which is a critical component of the radiative-transfer equation, see below) will be assumed to be a known physical quantity. A more detailed discussion concerning opacity calculations will be presented in Chapter 4.

3.3 Specific Intensity and Radiative Moments

The radiation field in stars can be characterized by a physical quantity I_ν called specific intensity. The specific intensity defines the directional value of the radiation field. Since the specific intensity is defined per unit of solid angle, this section will begin with a reminder of this mathematical quantity.

A solid angle is the two-dimensional equivalent to an angle and its units are steradians (sr). A solid angle can be used to define direction. In spherical coordinates, the solid angle subtended between the angles θ and $\theta + d\theta$, and φ and $\varphi + d\varphi$ (or in other words the solid angle subtended by the surface dS) is

$$d\Omega = \frac{dS}{r_0^2} = \sin\theta\, d\theta\, d\varphi \tag{3.5}$$

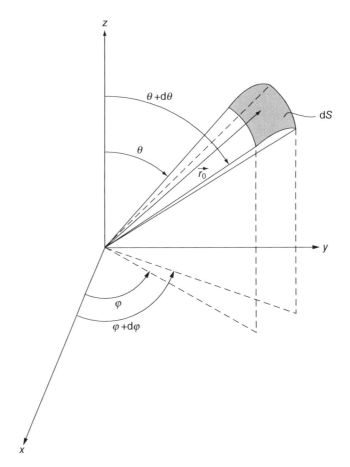

Figure 3.4 Illustration showing the solid angle subtended between the angles θ and $\theta + d\theta$, and φ and $\varphi + d\varphi$.

where dS is the surface defined on Figure 3.4. Here, r_0 is the distance from the origin of the axis to the surface dS. Contrarily to area, a solid angle is independent of r_0. In one dimension, the equivalent equation is $d\varphi = ds/r_0$, where ds is the arc length defined by the angle $d\varphi$ at radius r_0 from the origin of the axis. Please note that when $d\Omega$ is integrated over all solid angles (or directions) a value of 4π is obtained

$$\oint d\Omega = \int_0^{2\pi} d\varphi \int_0^{\pi} \sin\theta d\theta = 4\pi \tag{3.6}$$

Meanwhile, the one-dimensional equivalence of integrating $d\varphi$ gives 2π.

The specific intensity I_v (see Figure 3.5) represents the quantity of energy emanating from the surface dA (i.e. either being irradiated from or passing through this surface) per unit time in the spectral range between v and $v+dv$, travelling between the solid angles Ω and $\Omega + d\Omega$, crossing the surface $dA'=\cos\theta dA$ perpendicular to the direction defined

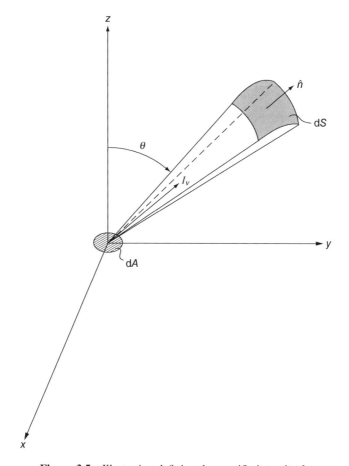

Figure 3.5   Illustration defining the specific intensity I_ν.

by the solid angle. Its units are erg/s/Hz/cm²/sr. The specific intensity defines the directional value of the radiation field. The most important example for the specific intensity is that of blackbody radiation. In this case, $I_\nu = B_\nu$, where B_ν is the specific intensity of blackbody radiation (see Chapter 1). This result is of critical importance for understanding the functioning of stars. The specific intensity per unit wavelength I_λ can also be defined.

The specific intensity $I_\nu(\vec{r}, \hat{n}, t)$ generally depends on the frequency, on the position $\vec{r}$ (i.e. the vector $\vec{r}$ may be defined as the vector between the centre of the star and the origin of the axis shown in Figure 3.5), on the direction of the unit vector $\hat{n}$ normal to the surface dS (also shown in Figure 3.5) and on time. The vector $\hat{n}$ defines the direction of the specific intensity that may also be defined by the angles θ and φ. The specific intensity can be seen as a beam of photons crossing the surface $dA'=\cos\theta dA$ in the frequency range between ν and $\nu+d\nu$ at point $\vec{r}$ going in the direction $\hat{n}$ within the solid angle $d\Omega$.

Since I_ν is a quantity given per unit solid angle, strictly speaking, it cannot be considered a vector. However, it is often useful to represent it as such in figures to more easily visu-

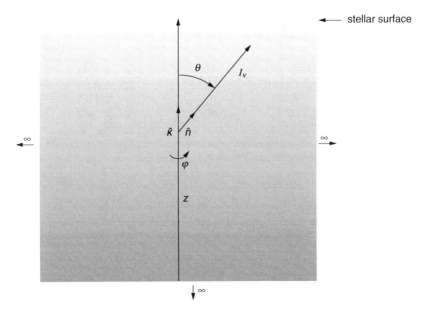

Figure 3.6   Diagram describing the plane-parallel approximation. The dimensions of the atmosphere in this approximation are infinite in all directions except $+z$ that ends at the stellar surface. In this approximation, the specific intensity is independent of φ. The grey scale represents the growing density in the atmosphere.

alise its physical character. Since the specific intensity is a quantity defined per unit solid angle, in the absence of opacity or radiative sources it is independent of distance.

Except for when a star is in a rapid phase of evolution, the specific intensity may be assumed independent of time. This will be the case for the applications presented here.

For the purposes here and in the next chapter, the specific intensity will be applied to stellar atmospheres. Since the curvature of a star can often be considered negligible in its exterior layers, the so-called plane-parallel approximation can be adopted. In this approximation, it can be supposed that the atmosphere is composed of plane-parallel layers of infinite extent in the x- and y-axis, and semi-infinite in the direction $-z$ (see Figure 3.6). All of the local physical properties (temperature, pressure, etc.) of the star then depend only on z. As will be seen later in this chapter, the radiation field at a certain depth will depend mostly on the physical properties of the layers relatively close it. A semi-infinite atmosphere is therefore a reasonable approximation of a real stellar atmosphere that evidently has a finite depth, since the very deep layers contribute negligibly to the radiation field near the surface (for more details see Sections 3.4 and 3.6).

The direction $\hat{n}$ shown in Figure 3.5 can be defined by the angles θ and φ. However, in the plane-parallel approximation and due to its symmetrical properties relative to φ, for a given value of θ the value of I_v is the same for all values of φ. The specific intensity can then be written $I_v(z, \theta)$. Another way of understanding this symmetry can be obtained by a simple mind experiment. An observer found in the atmosphere measuring the intensity of radiation for a given angle θ finds that it is independent of φ. The reason is that since the atmosphere is infinite in the *xy*-plane, the observer sees the same configuration for the atmospheric matter for any value of φ. Therefore, I_ν is independent of φ. Another variable $u = \cos\theta$ is often used instead of θ, and the specific intensity is then written as $I_\nu(z, u)$. In the next section, the radiative-transfer equation will be introduced, which when solved, gives the specific intensity. But beforehand, quantities of great importance for stars that depend on I_ν are discussed.

A physical quantity that often intervenes in stellar astrophysics is the average intensity J_ν. It is defined as the average of the specific intensity over all solid angles (or directions)

$$J_\nu(z) = \frac{1}{4\pi}\oint I_\nu(z, u)\,d\Omega = \frac{1}{4\pi}\int_0^{2\pi} d\varphi \int_0^\pi I_\nu(z, u)\sin\theta\,d\theta \tag{3.7}$$

By making the change of variable $u = \cos\theta$, and therefore $du = -\sin\theta\,d\theta$, the second integral over θ shown above becomes

$$\int_0^\pi I_\nu(z, u)\sin\theta\,d\theta = -\int_1^{-1} I_\nu(z, u)\,du = \int_{-1}^1 I_\nu(z, u)\,du \tag{3.8}$$

and since the first integral (over φ) equals 2π, the average intensity is then simply

$$J_\nu(z) = \frac{1}{2}\int_{-1}^1 I_\nu(z, u)\,du \tag{3.9}$$

The units of J_ν are the same as those of the specific intensity.

Another physical quantity that is of critical importance in stars and that can be calculated with the specific intensity is the monochromatic radiative flux F_ν, whose units are erg/s/Hz/cm² (see Sections 1.3 and 1.4). The flux is important since it is a quantity that can be measured by astronomers and can therefore be used as a quantifiable entity to study stars and other astronomical objects. The net monochromatic flux of energy across a plane-parallel surface in the outgoing direction in a star, depends on the projection of the specific intensity in the outgoing direction $\hat{k}$. The component of the specific intensity along the unitary vector $\hat{k}$ pointing in the direction $+z$, which is equal to $I_\nu\,\hat{n}\cdot\hat{k}$ must therefore be evaluated for all directions. The monochromatic flux is then given by the integration of this component of the specific intensity over all solid angles (or directions) and since the scalar product $\hat{n}\cdot\hat{k} = \cos\theta = u$ the monochromatic flux is given by the expression

$$F_\nu(z) = \oint I_\nu(z, u)\hat{n}\cdot\hat{k}\,d\Omega = \oint I_\nu(z, u)\cos\theta\,d\Omega = \int_0^{2\pi} d\varphi \int_0^\pi I_\nu(z, u)u\sin\theta\,d\theta = 2\pi\int_{-1}^1 I_\nu(z, u)u\,du \tag{3.10}$$

It should be noted that for the case when the specific intensity is isotropic, the radiative flux is nil since the quantity of radiative energy crossing a surface is the same for each opposing direction. More generally, when $I_\nu(z, u)$ is an even function with respect to u, the flux is zero since the function u found in the integral above is odd. The physical meaning of this result is that an equal amount of energy is going in the $+z$ and in the $-z$ directions. In a stellar atmosphere, the specific intensity is not isotropic since the temperature generally increases with depth. At a given depth, more energy is irradiated upwards from the layers below this depth than the energy irradiated downwards from layers above. This is due to the general increase of temperature with depth in stars. The specific intensity for $u > 0$ is then larger than for $u < 0$ and a net radiative flux exists in the outgoing direction.

Example 3.1: In Chapter 1, it was mentioned that the monochromatic flux emanating from the surface of a blackbody is $F_v = \pi B_v$. Prove it.

Answer:

By definition, the specific intensity at the surface of a blackbody is equal to B_v for $0 \le u \le 1$ and is nil for $-1 \le u < 0$. The monochromatic flux is then

$$F_v = 2\pi \int_{-1}^{1} I_v u \, du = 2\pi \int_{0}^{1} B_v u \, du = 2\pi B_v \left[\frac{u^2}{2} \right]_0^1 = \pi B_v \qquad (3.11)$$

Example 3.2: If a star possesses a specific intensity of the form $I_v (z, u) = a_v (z) + b_v (z)u$, where the functions $a_v(z)$ and $b_v(z)$ are independent of u, calculate $F_v(z)$.

Answer:

By definition, the monochromatic flux is

$$F_v(z) = 2\pi \int_{-1}^{1} I_v(z, u) u \, du = 2\pi \int_{-1}^{1} [a_v(z) + b_v(z)u] u \, du$$

$$= 2\pi \left[\frac{a_v(z) u^2}{2} + \frac{b_v(z) u^3}{3} \right]_{-1}^{1} = \frac{4\pi}{3} b_v(z) \qquad (3.12)$$

The flux in this case only depends on the second term of the expression for $I_v (z, u)$ because this term is an odd function with respect to u. The first term does not contribute because it is an even function with respect to u.

As seen in Chapter 1, and by definition of the effective temperature, the integrated radiative flux F at the surface of a star is

$$F = \int_{0}^{\infty} F_v \, dv = \sigma T_{\mathrm{eff}}^4 \qquad (3.13)$$

In the plane-parallel atmosphere and assuming that the energy is solely transported by radiation (and not by the other two modes of energy transport: convection and conduction), the integrated radiative flux is constant throughout the atmosphere. This is true assuming that the star has reached equilibrium so that each layer irradiates as much energy as it absorbs. It also assumes that there are no energy sources in the atmosphere (i.e. no nuclear fusion) which is the case since the temperature there is too low for nuclear burning.

As will be discussed in Section 3.7 and in the next chapter, the fact that the integrated radiative flux is constant as a function of depth in the case where all of the energy is transported by radiation in a plane-parallel atmosphere can be useful, especially when calculating model atmospheres. However, in spherical stellar models, the integrated flux first increases with depth since luminosity there is constant and is distributed on a smaller surface area (see Eq. 1.11 for instance). This is true up to the stellar core (where nuclear energy production takes place) since the luminosity decreases there as the centre is approached. More details concerning this property will be discussed in Chapter 5.

It should, however, be noted that even if the integrated radiative flux is constant with depth in the case mentioned above, the monochromatic flux is not. In deeper and therefore hotter layers, the monochromatic flux distribution changes with depth due to Wien's law. Also, since temperature varies with depth, so do ionisation fractions and the opacity spectrum. These changes lead to modifications in the frequency dependence of the monochromatic flux as a function of depth. The atomic lines in the spectra of various ions appear (or disappear) as the population of these ions change.

Once outside the star and supposing no more interaction with matter, the frequency distribution of the monochromatic flux will stay constant, but its intensity will decrease as $1/d^2$ due to geometrical dilution, where d is the distance from the observer to the star.

More generally, moments of order n of the radiative field $M_v(z, n)$ may also be defined as

$$M_v(z, n) = \frac{1}{2} \int_{-1}^{1} I_v(z, n) u^n du \tag{3.14}$$

and thus,

$$M_v(z, 0) = J_v(z) \tag{3.15}$$

$$M_v(z, 1) = H_v(z) = \frac{F_v(z)}{4\pi} \tag{3.16}$$

$$M_v(z, 2) = K_v(z) \tag{3.17}$$

The moment of order $n = 0$ is simply J_v. The moment of order $n = 1$ is called the Eddington flux H_v. A simple relation exists between the Eddington flux and the true flux

$$F_v = 4\pi H_v \tag{3.18}$$

The moment of order $n = 2$, represented by K_v, and sometimes called the K-integral, is related to radiative pressure and will be discussed in (the optional) Section 3.12.

Special Topic – Radiative Energy Density

With the knowledge of the specific intensity, it is possible to evaluate the energy density of the associated radiation field. The energy crossing a surface dA from the specific intensity arriving at an angle θ with respect to the normal of that surface (see Figure 3.7), within an infinitesimal solid angle $d\Omega$, time dt and frequency range $d\nu$ is

$$dE_\nu = I_\nu \cos\theta dA d\Omega d\nu dt \tag{3.19}$$

During the time dt, the volume dV occupied by the radiation from the beam under consideration is $dA\cos\theta cdt$ (see Figure 3.7). It is then trivial to write the equation above in terms of dV

$$dE_\nu = \frac{I_\nu dV d\Omega d\nu}{c} \tag{3.20}$$

When dV is infinitesimally small, the specific intensity inside this volume may be assumed to be constant. The energy density U_ν within the frequency interval $d\nu$ can therefore be written as the following (by using the definition of J_ν given in Eq. 3.7) where the sum from the radiation coming from all directions is undertaken (i.e. the same argument made to find the energy in an infinitesimal volume for a given angle can be repeated for all directions)

$$U_\nu = \frac{1}{c}\oint I_\nu d\Omega = \frac{4\pi}{c}J_\nu \tag{3.21}$$

The energy density U_ν at a given point is therefore proportional to the average intensity J_ν of the field there.

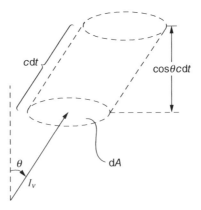

Figure 3.7 Illustration showing a beam of radiation crossing a surface dA. During the time interval dt, the volume $dV = dA\cos\theta cdt$ is filled with radiation from the specific intensity in the given direction that crosses the surface dA.

3.4 Radiative Transfer Equation

As seen in the previous section, to calculate the radiative flux of a star (Eq. 3.10), the specific intensity must be determined. To do this, one must solve a differential equation called the radiative-transfer equation. In this section, this equation will be introduced and each of its terms will be explained. Its demonstration, which is not of critical importance at this stage, is given in (the optional) Section 3.11. The reader who wishes to cover that section may read it before continuing.

For the discussion here, the plane-parallel stellar atmosphere approximation will be used. Figure 3.8 shows schematically a beam of photons specified by I_v as it crosses a slab of matter of thickness dz. As the beam crosses the layer in question, some photons are absorbed or scattered by the particles of matter present. Meanwhile, others photons are added to I_v since matter can also emit photons. The equation of radiative transfer is the following

$$\frac{u}{\rho} \frac{dI_v(z,u)}{dz} = \underset{\uparrow}{-k_v I_v(z,u)} + \underset{\uparrow}{j_v}$$

$$\underset{\text{variation of } I_v}{\uparrow} \qquad \qquad \underset{\text{radiation added to } I_v}{\uparrow}$$

(with "radiation subtracted from I_v" labelling the $-k_v I_v(z,u)$ term) (3.22)

The term on the left-hand side of this equation represents the variation to I_v as it crosses the slab of stellar plasma of thickness dz. The variable u is present because, in reality, the beam travels a distance dz/u. The density is present for the simple reason that the opacity per unit mass k_v is employed here. The first term on the right-hand side of the equation represents the radiation subtracted from the beam due to absorption or scattering processes, explaining the minus sign found there. It is proportional to I_v because the quantity of energy absorbed or scattered increases as the intensity of the beam increases. The last term j_v is called the emissivity of matter and represents the radiation added to I_v as it crosses the slab of matter. Its units are erg/Hz/s/ster/g.

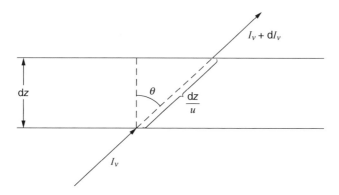

Figure 3.8 Illustration of a beam of photons characterized by I_v that crosses a slab of plasma of thickness dz at an angle θ and emerges as $I_v + dI_v$. The distance travelled through the slab is dz/u.

In order to better comprehend its importance, the emissivity can be approximated by the following expression

$$j_v \approx \kappa_v B_v + \sigma_v J_v \tag{3.23}$$

where κ_v is the opacity due to absorption and σ_v is the opacity due to scattering. The first term of this expression is related to the radiation that is emitted by matter, which is set equal to the quantity of energy absorbed by matter. This explains why it is equal to $\kappa_v B_v$, where for this approximation the radiation field is simply assumed to be related to the Planck distribution. Meanwhile, the second term is related to the quantity of energy scattered. It depends on the scattering cross section and on the average intensity. This is a simplified view of emissivity, but it is instructive and can lead to a better understanding of the interaction between matter and radiation. It will be used below to gain insight into another important physical quantity intervening in radiative transfer called the source function.

Example 3.3: Suppose that a single beam of radiation with specific intensity I_v^0 arrives on the surface of a cold cloud of gas with no emissivity (see Figure 3.9). The cloud has a density ρ and an opacity k_v both constant within it. Find $I_v(z)$ (since a single direction is studied here, u is omitted from the arguments of I_v).

Answer:

Since for the beam in question $u = 1$ and that no emissivity exists in the cloud under consideration ($j_v = 0$), the radiative-transfer equation becomes

$$\frac{dI_v}{dz} = -k_v \rho I_v \tag{3.24}$$

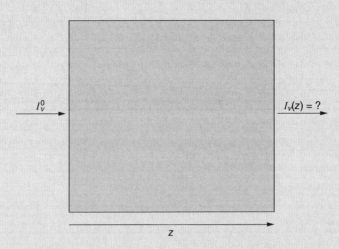

Figure 3.9   Illustration of the problem discussed in Example 3.3.

This equation may be directly integrated

$$\int_{I_\nu^0}^{I_\nu(z)} \frac{dI_\nu}{I_\nu} = -\int_0^z k_\nu \rho \, dz \tag{3.25}$$

and since k_ν and ρ are constants

$$\ln(I_\nu(z)) - \ln(I_\nu^0) = -k_\nu \rho z \tag{3.26}$$

which can be expressed as

$$I_\nu(z) = I_\nu^0 e^{-k_\nu \rho z} \tag{3.27}$$

This is an idealized situation since emissivity is always present in plasma. By assuming that Eq. (3.23) represents emissivity, it is negligible for very low temperatures or at frequencies where B_ν is very weak and when no scattering is present. However, the result found here is instructive to better understanding radiative transfer.

In this last example, the specific intensity was found to decrease (or to be extinct) exponentially with respect to z. This extinction increases as the opacity increases. The result found above is very useful to understand how radiation is transmitted through plasma. Of course, if emissivity would have been present, the dependence of the specific intensity as a function of depth would have been more complicated than the expression found above. This is due to the fact that in reality, as radiation is transported through a given layer in a star, photons are absorbed or scattered while others are emitted by the stellar plasma.

Writing the specific intensity as a function of z might at first seem more natural, but it isn't very useful. As seen in the last example, the specific intensity diminishes not only as a function of z, but rather as a function of $k_\nu \rho z$. A quantity defined by this product is a more natural variable to employ when studying radiative transfer in stars. The radiative-transfer equation may be rewritten by dividing it by $-k_\nu$

$$\frac{u}{-k_\nu \rho} \frac{dI_\nu(z, u)}{dz} = I_\nu(z, u) - \frac{j_\nu}{k_\nu} \tag{3.28}$$

A new variable τ_ν, called the optical depth, can be defined such that

$$d\tau_\nu = -k_\nu \rho \, dz \tag{3.29}$$

The radiative-transfer equation then becomes

$$u \frac{dI_\nu(\tau_\nu, u)}{d\tau_\nu} = I_\nu(\tau_\nu, u) - \frac{j_\nu}{k_\nu} \tag{3.30}$$

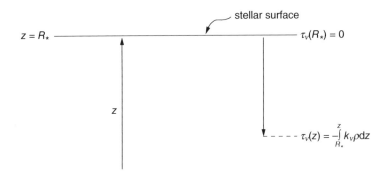

Figure 3.10   Illustration of the concept of optical depth with respect to the stellar surface. The integration to obtain $\tau_\nu(z)$ is performed from the surface to the depth z.

Note that the specific intensity is now written as a function of optical depth instead of z. Figure 3.10 illustrates the optical depth and its relation to z which is

$$\tau_\nu(z) = -\int_{R_*}^{z} k_\nu \rho \, dz \tag{3.31}$$

The optical depth is nil at the surface of the star, and increases with geometrical depth. However, equivalently to distance the optical depth between any two points in a star can also be calculated.

The optical depth is a measure of the opaqueness of a medium at a given frequency. It possesses no physical units. It is a variable with greater physical significance than z when studying radiative transfer. An optical depth scale may be constructed for any frequency. However, in stellar modelling it is customary to define a standard optical depth scale at a single wavelength (at 5000 Å for instance) or related to an average opacity (such as the Rosseland mean opacity that will be defined in Section 3.9). Each value of the standard optical depth scale (τ_{std}) chosen corresponds to a value of z. To obtain a representative monochromatic flux spectrum, the following radiative-transfer equation must be solved for a given number of frequencies and values of u

$$u \frac{dI_\nu(\tau_{std}, u)}{d\tau_{std}} = I_\nu(\tau_{std}, u) - \frac{j_\nu}{k_\nu} \tag{3.32}$$

More details surrounding such calculations will be given in Chapter 4.

Outer stellar regions with optical depths much smaller than unity are commonly called optically thin regions, while deeper regions with optical depths far greater than unity are called optically thick regions.

The solution found in Example 3.3 can then be written $I_\nu(z) = I_\nu^0 e^{-\tau_\nu}$ and as a consequence, the original signal is attenuated by a factor e each time the optical depth increases by one unit. From this result, the reader may now better comprehend why the radiation field emanating from stars comes from the external layers, i.e. those limited to optical depths of less than several units. This is also the reason why spectral classes seen in Chapter 1 depend on the surface temperature of stars. Photons coming from deep regions of the atmosphere are extinct before they can reach the surface. However, this does not mean that the energy is not being transported to the surface. The energy of the photons created by nuclear fusion in the central regions is transported to the surface from layer to layer by photons of generally decreasing frequencies (due to Wien's law) as the surface is approached. At a given layer, the frequency distribution of the radiation field is related to the temperatures of the surrounding medium and is thus affected by the transport of energy. In Section 3.6, the specific intensity at the surface of stars will be found to be a weighted sum of the source function $S_v(\tau_v)$ being defined as

$$S_v(\tau_v) = \frac{j_v}{k_v} \qquad (3.33)$$

Since neither the opacity nor the emissivity depends on direction, the source function is a function solely of depth (and of course frequency).

The radiative-transfer equation can then be written

$$u \frac{dI_v(\tau_v, u)}{d\tau_v} = I_v(\tau_v, u) - S_v(\tau_v) \qquad (3.34)$$

By using the approximation for the emissivity given above (Eq. 3.23), the source function can be approximated by

$$S_v \approx \frac{\kappa_v B_v}{\kappa_v + \sigma_v} + \frac{\sigma_v J_v}{\kappa_v + \sigma_v} \qquad (3.35)$$

Therefore, when scattering processes are negligible $S_v \approx B_v$. Such an approximation can be very useful to gain insight into the radiative-transport process in stars. The knowledge of the true source function requires complex calculations. Such calculations are outside the scope of this book and will not be discussed here. However, the physical importance of the source function will become clearer in Section 3.6, where the transfer equation will be solved.

3.5 Local Thermodynamic Equilibrium

In this section, the concept of local thermodynamic equilibrium (LTE) is discussed. The LTE concept comes in many forms. The first form and the weakest, simply supposes that the atomic energy levels and ionic populations are determined solely by collisions in the plasma. Since the temperature increases with depth in stars, it is then not trivial to gauge the relevance of the local temperature. If the mean free path of the particles in the plasma is small compared to the distance scale on which the temperature changes within the system, the local temperature can be directly associated to the thermodynamic properties of the local plasma. This local value for the temperature may then be used to obtain the velocity distribution of the particles via the Maxwell distribution. The atomic-energy-level and ionic populations can then be obtained, respectively, with the Boltzmann and Saha equations using the local thermodynamic conditions (T and n_e).

However, under certain conditions sometimes found in stars, radiative processes (photoexcitations and photoionisations) can become important and must be taken into account when calculating atomic-energy-level and ionic populations. Such calculations are commonly called non-LTE or NLTE. These computations are complicated since detailed rate equations taking into account the excitations and ionisations due to the radiation field come into play. NLTE effects can be important in the outer regions of stars, because collisional processes are weaker due to the relatively lower densities present there.

A second and more stringent definition of LTE consists of assuming that matter emits radiation such that the source function is $S_v = B_v$. As shown above, this result is also obtained when using Eq. (3.35) and when scattering processes are negligible. The assumption that the source function is equal to the Planck distribution is very useful for understanding radiative transfer in stars. For example, this approximation will be used in the next section for explaining the relative importance of the contribution of the various depths to the emerging radiative flux at the surface of stars.

A third and even stricter definition of LTE is assuming that the specific intensity is equal to the Planck distribution ($I_v = B_v$), where the Planck function is determined by using the local temperature. This approximation also leads to $S_v = B_v$.

It should be noted that the last two definition of LTE encompass the supposition made in the first definition described at the beginning of this section, i.e. that the various populations can be calculated with the use of the local physical conditions with the Boltzmann and Saha equations.

3.6   Solution of the Radiative-Transfer Equation

To evaluate the radiative flux present at a given depth inside a star and the flux emerging at its surface, one must solve the radiative-transfer equation in order to obtain the specific intensity. The aim of this section is to find a general solution to the radiative-transfer equation. To achieve this goal, the radiative-transfer equation can be multiplied by the integration factor $e^{-\frac{\tau_v}{u}}$

$$u\frac{dI_v(\tau_v,u)}{d\tau_v}e^{-\frac{\tau_v}{u}} = I_v(\tau_v,u)e^{-\frac{\tau_v}{u}} - S_v(\tau_v)e^{-\frac{\tau_v}{u}} \tag{3.36}$$

The terms found in this equation may be rearranged by dividing it by u and by transferring the first term on the right-hand side of the equation to the left-hand side

$$\frac{dI_v(\tau_v,u)}{d\tau_v}e^{-\frac{\tau_v}{u}} - \frac{I_v(\tau_v,u)e^{-\frac{\tau_v}{u}}}{u} = -\frac{S_v(\tau_v)e^{-\frac{\tau_v}{u}}}{u} \tag{3.37}$$

The two terms on the left-hand side may be written as a derivative of $I_v(\tau_v, u)e^{-\frac{\tau_v}{u}}$ with respect to τ_v

$$\frac{d\left(I_v(\tau_v, u)e^{-\frac{\tau_v}{u}}\right)}{d\tau_v} = -\frac{S_v(\tau_v)e^{-\frac{\tau_v}{u}}}{u} \tag{3.38}$$

From a purely mathematical point of view, this equation may be integrated between two points (or depths) within the star, $\tau_{v,1}$ and $\tau_{v,2}$, where the subscripts 1 and 2 define these two depths

$$\int_{\tau_{v,1}}^{\tau_{v,2}} d\left(I_v(t, u)e^{-\frac{t}{u}}\right) = -\int_{\tau_{v,1}}^{\tau_{v,2}} S_v(t)e^{-\frac{t}{u}}\frac{dt}{u} \tag{3.39}$$

In order to avoid confusion, the mute variable now employed in the integral is t instead of τ_v. Integration of the left-hand side of this equation leads to

$$I_v(\tau_{v,2}, u)e^{-\frac{\tau_{v,2}}{u}} - I_v(\tau_{v,1}, u)e^{-\frac{\tau_{v,1}}{u}} = -\int_{\tau_{v,1}}^{\tau_{v,2}} S_v(t)e^{-\frac{t}{u}}\frac{dt}{u} \tag{3.40}$$

and by rearranging the various terms this equation becomes

$$I_v(\tau_{v,1}, u) = I_v(\tau_{v,2}, u)e^{\frac{\tau_{v,1}-\tau_{v,2}}{u}} + \int_{\tau_{v,1}}^{\tau_{v,2}} S_v(t)e^{\frac{\tau_{v,1}-t}{u}}\frac{dt}{u} \tag{3.41}$$

Figure 3.11 will aid in the physical interpretation of this general solution. The solutions for the outgoing and ingoing directions will be discussed separately. For the outgoing $(u \geq 0)$ directions (see Figure 3.11(a)), the term $I_v(\tau_{v,2}, u)e^{\frac{\tau_{v,1}-\tau_{v,2}}{u}}$ is the residual value of

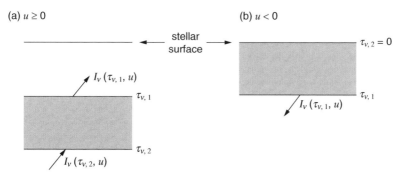

(a) $u \geq 0$ (b) $u < 0$

Figure 3.11 Illustrative explanation of the general solution for the radiative-transfer equation for the outgoing (a) and ingoing (b) directions. The shaded areas represent the portion of the star that contributes to the intensity at depth $\tau_{v,1}$.

the specific intensity emanating from the depth $\tau_{v,2}$ (i.e. the intensity due to depths larger than $\tau_{v,2}$) that remains at $\tau_{v,1}$ following its partial extinction while travelling between the two depths in question, which explains the term $e^{\frac{\tau_{v,1}-\tau_{v,2}}{u}}$. The integral in this equation, which is a weighed sum of the source function, is related to the radiation emitted by the medium found between the depths $\tau_{v,1}$ and $\tau_{v,2}$ (the shaded area in Figure 3.11(a)). In other words, the radiation travelling in the outgoing directions ($u \geq 0$) at depth $\tau_{v,1}$ is due to the radiation coming from the layers below this depth. For a semi-infinite atmosphere, $\tau_{v,2} = \infty$ and the value of the intensity at $\tau_{v,2}$ is therefore not necessary since $e^{\frac{\tau_{v,1}-\tau_{v,2}}{u}} \to 0$. The specific intensity in the outgoing direction ($u \geq 0$) can then be written (see Figure 3.11(a))

$$I_v(\tau_v, u) = \int_{\tau_v}^{\infty} S_v(t) e^{\frac{\tau_v - t}{u}} \frac{dt}{u} \quad \text{(for } u \geq 0\text{)} \tag{3.42}$$

For the ingoing ($u < 0$) specific intensity, the radiation is due to the matter above $\tau_{v,1}$ (the shaded area in Figure 3.11(b)). For this case (Eq. 3.41), $\tau_{v,2} = 0$ and $I_v(\tau_{v,2} u) = 0$ when assuming that there is no incoming radiation on the stellar surface from exterior sources such as other stars.[2] The specific intensity for this case is therefore

$$I_v(\tau_v, u) = -\int_{0}^{\tau_v} S_v(t) e^{\frac{\tau_v - t}{u}} \frac{dt}{u} \quad \text{(for } u < 0\text{)} \tag{3.43}$$

The minus sign in front of the integral above assures that the specific intensity is positive since u is negative. These last two results enable the calculation of the specific intensity (and thus the flux) at all depths, providing the source function is known.

However, it should be noted that the resolution of the two equations shown above in a real star can be very complex since the source function depends on the emissivity of matter, which itself depends on the radiative field. The solution of these equations is therefore far more complicated than a simple integration. More insight may be obtained about this complexity in the (advanced optional) Section 3.10.

An important application of the equations found above is related to the radiation field at the surface of stars since it offers a lot of physical insight about the stellar layers that contribute to the formation of this field. For a semi-infinite atmosphere, the specific intensity at the surface can be obtained from Eq. (3.42) by knowing that for this case $\tau_v = 0$. This leads to the following expression

$$I_v(0, u) = \int_{0}^{\infty} S_v(t) e^{-\frac{t}{u}} \frac{dt}{u} \quad (u \geq 0) \tag{3.44}$$

---

[2]In a binary star system for instance, this approximation may not be valid since incoming radiation from the companion is not necessarily negligible. Such a process may cause heating on the portion of the star facing its companion and along with rotation can lead to it becoming a variable star (see Chapter 5).

The specific intensity at the surface is thus a weighted sum of the source function through the atmosphere. If the source function is approximated by the Planck function ($S_v \approx B_v$), the specific intensity at the surface then becomes a weighted sum of the Planck functions coming from the various depths. Also, since the term inside the integral above is weighted by the factor $e^{-\frac{\tau_v}{u}}$, only the superficial layers (those with optical depths of less than or equal to the order of unity) contribute significantly to the radiation field emanating from the stellar surface. That fact is fundamental as to why it is solely the temperatures of the surface layers that determine the spectral type of stars: the photons exiting a star come from layers with optical depths of less than a few units.

Since the opacity depends on frequency, radiation at frequencies with relatively large opacities will come from shallower layers (in z) than radiation at frequencies with lower opacities. For instance, the radiation field at the central (or natural) frequency of an atomic line, will be formed in (or will come from) shallower layers than the photons found in the wings of the line. The wings of an atomic line are defined as the parts of the line profile found farther away from, and on either side of, the natural frequency.

Also, note that even though there is a u^{-1} term in the integral above, it does not diverge for $u = 0$ (or $\theta = \pi/2$). The reason being that when u tends toward zero, the term $e^{-\frac{t}{u}}$ tends towards zero faster than the u^{-1} term diverges to infinity.

Example 3.4: Calculate the specific intensity at the surface of an isothermal gas cloud (with known temperature) with the assumption $S_v = B_v$, where B_v is the Planck function associated to the temperature of the cloud. Use the semi-infinite atmosphere approximation and assume that the cloud has a very large optical depth.

Answer:

Assuming no radiation source exists outside this cloud, the specific intensity is nil for the ingoing direction. For the outgoing direction the intensity is formally

$$I_v(0, u) = \int_0^\infty S_v(t) e^{-\frac{t}{u}} \frac{dt}{u} \quad (u \geq 0) \tag{3.45}$$

Since the cloud is isothermal, the source function is independent of depth and it may be taken out of the integral above

$$I_v(0, u) = B_v \int_0^\infty e^{-\frac{t}{u}} \frac{dt}{u} = B_v \left[-e^{-\frac{t}{u}} \right]_0^\infty = B_v \quad (u \geq 0) \tag{3.46}$$

Here, since the cloud is optically thick it was assumed that the integral above could be summed to infinity. The specific intensity for an optically thick isothermal gas is therefore simply equal to the Planck function.

Example 3.5: Assuming a given stellar atmosphere possesses a source function $S_\nu(\tau_\nu) = a_\nu + b_\nu\tau_\nu$, where a_ν and b_ν are functions of frequency, calculate the specific intensity at its surface.

Answer:

Assuming no radiation source exists outside this star, the specific intensity is nil for the ingoing direction. For the outgoing direction the intensity is formally

$$I_\nu(0,u) = \int_0^\infty S_\nu(t)e^{-\frac{t}{u}}\frac{dt}{u} \quad (u \ge 0) \tag{3.47}$$

then

$$I_\nu(0,u) = \int_0^\infty a_\nu e^{-\frac{t}{u}}\frac{dt}{u} + \int_0^\infty b_\nu t e^{-\frac{t}{u}}\frac{dt}{u} \tag{3.48}$$

$$I_\nu(0,u) = \frac{a_\nu}{u}\left[-ue^{-\frac{t}{u}}\right]_0^\infty + \frac{b_\nu}{u}\left[-ue^{-\frac{t}{u}}(t+u)\right]_0^\infty = a_\nu + b_\nu u \tag{3.49}$$

Special Topic – The Sun's Corona and the Solar Cycle

The solar corona is an exterior region of the Sun that reaches several million kilometres outside the surface (the surface of the Sun for the purposes here may be defined as where $\tau \approx 1$). It is composed of a very tenuous and hot gas that has a temperature of up to one to two million degrees. This gas is heated by nonthermal processes. A possible origin for heating the corona comes from a phenomenon called magnetic-line reconnection. Magnetic lines at the surface of the sun are continuously rearranged so as to avoid physically forbidden magnetic line intersections, and this process is believed to heat the corona.

One could think that such a hot gas would dominate the radiative flux emerging from the Sun. However, since the density of the gas composing the solar corona is extremely weak (and so is its optical depth) this component of the Sun does not contribute much to the solar flux (except at very high frequencies such as X-rays). Equation (3.44) shows that the contribution from very weak optical depths are not as important as those from optical depths of values on the order of unity (i.e. the atmosphere of the Sun). The intensity of the radiation coming from the corona is negligible as compared to the radiation coming from the solar disk, which renders the corona invisible in most wavelengths. However, the corona can be seen during

Figure 3.12   The solar corona as seen during a total solar eclipse (NASA/courtesy of nasaimages.org). (see colour plate.)

a total solar eclipse (see Figure 3.12), since the radiation coming from the photosphere is blocked by the Moon. The corona then becomes visible as compared to the dark sky.

The corona can also be observed with an instrument conceived by the French astronomer Bernard Lyot (1897–1952) called a coronagraph. A coronagraph is an optical system in which a disk obscures the solar surface and thus mimics a solar eclipse. Figure 3.13 shows a picture taken while using a coronagraph to which is superimposed a photo of the Sun's surface. Solar prominences can be observed near the surface. These structures are due to magnetic field lines (or loops) that extend outside the solar surface. They are formed of charge particles (mainly hydrogen) following these extruding magnetic lines.

Dark spots, called sunspots, can also be observed on the surface of the Sun (see Figure 3.14). Sunspots are related to regions of more intense magnetic fields. They are darker than the surrounding regions due to their lower temperature.[3] This temperature drop (up to approximately 1500 K) is related to the presence of large magnetic fields in sunspots. Magnetic fields can contribute to the total pressure via the so-called magnetic pressure. Therefore, the contribution of the gas pressure to the total pressure diminishes and leads to a lower temperature. Sunspots are transient features that last from approximately a day up to several months. They also appear in pairs (of opposite magnetic polarity) caused by the ingoing and outgoing magnetic lines from the solar surface.

---

[3] In reality, sunspots are not dark since their temperature is approximately 4500 K or more. They appear dark at the surface of the Sun because of the contrast with the brighter (hotter) surrounding regions.

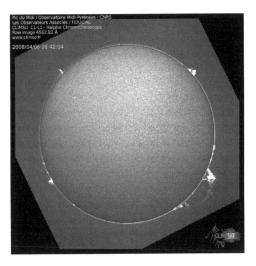

Figure 3.13 A picture of the solar corona taken with a coronagraph, superimposed on a picture of the solar surface, using a H_α filter (photo courtesy of Jacques-Clair Noëns and the Associated Observers on the instrument CLIMSO from the Observatoire Midi-Pyrénées).

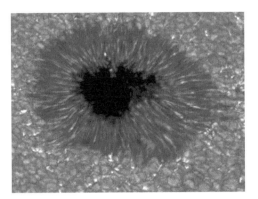

Figure 3.14 Photo of a sunspot on the solar surface. The darker central region is called the umbra, while the lighter region is called the penumbra where the magnetic field is relatively smaller than in the umbra. Also seen on this picture are hundreds of granules. These granules measure about 1000 km and are manifestations of convective cells of matter related to convective energy transport (NASA/courtesy of nasaimages.org).

The number of sunspots and their position relative to the equator is a periodic phenomenon with a period of approximately 11 years that is commonly called the solar cycle. This is due to the inversion of the polarity of the Sun's magnetic field over that period. During the solar cycle, the sunspots migrate towards the solar equator (see Figure 3.15) and the number of sunspots on the solar surface also varies

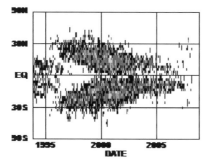

Figure 3.15   Figure showing the migration of sunspots towards the equator during the solar cycle (NASA/courtesy of nasaimages.org).

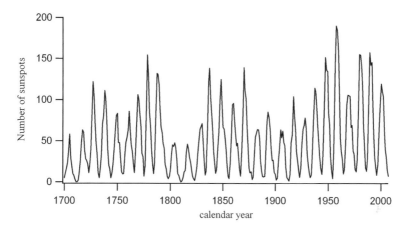

Figure 3.16   The number of sunspots observed over the last three centuries.

with time with a period of 11 years (see Figure 3.16). The sunspots follow the magnetic lines that are distorted by the differential rotation of the Sun (i.e. the equator turns more rapidly than upper latitudes). The generally dipolar magnetic field present at the beginning of a solar cycle is progressively transformed into a toroidal field to eventually regenerate into a generally dipolar field of opposite polarity. Figures showing the migration of the sunspots over time such as Figure 3.15 are often called butterfly diagrams since the shape of the sunspot distribution in these figures resembles the wings of a butterfly. They are also sometimes called Maunder's butterfly diagrams that are named after the British astronomer Edward W. Maunder (1851–1928). The period for a full cycle during which the solar magnetic field polarity returns to its original configuration is approximately 22 years.

   Another structure that is visible in Figure 3.14 is that of granules. Granules are visual manifestations of the presence of rising and sinking cells of plasma related to the convective transport process taking place in the outer region of the Sun.

3.7 Radiative Equilibrium

In Chapter 1, three modes of energy transport were briefly discussed: radiative, convective and conductive transport. Radiative equilibrium is a state defined as when all of the energy is transported by radiation. Of course, this state is not achieved in all regions or at all evolutionary stages of stars, since for example convection can sometimes be present (i.e. as in certain parts of the Sun). It should be mentioned that in stellar atmospheres, conductive transport is never important. Convection and conduction will be seen in more detail when discussing stellar interiors in Chapter 5.

Since there exists no energy sources or sinks in a stellar atmosphere, the amount of energy absorbed by a given layer is equal to the energy that it emits. In the plane-parallel approximation, this leads to a constant integrated flux throughout the atmosphere since the energy emitted is distributed over the same area independently of depth. By definition of the effective temperature, the integrated flux is $F = \sigma T_{\text{eff}}^4$ for all depths (see Section 1.4).

To extract information for an atmosphere in the state of radiative equilibrium, the radiative-transfer equation

$$\frac{u}{\rho}\frac{dI_v}{dz} = -k_v I_v + j_v \tag{3.50}$$

may be integrated over all frequencies and solid angles

$$\frac{1}{\rho}\frac{d\left[\int_0^\infty dv \oint u I_v d\Omega\right]}{dz} = \int_0^\infty dv \oint (-k_v I_v + j_v) d\Omega \tag{3.51}$$

Since the integrated flux

$$F = \int_0^\infty dv \oint u I_v d\Omega \tag{3.52}$$

is constant throughout the atmosphere, the left hand-side of Eq. (3.51) in nil, and this equation may then be written

$$\int_0^\infty dv \oint k_v I_v d\Omega = \int_0^\infty dv \oint j_v d\Omega \tag{3.53}$$

The left-hand side of this equation represents the total energy absorbed or scattered by matter and the right-hand side represents the energy that it radiates. A simple integration over all solid angles, and using the definitions of J_v and S_v given earlier, leads to the following equation

$$\int_0^\infty k_v J_v dv = \int_0^\infty k_v S_v dv \tag{3.54}$$

If the source function is approximated by the following expression discussed in Section 3.4

$$S_v \approx \frac{\kappa_v B_v}{\kappa_v + \sigma_v} + \frac{\sigma_v J_v}{\kappa_v + \sigma_v} \tag{3.55}$$

the integral above can be simplified to give

$$\int_0^\infty \kappa_v J_v dv = \int_0^\infty \kappa_v B_v dv \tag{3.56}$$

or

$$\int_0^\infty \kappa_v (J_v - B_v) dv = 0 \tag{3.57}$$

Only the absorption opacities (κ_v) are present in these last two equations. Scattering processes (σ_v) are absent because no photons are destroyed or created by this process. This equation shows that the quantity of energy absorbed by the medium is equal to the amount it emits. But, it also has a deeper physical significance. This equation demonstrates that the local properties of the stellar plasma, defined by the local Planck function B_v, depend on more global (or nonlocal) properties defined by the average intensity J_v. The value of J_v is the directional average of the specific intensity, which itself is a weighted sum of the source functions of the surrounding area and therefore J_v is nonlocal in nature. As will be seen in (the advanced optional) Section 4.5.4, this equation can lead to a method for correcting the temperature during the modelling of a stellar atmosphere.

3.8 Radiative Transfer at Large Optical Depths

The aim of this section is to estimate the specific intensity and its moments at large optical depths ($\tau_v \gg 1$), or in other words, in the interior regions of stars. The results found here will be helpful to better understand radiative-energy transport in stellar interiors. This topic will be discussed in Chapter 5.

At large optical depths, the mean free path of photons is relatively small and therefore the radiative field depends mostly on the local conditions. In this context, the source function at each depth can be approximated by the Planck function calculated at the local temperature, i.e. $S_v(\tau_v) \approx B_v(\tau_v)$. The source function at a depth t can then be written as a Taylor series of the function B_v around the depth τ_v

$$S_v(t) = B_v(\tau_v) + (t - \tau_v)\frac{dB_v}{d\tau_v}\bigg|_{\tau_v} + \frac{(t - \tau_v)^2}{2!}\frac{d^2 B_v}{d\tau_v^2}\bigg|_{\tau_v} + \dots = \sum_{n=0}^\infty \frac{(t - \tau_v)^n}{n!}\frac{d^n B_v}{d\tau_v^n}\bigg|_{\tau_v} \tag{3.58}$$

Using this expression for the source function leads to simple equations for I_v, J_v and H_v at large optical depths. Such simple expressions are very insightful and help in better understanding radiative transfer in stars. These expressions can also be very useful to undertake calculations in stellar interiors.

In Section 3.6, it was seen that for $u \geq 0$, the specific intensity at depth τ_v is

$$I_v(\tau_v, u) = \int_{\tau_v}^{\infty} S_v(t) e^{\frac{\tau_v - t}{u}} \frac{dt}{u} \quad \text{(for } u \geq 0\text{)} \tag{3.59}$$

By inserting the expression for source function given in Eq. (3.58), the specific intensity becomes

$$I_v(\tau_v, u) = \sum_{n=0}^{\infty} \frac{1}{n!} \frac{d^n B_v}{d\tau_v^n}\bigg|_{\tau_v} \int_{\tau_v}^{\infty} (t - \tau_v)^n e^{\frac{\tau_v - t}{u}} \frac{dt}{u} \quad \text{(for } u \geq 0\text{)} \tag{3.60}$$

By applying the following change of variable $x = \dfrac{t - \tau_v}{u}$ (which also leads to $dx = \dfrac{dt}{u}$), this equation can be written

$$I_v(\tau_v, u) = \sum_{n=0}^{\infty} \frac{1}{n!} \frac{d^n B_v}{d\tau_v^n}\bigg|_{\tau_v} u^n \int_0^{\infty} x^n e^{-x} dx \quad \text{(for } u \geq 0\text{)} \tag{3.61}$$

and since $\int_0^{\infty} x^n e^{-x} dx = n!$, the specific intensity finally becomes

$$I_v(\tau_v, u) = \sum_{n=0}^{\infty} u^n \frac{d^n B_v}{d\tau_v^n}\bigg|_{\tau_v} \quad \text{(for } u \geq 0\text{)} \tag{3.62}$$

It can be shown (see Exercise 3.9) that the exact same result is found for $u < 0$, therefore

$$I_v(\tau_v, u) = \sum_{n=0}^{\infty} u^n \frac{d^n B_v}{d\tau_v^n}\bigg|_{\tau_v} = B_v + u \frac{dB_v}{d\tau_v} + u^2 \frac{d^2 B_v}{d\tau_v^2} + \dots \quad (-1 \leq u \leq 1) \tag{3.63}$$

It is then trivial to show that at large optical depths, the average intensity is

$$J_v(\tau_v) = \frac{1}{2} \int_{-1}^{1} I_v(\tau_v, u) du = B_v + \frac{1}{3} \frac{d^2 B_v}{d\tau_v^2} + \dots \tag{3.64}$$

The flux and the K-integral can also be easily obtained

$$H_v(\tau_v) = \frac{1}{2} \int_{-1}^{1} I_v(\tau_v, u) u \, du = \frac{1}{3} \frac{dB_v}{d\tau_v} + \frac{1}{5} \frac{d^3 B_v}{d\tau_v^3} + \dots \tag{3.65}$$

$$K_v(\tau_v) = \frac{1}{2} \int_{-1}^{1} I_v(\tau_v, u) u^2 \, du = \frac{B_v}{3} + \frac{1}{5} \frac{d^2 B_v}{d\tau_v^2} + \dots \tag{3.66}$$

By developing a Taylor series for B_v, it can be shown (see Exercise 3.10) that its derivatives vary as

$$\frac{\left(\dfrac{d^{n+2} B_v}{d\tau_v^{n+2}} \right)}{\left(\dfrac{d^n B_v}{d\tau_v^n} \right)} \approx \frac{1}{\tau_v^2} \tag{3.67}$$

Therefore, the terms containing higher-order derivatives of B_ν can be neglected at large optical depths and approximate values for the specific intensity, the average intensity, the radiative flux and the K-integral are given by the following expressions

$$I_\nu \approx B_\nu + u \frac{dB_\nu}{d\tau_\nu} \tag{3.68}$$

$$J_\nu \approx B_\nu \tag{3.69}$$

$$H_\nu \approx \frac{1}{3} \frac{dB_\nu}{d\tau_\nu} \tag{3.70}$$

$$K_\nu \approx \frac{B_\nu}{3} \tag{3.71}$$

The first term on the left-hand side of the equation for the specific intensity at large optical depths is isotropic and it does not contribute to the flux. In this approximation, solely the second term is responsible for the flux. The relative value of the isotropic term of I_ν as compared to its anisotropic term at large optical depths may be estimated by using the results found above (see Eq. 3.68 and 3.70)

$$\frac{\frac{dB_\nu}{d\tau_\nu}}{B_\nu} \approx \frac{3H_\nu}{B_\nu} \approx \frac{T_{eff}^4}{T^4} \ll 1 \tag{3.72}$$

Here, the fact that the frequency integrals of H_ν and B_ν vary, respectively, as T_{eff}^4 and T^4 was used to estimate the dependence of $\frac{H_\nu}{B_\nu}$ with respect to temperature (or depth). Since in stellar interiors $T \gg T_{eff}$, this result shows that the anisotropic term in I_ν is extremely small there. In other words, there is, relatively speaking, only a little more energy flowing in the outgoing ($u > 0$) direction than in the ingoing direction ($u < 0$).

Since $d\tau_\nu = -k_\nu \rho dz$, the flux may also be written as a function of the temperature gradient as follows

$$H_\nu \approx \frac{1}{3} \frac{dB_\nu}{d\tau_\nu} = -\frac{1}{3k_\nu \rho} \frac{dB_\nu}{dz} = -\frac{1}{3k_\nu \rho} \frac{dB_\nu}{dT} \frac{dT}{dz} \tag{3.73}$$

This equation is very rich in physical insight. First, and assuming all else equal, at frequencies with large opacities the monochromatic flux is weaker. The radiative flux that is transported through the star preferably passes through frequencies that are not too opaque. Secondly, the flux is proportional to $-dT/dz$, so the flux increases proportionally to the temperature gradient. This is a well-known result of thermodynamics. Finally, it should be mentioned that the monochromatic flux is not proportional to the local Planck function, but rather to its derivative with respect to temperature.

This approximation for the flux is often used in numerical models that compute stellar structure since high precision for the radiative field is generally not warranted there. The equation above is commonly called the diffusion approximation since it has the mathematical form of a typical diffusion equation, i.e. the flux being proportional to a gradient (historically called Fick's law). With the knowledge of the opacity and the stellar density and temperature profiles (i.e. their dependence with depth), this equation enables the direct calculation of the monochromatic flux without having to solve the radiative-transfer equation. This approximation of the radiative flux at large optical depths is one of the so-called equations of stellar structure that will be discussed in Chapter 5. However, in stellar atmospheres, the radiative-transfer equation must be solved in detail since the diffusion approximation is not valid at small optical depths. This topic will be discussed in more detail in Chapter 4.

3.9   Rosseland and Other Mean Opacities

As seen earlier (see Figures 3.2 and 3.3 for example), the monochromatic opacity in stellar plasma has a complex frequency dependence. It is thus difficult to draw general conclusions from the opacity at a single frequency. The definition of an average opacity could then be helpful. Several average opacities are commonly used in stellar astrophysics, but the most prevalent one is the Rosseland opacity, named after the Norwegian astronomer Svein Rosseland (1894–1985) who first introduced it.

The Rosseland opacity (k_R) is defined such that the total integrated Eddington flux H is given by an equation similar to the one found for the monochromatic flux (Eq. 3.73) in stellar interiors (i.e. at large optical depths)

$$H = -\frac{1}{3k_R \rho} \frac{dB}{dT} \frac{dT}{dz} \tag{3.74}$$

where B is the integrated Planck function and where the monochromatic opacity found in Eq. (3.73) is replaced by k_R. An expression for the Rosseland opacity may be found by integrating the flux given by Eq. (3.73) over all frequencies and by equating it to the expression given just above

$$H = \int_0^\infty H_\nu d\nu = -\frac{1}{3\rho} \frac{dT}{dz} \int_0^\infty \frac{1}{k_\nu} \frac{dB_\nu}{dT} d\nu = -\frac{1}{3k_R \rho} \frac{dB}{dT} \frac{dT}{dz} \tag{3.75}$$

and this leads to the Rosseland mean opacity

$$\frac{1}{k_R} = \frac{\displaystyle\int_0^\infty \frac{1}{k_\nu} \frac{dB_\nu}{dT} d\nu}{\displaystyle\frac{dB}{dT}} \tag{3.76}$$

Since $B = \dfrac{\sigma T^4}{\pi}$, the Rosseland mean may also be written as follows

$$\frac{1}{k_R} = \frac{\pi}{4\sigma T^3} \int_0^\infty \frac{1}{k_\nu} \frac{dB_\nu}{dT} d\nu \tag{3.77}$$

Since the Rosseland opacity depends on an integration of the inverse of k_ν, care must be taken to properly calculate this physical quantity. For instance, if missing opacity sources makes it such that k_ν is nil or unduly small, it can skew the value obtained for the Rosseland opacity.

The Rosseland mean opacity has properties similar to the monochromatic opacity. It depends on both the local physical conditions of the plasma and on the abundances of the elements present there. Also, when the monochromatic opacity is smaller or larger, so is the Rosseland mean. By Eq. (3.74), if for certain reasons or under certain conditions the Rosseland mean becomes large, the temperature gradient has to increase (assuming the same integrated flux). This relates to the fact that when the opacity increases, there exists more resistance to radiation transport, and a larger temperature gradient is needed to sustain radiative-energy transport. As will be seen in Chapter 5, this property of radiative-energy transport is critical for properly understanding stellar interiors. It will be shown that stellar regions with large opacities are prone to develop convective transport since radiation cannot by itself furnish the necessary flux to transport the energy.

Since Rosseland opacity calculations are time consuming, grids of this physical quantity can be pretabulated in order to save computing time when numerically modelling stars. Rosseland opacity grids are usually built on a given temperature-density grid (or variations thereof, see discussion below) and for a number of element mixtures (i.e. different values for X, Y and Z). Instead of showing a three-dimensional variation of the Rosseland opacity versus temperature and density, a two-dimensional cut of such data will be shown here. Figure 3.17 shows the Rosseland opacity as a function of temperature from the results of the Opacity Project (OP), which consists of a database of atomic data, radiative opacities and other astrophysical data made available for use by the scientific community. The OP opacities include H, He and 15 metals (C, N, O, Ne, Na, Mg, Al, Si, S, Ar, Ca, Cr, Mn, Fe and Ni). The results shown in that figure are for densities satisfying the following relation

$$\log R = \log \frac{\rho}{T_6^3} = -3 \qquad (3.78)$$

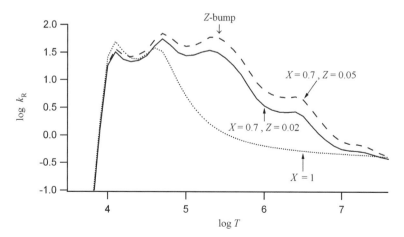

Figure 3.17 Rosseland mean opacity as a function of temperature from the Opacity Project data. The densities used are those for $\log R = -3$. Three curves are shown for different abundances that are defined in the figure. Also identified in the figure is the position of the Z-bump.

where T_6 is temperature in units of 10^6 K and R is a variable defined as $\dfrac{\rho}{T_6^3}$ that serves as the second dimension for the opacity tables (the first dimension being temperature). The value of $\log R = -3$ is chosen because it is close to corresponding values found inside typical stars. The grids on which the opacities are calculated use grid variables such as $\log R$ instead of density because a T–R grid has a larger number of points of interest for the conditions encountered in stars than does a T–ρ grid. In other words, choosing a rectangular T–ρ grid leads to a larger number of useless grid points (i.e. that are of no interest for applications in stars).

Several curves are shown in Figure 3.17. First, the curve (dotted curve) shown for a pure hydrogen ($X = 1$) medium will be discussed. Two maxima can be seen between $\log T = 4$ and 5. The first maximum (found at the lower temperature) is due to the opacity from hydrogen found in excited states (for example the opacity due to Balmer lines and the Balmer jump). At even lower temperatures, H^- opacity dominates. The second maximum is due to transitions from the fundamental state of hydrogen ($n = 1$). Lyman transitions are only of importance at these higher temperatures since the radiation field is more intense in the ultraviolet part of the spectrum in these hotter regions. At even higher temperatures, the Rosseland opacity decreases due to hydrogen ionisation. Opacity there is mostly due to electron scattering.

Two other curves are shown in Figure 3.17. The first of these two curves (solid curve) supposes the presence of hydrogen ($X = 0.7$), helium ($Y = 0.28$) and metals ($Z = 0.02$), while the other curve (dashed curved) shows the Rosseland mean for a higher metallicity ($Z = 0.05$). The reader is reminded that the metallicity for the Sun is approximately $Z = 0.017$. Not surprisingly, as the metallicity increases, so does the Rosseland mean. An interesting feature that is identified in this figure is that of the so-called Z-bump. This bump is defined by a particularly large increase of the opacity beginning near $\log T \approx 5.2$ due to the opacity of the metals. It is believed that this feature might be responsible for the pulsations of certain types of stars (this aspect of stellar astrophysics will be discussed further in Chapter 5). For more detailed information about Rosseland mean opacities and the Opacity Project calculations, the reader is referred to M.J. Seaton *et al.*, *Monthly Notices of the Royal Astronomical Society*, 266, 805 (1994).

Other mean opacities are also occasionally used in astrophysics. Two such mean opacities are the flux mean

$$k_{\mathrm{F}} = \frac{\displaystyle\int_0^\infty k_\nu F_\nu \mathrm{d}\nu}{\displaystyle\int_0^\infty F_\nu \mathrm{d}\nu} \tag{3.79}$$

and the Planck mean

$$k_{\mathrm{P}} = \frac{\displaystyle\int_0^\infty k_\nu B_\nu \mathrm{d}\nu}{\displaystyle\int_0^\infty B_\nu \mathrm{d}\nu} \tag{3.80}$$

However, since these average opacities are much less prevalent and useful than the Rosseland mean, they will not be discussed any further in this book.

3.10   Schwarzschild–Milne Equations[††]

In Section 3.6, the solution for the specific intensity was found. In this section, these results will be used to find general equations for the average intensity and the flux. These expressions were, respectively, found by the astrophysicists Karl Schwarzschild (1873–1916) and Edward Arthur Milne (1896–1950) and are called the Schwarzschild–Milne equations.

First, when the general solution found for the specific intensity (Eq. 3.42 and 3.43) is inserted in the expression for the average intensity expression, it becomes

$$J_v(\tau_v) = \frac{1}{2}\int_{-1}^{1} I_v(\tau_v, u)\,du = -\frac{1}{2}\int_{-1}^{0} du \int_{0}^{\tau_v} S_v(t)\,e^{\frac{\tau_v - t}{u}}\,\frac{dt}{u} + \frac{1}{2}\int_{0}^{\infty} du \int_{\tau_v}^{\infty} S_v(t)\,e^{\frac{\tau_v - t}{u}}\,\frac{dt}{u} \qquad (3.81)$$

The two integrals found on the right-hand side of this equation represent the contribution to the average intensity from the layers respectively found above and below the optical depth τ_v. This equation may be simplified by applying a change of variable for each of the two integrals found on its right-hand side. For the first of these two integrals (where $u \leq 0$), the variable $y = -1/u$ is defined and thus $du = dy/y^2$. For the second integral (where $u \geq 0$), the following change of variable is used: $y = 1/u$ and thus $du = -dy/y^2$. The average intensity then becomes

$$J_v(\tau_v) = \frac{1}{2}\int_{0}^{\tau_v} S_v(t)\,dt \int_{1}^{\infty} e^{-y(\tau_v - t)}\,\frac{dy}{y} + \frac{1}{2}\int_{\tau_v}^{\infty} S_v(t)\,dt \int_{1}^{\infty} e^{-y(t - \tau_v)}\,\frac{dy}{y} \qquad (3.82)$$

This expression can be simplified by using mathematical functions called exponential integrals E_n defined as

$$E_n(x) = \int_{1}^{\infty} e^{-xy}\,\frac{dy}{y^n} \qquad (3.83)$$

Special Topic – Properties of the Exponential Functions

Here are two very useful properties for $E_n(x)$:

$$E_n(0) = \int_{1}^{\infty} \frac{dy}{y^n} = \frac{1}{n-1} \qquad (3.84)$$

$$\frac{dE_n(x)}{dx} = -\int_{1}^{\infty} e^{-xy}\,\frac{dy}{y^{n-1}} = -E_{n-1}(x) \qquad (3.85)$$

The average intensity can then be written

$$J_\nu(\tau_\nu) = \frac{1}{2}\int_0^{\tau_\nu} S_\nu(t)E_1(\tau_\nu - t)dt + \frac{1}{2}\int_{\tau_\nu}^\infty S_\nu(t)E_1(t - \tau_\nu)dt \tag{3.86}$$

Since the argument of the exponential integral is positive for both of these integrals, namely $\tau_\nu - t \geq 0$ for the first integral and $t - \tau_\nu \geq 0$ for the second term, this expression may be written

$$J_\nu(\tau_\nu) = \frac{1}{2}\int_0^\infty S_\nu(t)E_1(|t - \tau_\nu|)dt \tag{3.87}$$

which is called the Schwarzschild equation.

An operator Λ, called the lambda operator, may be defined such as

$$\Lambda_\nu(f(t)) = \frac{1}{2}\int_0^\infty f(t)E_1(|t - \tau_\nu|)dt \tag{3.88}$$

and the average intensity can therefore be written in terms of the lambda operator

$$J_\nu(\tau_\nu) = \Lambda_\nu(S_\nu(t)) \tag{3.89}$$

A similar expression can also be found for the Eddington flux, called the Milne equation

$$H_\nu(\tau_\nu) = \frac{1}{2}\int_0^{\tau_\nu} S_\nu(t)E_2(\tau_\nu - t)dt - \frac{1}{2}\int_{\tau_\nu}^\infty S_\nu(t)E_2(t - \tau_\nu)dt = \frac{1}{2}\int_0^\infty S_\nu(t)E_2(|t - \tau_\nu|)dt \tag{3.90}$$

or

$$H_\nu(\tau_\nu) = \Phi_\nu(S_\nu(t)) \tag{3.91}$$

where the phi operator Φ_ν is defined as

$$\Phi_\nu(f(t)) = \frac{1}{2}\int_0^\infty f(t)E_2(|t - \tau_\nu|)dt \tag{3.92}$$

It can also be shown that the K-integral is given by the following expression

$$K_\nu(\tau_\nu) = \frac{1}{2}\int_0^\infty S_\nu(t)E_3(|t - \tau_\nu|)dt \tag{3.93}$$

As mentioned previously for the case of the solution of the radiative-transfer equation found in Sections 3.6, the solution of the Schwarzschild–Milne equations are not as direct as it may seem. The solutions of these equations depend on the source function that in turn depends on the radiation field (Eq. 3.33 and 3.35). There exists a nonlocal coupling for the radiation field since the radiation field at a given depth depends on the field found in other layers. In general, the radiative-transfer problem is therefore quite complex. For more information about such advanced concepts of radiative transfer, the reader is referred to Mihalas, D., *Stellar Atmospheres*, W.H. Freeman and Company, San Francisco (1970 and 1978) and Chandrasekhar, S., *Radiative Transfer*, Dover, New York (1960).

3.11 Demonstration of the Radiative-Transfer Equation[†]

Figure 3.18 shows a beam of radiation, characterized by the specific intensity that crosses an element of matter of area dA and thickness ds. The orientation of the surface dA was chosen so that it is perpendicular to the direction of the specific intensity. From this figure, it is clear that $ds = dz/u$. This result will be used below to write the radiative-transfer equation as a function of z instead of s. The monochromatic opacity k_v is defined such that the energy taken away (either by absorption or scattering processes) from I_v between the solid angles Ω and $\Omega + d\Omega$, during the time dt, by the mass within the volume $dAds$, between the frequencies v and $v+dv$ is

$$dE_v = k_v \rho I_v d\Omega dv dt dA ds \tag{3.94}$$

Since matter can also add energy to the radiation field, we may define a term called the emissivity j_v such that the energy added by the mass within the volume $dAds$, between the frequencies v and $v+dv$ and between the solid angles Ω and $\Omega + d\Omega$ is

$$dE_v = j_v \rho d\Omega dv dt dA ds \tag{3.95}$$

Since the emission of radiation by unit mass depends solely on the properties of matter and on the local physical conditions, the specific intensity is absent in this last expression.

When the two equations above are subtracted from one another, it gives the global amount of energy either added or taken away from the specific intensity, depending on the relative strength of these two terms. This leads to a variation dI_v of the specific intensity that may be written as follows

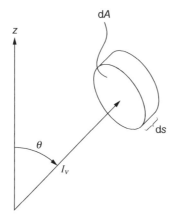

Figure 3.18 Illustration of a beam of radiation defined by the specific intensity I_v arriving perpendicular to the surface dA of a mass element of volume $dAds$.

$$dI_v d\Omega dv dt dA = -k_v \rho I_v d\Omega dv dt dA ds + j_v \rho d\Omega dv dt dA ds \qquad (3.96)$$

The differential ds is absent from the left-hand side of the equation by virtue of the definition of I_v.

As mentioned previously, since ds = dz/u, the equation above may be written as a function of z instead of s and rearranged to give

$$\frac{u}{\rho} \frac{dI_v}{dz} = -k_v I_v + j_v \qquad (3.97)$$

which is the well-known radiative-transfer equation.

3.12   Radiative Acceleration of Matter and Radiative Pressure[†]

3.12.1   Radiative Acceleration of Matter

Since photons possess momentum, when radiation interacts with matter, momentum is transferred from the photons to atoms. An element of matter found in a star absorbs and scatters photons coming from all directions. However, since a net flux of radiative energy exists in the outgoing direction, momentum is transferred to the mass element in this direction. This transfer of momentum may be expressed as an acceleration of matter, which is commonly called radiative acceleration. In a star, this force pushes matter outwards because of the outgoing radiative flux, in other words, because of assumed spherical symmetry, the only direction in which there can be a net radiative acceleration is the outgoing direction. This affects the pressure gradient within stars (see below).

In the previous section, the energy transmitted to an element of matter of volume dAds (see Figure 3.18) from a beam of radiation travelling in the direction normal to the surface dA was found to be

$$dE_v = k_v \rho I_v d\Omega dv dt dA ds \qquad (3.98)$$

For radiation, the relation between momentum and energy is $p = E/c$. In a star, the component of momentum dp_v transferred to the element of mass in the direction $\hat{z}$ from photons with frequencies between v and $v+dv$ contained in the specific intensity I_v is

$$dp_v = \frac{u dE_v}{c} = \frac{u}{c} k_v \rho I_v d\Omega dv dt dA ds \qquad (3.99)$$

The momentum due to photons arriving from all directions can be obtained by integrating over all solid angles, and by using the definition of the Eddington flux, this gives

$$dp_v = \frac{4\pi}{c} k_v \rho H_v dv dt dA ds \qquad (3.100)$$

By integrating over the entire spectrum, the integrated momentum dp transferred to the mass element is found to be

$$dp = \frac{4\pi}{c} \rho dt dA ds \int_0^\infty k_\nu H_\nu d\nu \tag{3.101}$$

Since force is equal to the derivative of momentum with respect to time, for the case under consideration here, the radiative force F_{rad} can be written

$$F_{rad} = \frac{dp}{dt} = \frac{4\pi}{c} \rho dA ds \int_0^\infty k_\nu H_\nu d\nu = \rho dA ds g_{rad} \tag{3.102}$$

where $\rho dA ds$ is the mass of the matter element under consideration and where an acceleration g_{rad} is introduced and is defined as the radiative acceleration of matter. From the equation above, the following expression is found for the radiative acceleration

$$g_{rad} = \frac{4\pi}{c} \int_0^\infty k_\nu H_\nu d\nu \tag{3.103}$$

Radiative acceleration is proportional to both opacity and flux. At higher opacities, more photons are absorbed or scattered and this contributes to radiative acceleration. It might at first seem strange that scattering processes intervene in g_{rad} since these photons are not absorbed. However, during the scattering process, momentum is transferred to matter. For example, when atoms absorb photons to re-emit them soon after (which is a scattering process, see Figure 3.1(a)), momentum is gained during the absorption. And since the emission of the photons is isotropic, no net momentum is lost by matter thereafter. Therefore, momentum is gained by the atom during such an interaction. Also, g_{rad} is proportional to the flux (instead of other moments of the radiation field for instance) because only the anisotropic portion of the radiative field contributes to radiative acceleration. For example, for an isotropic radiation field, the same amount of force is exerted on the opposing sides of a given stellar plasma mass element and no net acceleration ensues.

The expression for radiative acceleration given above is the total radiative acceleration of matter. In reality, each species is accelerated by radiation and the intensity of this acceleration depends on its opacity spectrum. Therefore, the various species have different radiative accelerations. This leads to the relative diffusion of the species within a star that can lead to structural changes. This subject will be discussed in Chapter 7.

The hydrostatic equilibrium equation must therefore include the force exerted by radiation. The effective acceleration on the medium is therefore equal to the vectorial sum of the gravitational and the radiative acceleration. The gradient of the gas pressure P is thus equal to

$$\frac{dP(r)}{dr} = -\rho(r)[g(r) - g_{rad}(r)] \tag{3.104}$$

This equation leads to a very interesting and important result. In massive stars, the surface temperature can be large enough so that $g_{rad} > g$ due to the large radiative flux present at their surface. In this case, matter is pushed out of the star and radiative acceleration of matter imposes an upper limit to stellar masses. The following example illustrates this property.

Example 3.6: Using the concept of radiative acceleration, estimate the upper limit for the effective temperature for stars while assuming that the opacity is dominated by electron scattering (assume a star with surface gravity log $g = 4$).

Answer:

The outer regions of a star with a surface gravity g stay bound to the star up to the point where radiative acceleration equals gravitational acceleration

$$g_{rad} = \frac{4\pi}{c} \int_0^\infty k_\nu H_\nu d\nu = g \tag{3.105}$$

where

$$k_\nu = \frac{n_e \sigma_T}{\rho} \tag{3.106}$$

since only electron scattering is considered here (see Eq. 3.4). For stars with suffi-ciently large T_{eff} (or mass) such as to lead to $g_{rad} > g$, these outer layers are pushed out by radiation pressure.

By inserting the electron scattering opacity in the g_{rad} equation and equating it to the surface gravity of the star under consideration, the following relation ensues

$$g_{rad} = \frac{4\pi}{c} \frac{n_e \sigma_T}{\rho} \int_0^\infty H_\nu d\nu = \frac{\sigma T_{eff}^4}{c} \frac{n_e \sigma_T}{\rho} = 10^4 \text{ cm/s}^2 \tag{3.107}$$

Since the temperature at the surface is relatively large for massive stars, the assump-tion that all of the hydrogen is ionised can be made. To estimate the maximal effective temperature, a star made of pure hydrogen may be considered. The density is then

$$\rho = n_p m_p + n_e m_e \approx n_p m_p = n_e m_p \tag{3.108}$$

and consequently

$$\frac{\sigma T_{eff}^4}{c} \frac{\sigma_T}{m_p} = 10^4 \text{ cm/s}^2 \tag{3.109}$$

The maximal effective temperature is

$$T_{eff} = \left[\frac{c m_p}{\sigma \sigma_T} 10^4 \right]^{1/4} \approx 60\,000\,\text{K} \tag{3.110}$$

This value is a reasonably good approximation as compared to the maximum observed stellar effective temperatures.

3.12.2   Radiative Pressure

As previously discussed, photons can transfer their momentum to matter and they can therefore participate in the total pressure in stellar plasma. The pressure caused by photons is called radiative pressure. In the previous section, the hydrostatic equilibrium equation found (Eq. 3.104) related the gas pressure P to the radiative and gravitational accelerations. A total pressure P_{tot} may also be defined as the pressure due to both gas and radiation

$$P_{tot} = P + P_{rad} \tag{3.111}$$

where P_{rad} is the pressure due to radiation. Therefore, the pressure exerted by the weight of the mass above a certain point r must be counterbalanced by the total pressure

$$\frac{dP_{tot}(r)}{dr} = -\rho(r)g(r) \tag{3.112}$$

and thus

$$\frac{dP(r)}{dr} + \frac{dP_{rad}(r)}{dr} = -\rho(r)g(r) \tag{3.113}$$

With Eq. (3.103), the radiative-pressure gradient may be related to radiative acceleration on the medium via the expression

$$\frac{dP_{rad}(r)}{dr} = -\rho(r)g_{rad}(r) = -\rho(r)\frac{4\pi}{c}\int_0^\infty k_\nu H_\nu d\nu \tag{3.114}$$

Using this equation, it is possible to find an expression for P_{rad}. To do this, the results of Eq. (3.70) and (3.71) may be combined to show that at large optical depths, the following relation exists between the flux and the K-integral

$$H_\nu = \frac{dK_\nu}{d\tau_\nu} \tag{3.115}$$

Using this result and the definition for the optical depth (Eq. 3.29), the radiation-pressure gradient may be written as follows

$$\frac{dP_{rad}}{dr} = -\rho\frac{4\pi}{c}\int_0^\infty k_\nu \frac{dK_\nu}{d\tau_\nu} d\nu = \rho\frac{4\pi}{c}\int_0^\infty k_\nu \frac{dK_\nu}{k_\nu \rho dr} d\nu = \frac{d}{dr}\left[\frac{4\pi}{c}\int_0^\infty K_\nu d\nu\right] \tag{3.116}$$

and consequently, radiative pressure is given by the expression

$$P_{rad} = \frac{4\pi}{c}\int_0^\infty K_\nu d\nu = \frac{4\pi}{c}K \tag{3.117}$$

A monochromatic radiation pressure $P_{rad}(\nu)$ can also be defined as

$$P_{rad}(\nu) = \frac{4\pi}{c}K_\nu \tag{3.118}$$

Since at large optical depths (Eq. 3.71), the K-integral is

$$K_v \approx \frac{B_v}{3} \tag{3.119}$$

it is easy to show that this leads to

$$P_{rad} = \frac{4\sigma T^4}{3c} \tag{3.120}$$

This equation is only valid at large optical depths. In the outer atmosphere of a star, the radiation pressure is more complex since the K-integral cannot be approximated by the simple relation used above.

3.13 Summary

Types of opacities: bound–bound, bound–free, free–free and scattering

Solid angle: $d\Omega = \sin\theta d\theta d\varphi$ $\tag{3.121}$

Average intensity: $J_v(z) = \dfrac{1}{2}\displaystyle\int_{-1}^{1} I_v(z, u)\,du$ $\tag{3.122}$

Monochromatic flux: $F_v(z) = 2\pi\displaystyle\int_{-1}^{1} I_v(z,u)u\,du$ $\tag{3.123}$

Eddington flux: $H_v = \dfrac{F_v}{4\pi}$ $\tag{3.124}$

Radiative energy density: $U_v = \dfrac{4\pi}{c}J_v$ $\tag{3.125}$

Optical depth: $d\tau_v = -k_v\rho dz$ $\tag{3.126}$

Radiative-transfer equation: $\dfrac{u}{\rho}\dfrac{dI_v(z, u)}{dz} = -k_v I_v(z, u) + j_v$

$\text{or } u\dfrac{dI_v(\tau_v, u)}{d\tau_v} = I_v(\tau_v, u) - S_v(\tau_v)$ $\tag{3.127}$

Source function: $S_v(\tau_v) = \dfrac{j_v}{k_v} \approx \dfrac{\kappa_v B_v}{\kappa_v + \sigma_v} + \dfrac{\sigma_v J_v}{\kappa_v + \sigma_v}$ $\tag{3.128}$

Specific intensity at the surface of stars: $I_v(0,u) = \displaystyle\int_{0}^{\infty} S_v(t)e^{-\frac{t}{u}}\dfrac{dt}{u} \quad (u \geq 0)$ $\tag{3.129}$

Large optical depth approximations: $I_v \approx B_v + u\dfrac{dB_v}{d\tau_v}$ $\tag{3.130}$

$$J_v \approx B_v \tag{3.131}$$

$$H_v \approx \frac{1}{3}\frac{dB_v}{d\tau_v} = -\frac{1}{3k_v\rho}\frac{dB_v}{dT}\frac{dT}{dz} \tag{3.132}$$

$$K_v \approx \frac{B_v}{3} \tag{3.133}$$

Rosseland mean opacity: $\dfrac{1}{k_R} = \dfrac{\pi}{4\sigma T^3}\displaystyle\int_0^\infty \dfrac{1}{k_v}\dfrac{dB_v}{dT}\,dv$ $\tag{3.134}$

3.14   Exercises

3.1   Assuming the specific intensity $I_v(\tau, u) = a_v(\tau)u + b_v(\tau)u^2$, calculate $J_v(\tau)$ and $H_v(\tau)$.

3.2   If the source function inside a star is $S_v(\tau_v) = a_v + b_v\tau_v$, where a_v and b_v are functions of v, calculate the specific intensity $I_v(0, u)$ at the surface for the outgoing directions ($u \geq 0$).

3.3   By using its definition (Eq. 3.21), calculate the radiative energy density at large optical depths where $I_v \approx B_v + u\dfrac{dB_v}{d\tau_v}$. Please interpret the answer.

3.4   Assume that at a given frequency the opacity of the stellar plasma at the surface of a given star is dominated by Thomson scattering. Estimate the depth (in cm) of the atmosphere that is visible to an outside observer at that frequency, while assuming that the electron density in this region is constant and equal to $10^{16}\,\text{cm}^{-3}$. What percentage of the solar radius is this value?

3.5   Assuming an interstellar cloud with a known thickness z_0 and a density profile
$$\rho(z) = \rho_0\left(\frac{z}{z_0}\right)^2 + \rho_0\left(\frac{z}{z_0}\right)$$
(see Figure 3.19) and that the opacity k_v within the cloud is constant and known, calculate the relation for the optical depth (with respect to $z = 0$) for $z \leq z_0$. Calculate the optical depth at $z = z_0$ and physically interpret each quantity (including any constants) in this result.

3.6   Assume two adjacent interstellar gas clouds with known thickness d_1 and d_2 found in front of an astronomical light source (see Figure 3.20). Suppose that this source emits a beam of radiation with known specific intensity I_v^0 as it enters the first cloud of interstellar gas. Calculate the specific intensity I_v immediately after this beam has traversed the two clouds of gas. Assume that the densities (ρ_1 and ρ_2) and opacities (k_{v1} and k_{v2}) are constant within the clouds and are known quantities and that these clouds possess no emissivity.

3.7   A beam of radiation with initial intensity I_v^0 crosses an interstellar cloud of unknown thickness z_0 (see Figure 3.21). The intensity on the other side of the cloud is one thousandth

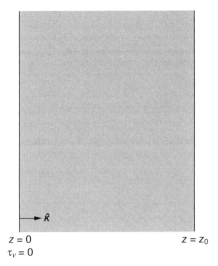

Figure 3.19   Illustration for Exercise 3.5.

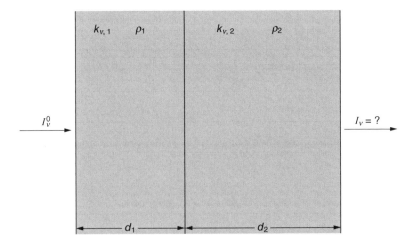

Figure 3.20   Illustration for Exercise 3.6.

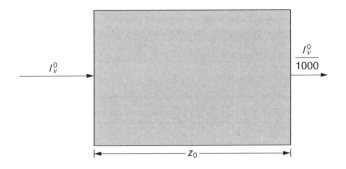

Figure 3.21   Illustration for Exercise 3.7.

$z = 0$

Figure 3.22   Illustration for Exercise 3.8.

that of the initial intensity. Assuming that the cloud possesses no emissivity and that the opacity respects the following relation $k_\nu \rho = \dfrac{\sqrt{z}}{4\xi}$ within the gas where ξ is a known constant, calculate the thickness z_0 of the cloud.

3.8   A beam of radiation with initial intensity I_ν^0 enters an interstellar cloud with known density (ρ), opacity (k_ν) and emissivity (j_ν) (see Figure 3.22) that are constant throughout the cloud. Calculate $I_\nu(z)$. What is its value for large values of z (or optical depth)? Physically interpret this result.

3.9   Similarly to the development in Section 3.8 show that for $u < 0$

$$I_\nu(\tau_\nu, u) = \sum_{n=0}^{\infty} u^n \left.\frac{d^n B_\nu}{d\tau_\nu^n}\right|_{\tau_\nu} = B_\nu + u\frac{dB_\nu}{d\tau_\nu} + u^2 \frac{d^2 B_\nu}{d\tau_\nu^2} + \dots \tag{3.135}$$

3.10   As mentioned in Section 3.8, show that at large optical depths

$$\frac{\left(\dfrac{d^{n+2} B_\nu}{d\tau_\nu^{n+2}}\right)}{\left(\dfrac{d^n B_\nu}{d\tau_\nu^n}\right)} \approx \frac{1}{\tau_\nu^2} \tag{3.136}$$

3.11[††]   Similar to the calculation shown in Section 3.10 for the average intensity, show that the flux can be written

$$H_\nu(\tau_\nu) = \frac{1}{2}\int_0^{\tau_\nu} S_\nu(t) E_2(\tau_\nu - t)\,dt - \frac{1}{2}\int_{\tau_\nu}^{\infty} S_\nu(t) E_2(t - \tau_\nu)\,dt \tag{3.137}$$

which is called the Milne equation.

4

Stellar Atmospheres

4.1 Introduction

The atmosphere of a star is defined as its outer regions that determine the properties of the radiative flux that emanates from its surface. The physical properties of these outer regions are therefore of critical importance to properly interpret astronomical observations. The principal aims of this chapter are to describe the physical processes in stellar atmospheres and to discuss their numerical calculations.

The purpose of stellar atmospheric modelling is to calculate, as accurately as possible, the physical structure of the outer regions of stars. The depth dependence of the various physical variables (T, ρ, n_e, etc.) defines the structure of a stellar atmosphere. With the physical structure of the outer regions, the monochromatic flux at the stellar surface may then be calculated in the theoretical framework seen in the previous chapter (assuming the knowledge of the required atomic data intervening in the interaction between the stellar plasma and the radiation field). Once the monochromatic flux is obtained, photometric colours can be calculated and compared to astronomical observations. High-resolution monochromatic radiative flux can also be computed and compared to observed stellar spectra. Such comparisons between theoretical models and observations are invaluable for better understanding the physical processes taking place in the atmospheres of stars.

Typically, the depth chosen for the bottom of a stellar atmosphere during its numerical modelling is on the order of $\tau \approx 10^2$ to 10^3, where τ is an optical depth scale of an average opacity or of an opacity at a frequency in the continuum (i.e. outside any atomic line). This depth corresponds to a geometrical depth of the order of one per cent of the stellar radius and is sufficient to assure that deeper regions do not contribute significantly to the surface radiative flux.

The physical structure of a stellar atmosphere has to respect the various physical equations that intervene there (i.e. hydrostatic equilibrium, radiative transfer, Saha and Boltzmann equations, etc.). Of course, as for any physical model (or simulation), some

An Introduction to Stellar Astrophysics Francis LeBlanc
© 2010 John Wiley & Sons, Ltd

approximations are required to render the numerical problem in question feasible. The resolution of these equations and the approximations usually made during atmospheric modelling will be discussed in Section 4.5. Such numerical simulations necessitate considerable computational resources. Beforehand, analytical results while supposing that the opacity is independent of frequency (or the so-called grey-atmosphere approximation) will be shown. Thereafter, the physics related to atomic lines and their importance in stars will be discussed. A discussion on the formation of atomic lines will then be presented. Finally, and as mentioned above, a section pertaining to the intricacies surrounding the calculation of stellar atmosphere models will be presented. Results from detailed stellar atmosphere modelling will also be shown.

4.2   The Grey Atmosphere

In Section 4.5 details concerning the calculation of the structure of a stellar atmosphere will be discussed. Such calculations necessitate the knowledge of the detailed monochromatic opacity spectrum and can only be undertaken with powerful numerical capabilities such as those provided by modern computers. However, by making appropriate approximations, certain interesting results can be obtained analytically. For example, by supposing that the opacity is independent of frequency (i.e. $k_v = k$), some illustrative and insightful results can be found. A stellar atmosphere while assuming such an opacity spectrum is called a grey atmosphere. Its name relates to the fact that in this approximation all wavelengths possess the same opacity and therefore the opacity spectrum can be considered 'colourless' or grey. Nevertheless, the radiation field present in such an atmosphere will possess a frequency distribution since matter emits a radiative spectrum associated to its temperature. In this section, an expression for the temperature profile (i.e. its dependence with respect to depth) and the flux emanating from a grey atmosphere will be found. But beforehand, some useful results obtained from the radiative-transfer equation for a grey atmosphere are found.

For a grey atmosphere, the frequency dependence for the optical depth found at the denominator on the left-hand side of the radiative-transfer equation (Eq. 3.34) disappears and this equation can then be written

$$u\frac{dI_v}{d\tau} = I_v - S_v \tag{4.1}$$

By integrating this equation over all frequencies, the radiative-transfer equation becomes

$$u\frac{dI}{d\tau} = I - S \tag{4.2}$$

where

$$I = \int_0^\infty I_v dv \quad \text{and} \quad S = \int_0^\infty S_v dv \tag{4.3}$$

By assuming radiative equilibrium, and since the opacity is independent of frequency, Eq. (3.54) becomes

$$k\int_0^\infty J_\nu d\nu = k\int_0^\infty S_\nu d\nu \tag{4.4}$$

which leads to $J = S$. Now, if the source function is approximated by the Planck distribution, a simple and useful expression is found for the integrated values of the average intensity and the source function as a function of the local temperature $T(\tau)$ in the atmosphere

$$J(\tau) = S(\tau) \approx B(\tau) = \int_0^\infty B_\nu(\tau) d\nu = \frac{\sigma T^4(\tau)}{\pi} \tag{4.5}$$

These results will be applied in the next section that is dedicated to the specification of the temperature profile $T(\tau)$ in a grey atmosphere.

4.2.1   The Temperature Profile in a Grey Atmosphere

Equation (4.2) may be integrated over all directions and this gives the moment of order 0 (see Section 3.3) of this equation and leads to the following expression

$$\oint u \frac{dI}{d\tau} d\Omega = \oint I d\Omega - \oint S d\Omega \tag{4.6}$$

Since the derivative on the left-hand side is independent of direction, it may be taken out of the integral. Then, by multiplying each side by $1/4\pi$, the equation above may be written

$$\frac{1}{4\pi}\frac{d}{d\tau}\Big[\oint uI d\Omega\Big] = \frac{1}{4\pi}\Big[\oint I d\Omega - \oint S d\Omega\Big] = J - S \tag{4.7}$$

and by the definition of the Eddington flux (Eq. 3.16), this equation gives

$$\frac{dH}{d\tau} = J - S \tag{4.8}$$

Since in radiative equilibrium, $J = S$, this equation shows that the integrated flux is constant throughout the atmosphere. By definition of the effective temperature, the value of H at the surface of a star (and consequently for all depths for the case under consideration) is equal to

$$H = \frac{\sigma T_{eff}^4}{4\pi} \tag{4.9}$$

A relation between H and K may be obtained with the moment of order 1 of Eq. (4.2). This moment is calculated by multiplying each side of this equation by u and integrating over all directions

$$\frac{1}{4\pi}\frac{\mathrm{d}}{\mathrm{d}\tau}\Big[\oint u^2 I \mathrm{d}\Omega\Big] = \frac{1}{4\pi}\Big[\oint uI\mathrm{d}\Omega - \oint uS\mathrm{d}\Omega\Big] \tag{4.10}$$

The integral on the left-hand side gives the integrated value of the K-integral. The integrals on the right-hand side, respectively, give H and zero. The last integral in the equation above is nil because S is isotropic. All these results lead to the following equation

$$\frac{\mathrm{d}K}{\mathrm{d}\tau} = H \tag{4.11}$$

Since H is a constant, this equation may be integrated to give

$$K(\tau) = H\tau + C \tag{4.12}$$

where C is an integration constant.

From Eqs. (3.69) and (3.71) (see Section 3.8), the relation between J and K at large optical depths is found to be $J = 3K$ and therefore, the average intensity J at large optical depths is

$$J(\tau) = 3[H\tau + C] \tag{4.13}$$

The approximation where $J = 3K$ is also valid at small optical depths for several simplified cases for the specific intensity (see Example 4.1 and Exercise 4.1, for example). This approximation, called the Eddington approximation, will be assumed valid throughout the grey atmosphere.

Example 4.1: Considering a two-stream model (sometime called Schuster's model) for the specific intensity, where $I(\tau,u) = I_{out}$ for $u \geq 0$, and $I(\tau,u) = I_{in}$ for $u < 0$, show that $J = 3K$.

Answer:

By definition

$$J(\tau) = \frac{1}{2}\int_{-1}^{1} I(\tau,u)\,\mathrm{d}u = \frac{1}{2}\Big[\int_{0}^{1} I_{out}\,\mathrm{d}u + \int_{-1}^{0} I_{in}\,\mathrm{d}u\Big] = \frac{I_{out} + I_{in}}{2} \tag{4.14}$$

and

$$K(\tau) = \frac{1}{2}\int_{-1}^{1} I(\tau,u)u^2\,\mathrm{d}u = \frac{1}{2}\Big[\int_{0}^{1} I_{out}u^2\,\mathrm{d}u + \int_{-1}^{0} I_{in}u^2\,\mathrm{d}u\Big] = \frac{I_{out} + I_{in}}{6} \tag{4.15}$$

Therefore, with these two results, it is trivial to show that $J = 3K$.

To simplify the equation above for the average intensity, the integration constant can be written as $C = yH$, thus giving

$$J(\tau) = 3H[\tau + y] \qquad (4.16)$$

and since

$$J(\tau) = \frac{\sigma T^4(\tau)}{\pi} \qquad (4.17)$$

and

$$H = \frac{\sigma T_{\text{eff}}^4}{4\pi} \qquad (4.18)$$

A relation may then be found for the temperature profile in a grey atmosphere, namely

$$T(\tau) = T_{\text{eff}} \left[\frac{3}{4}(\tau + y) \right]^{1/4} \qquad (4.19)$$

The value of constant y may be estimated by physical reasoning. From the discussion in Chapters 1 and 3 surrounding spectral types, the radiation field exiting a star is associated to layers with $T \approx T_{\text{eff}}$. The photons arising from the surface of stars come from layers that become transparent to radiation. It can be assumed that the layers where photons have a probability of 50 % of reaching the surface, before being absorbed or scattered, are those layers from where most of the radiation exiting a star emanates. Therefore, the optical depth of these layers is such that the attenuation of the radiation is $e^{-\tau} \approx 1/2$, thus $\tau \approx 2/3$. It might then be assumed that the temperature at $\tau = 2/3$ equals T_{eff} (i.e. $T(\tau = 2/3) = T_{\text{eff}}$). From the equation above, it is trivial to show that under this assumption $y = 2/3$. The average intensity and the temperature as a function of depth are, respectively,

$$J(\tau) = 3H \left[\tau + \frac{2}{3} \right] \qquad (4.20)$$

$$T(\tau) = T_{\text{eff}} \left[\frac{3}{4} \left(\tau + \frac{2}{3} \right) \right]^{1/4} \qquad (4.21)$$

A detailed calculation of the integration constant found in Eq. (4.19) is given below in an advanced special topics section and confirms the physical reasoning employed above that leads to these last two equations.

It is notable to realise that with simple physical considerations and by using basic physical equations related to the radiation field, an approximate relation giving the temperature of the exterior regions of stars as a function of depth is found. Even though the temperature profile found above is not very accurate when compared to detailed numerical models, it can still be quite useful and instructive. For example (see Section 4.5), this profile can be used as a first approximation when calculating detailed numerical model atmospheres. Also, this equation shows that the temperature at nil optical depth is $T\,(\tau = 0) = 0.84T_{eff}$. This result, as others seen previously, explains why the radiative flux at the surface of stars (and thus their spectral type) depends strongly on T_{eff}. This is due to the fact that the contributing layers to the flux have a temperature value close to T_{eff}.

---

[1]**Special Topic – Details Surrounding the Temperature Profile for a Grey Atmosphere**[††]

The integration constant C found in Eq. (4.13) can be obtained by calculating the flux at the stellar surface. As seen in (the optional) Section 3.10, at an optical depth τ the monochromatic Eddington flux is

$$H_v(\tau) = \frac{1}{2}\int_0^\infty S_v(t)E_2(|t-\tau|)dt \tag{4.22}$$

Therefore, a similar equation can be written for the integrated flux, which at the surface is equal to

$$H(0) = \frac{1}{2}\int_0^\infty S(t)E_2(t)dt = H \tag{4.23}$$

Since it was shown (Eq. 4.5 and 4.13) that

$$S(t) = J(t) = 3[Ht + C] \tag{4.24}$$

the integrated flux can be written

$$H = \frac{1}{2}\int_0^\infty 3[Ht + C]E_2(t)dt = \frac{3H}{2}\int_0^\infty tE_2(t)dt + \frac{3C}{2}\int_0^\infty E_2(t)dt \tag{4.25}$$

As seen in Section 3.10, the exponential integral has the property

$$\frac{dE_n(x)}{dx} = -E_{n-1}(x) \tag{4.26}$$

Therefore, the second integral on the right-hand side of Eq. (4.25) becomes (while neglecting the constant $3C/2$)

$$\int_0^\infty E_2(t)dt = -E_3(t)\big|_0^\infty = -E_3(\infty) + E_3(0) = \frac{1}{2} \tag{4.27}$$

(please note that $E_3(\infty) = 0$ and $E_3(0) = 1/2$, see Section 3.10).

---

[1]This special topic requires the content seen in (the optional) Section 3.10.

Meanwhile, the first integral on the right-hand side of Eq. (4.25) may be integrated by parts as follows (again while neglecting the constant $3H/2$)

$$\int_0^\infty tE_2(t)\,dt = tE_3(t)\big|_0^\infty + \int_0^\infty E_3(t)\,dt \tag{4.28}$$

The first term on the right-hand side is nil, while the second term gives $1/3$. Equation (4.25) then gives

$$H = \frac{H}{2} + \frac{3C}{4} \tag{4.29}$$

and consequently the integration constant is $C = \dfrac{2H}{3}$

This is the same result as the one found above by assuming that the optical depth at which the radiation exiting a star is formed is $\tau \approx 2/3$.

Figure 4.1 compares the temperature profile of a grey atmosphere to a detailed model atmosphere with $T_{\text{eff}} = 10\,000\,\text{K}$, $\log g = 4.0$, while assuming solar abundances. As expected, the grey-atmosphere temperature profile is different from the one from a comprehensive numerical model due to the various approximations made within the framework of the grey atmosphere.

The grey-atmosphere temperature profile has several drawbacks. For example, it does not depend on the atomic abundances present in the atmosphere. It is also independent

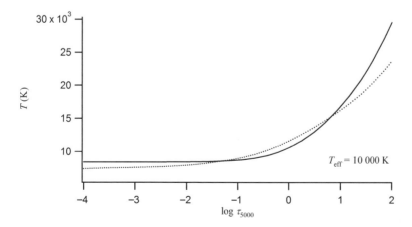

Figure 4.1 Temperature profile of a detailed atmospheric model with $T_{\text{eff}} = 10\,000\,\text{K}$, $\log g = 4.0$ and solar abundances (solid line) and the one for a grey atmosphere (dotted line), as a function of the optical depth calculated at $5000\,\text{Å}$.

of the surface gravity of the star that can in reality strongly affect the temperature profile in stellar atmospheres. Also, it is not clear which optical depth should be chosen when using Eq. (4.21), since no singular optical depth exists in a real stellar atmosphere. In Figure 4.1, the optical depth calculated at $5000\,\text{Å}$ was employed. However, if another optical depth had been used, the comparison between the grey-atmosphere temperature profile and the one of the detailed atmospheric model could have been much different.

Special Topic – Limb Darkening

Observations show that the radiation intensity at the centre of the solar disc is larger than the corresponding intensity detected near its limb. This is called the limb darkening effect. The physical reason for this effect is illustrated in Figure 4.2. Since the local temperature in the atmosphere (and thus the source function) increases with depth and the radiation leaving the star comes from layers with optical depth approximately equal to 2/3, the layers responsible for this radiation are hotter in the centre than near the limb. The limb-darkening effect can be estimated with the results found for the grey atmosphere. As seen in Chapter 3 (see Eq. 3.44), the specific intensity at the surface of stars is

$$I_v(0,u) = \int_0^\infty S_v(t) e^{-\frac{t}{u}} \frac{dt}{u} \tag{4.30}$$

The integrated value of the specific intensity at the surface of a grey atmosphere is given by the following expression

$$I(0,u) = \int_0^\infty S(t) e^{-\frac{t}{u}} \frac{dt}{u} = \int_0^\infty 3H\left[t+\frac{2}{3}\right] e^{-\frac{t}{u}} \frac{dt}{u} = 3H\left[u+\frac{2}{3}\right] \tag{4.31}$$

The direction of the intensity coming from the centre of the disc defines the angle $\theta = 0$ or $u = 1$ (see Figure 4.2), while the intensity near the limb of the disc is related to values of $u < 1$. The ratio of these intensities is

$$\frac{I(0,u)}{I(0,1)} = \frac{3}{5}\left[u+\frac{2}{3}\right] \tag{4.32}$$

At the limb ($u = 0$) this ratio gives 0.4. Therefore, this demonstrates that the intensity for the grey-atmosphere approximation at the limb is 40 % of the value of that at the stellar disc's centre.

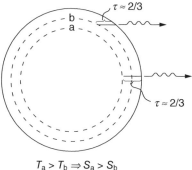

$$T_a > T_b \Rightarrow S_a > S_b$$

Figure 4.2 Illustration of the limb-darkening effect. The radiation from the disc's centre (here the observer is on the right side of the star) comes from hotter layers (with a larger value for the source function) than the radiation coming from the portions of the disc near the limb.

4.2.2[2] Radiative Flux in a Grey Atmosphere[††]

In the previous section, it was found that the integrated average intensity and source function for a grey atmosphere can be approximated by

$$J(\tau) = S(\tau) = 3H\left[\tau + \frac{2}{3}\right] \tag{4.33}$$

In reality, this relation gives the correct asymptotic value for the source function at large optical depths but is not very precise near the surface of stars. To better estimate the integrated average intensity and source function near the surface, they can be written as a function of $q(\tau)$ called the Hopf function such as

$$J(\tau) = S(\tau) = 3H[\tau + q(\tau)] \tag{4.34}$$

and consequently the temperature profile is

$$T(\tau) = T_{\text{eff}}\left[\frac{3}{4}(\tau + q(\tau))\right]^{1/4} \tag{4.35}$$

where $q(\infty) \approx 2/3$. The Hopf function was calculated in detail by the Indian astrophysicist S. Chandrasekhar (1910–1995). With these results, he obtained the flux distribution for a grey atmosphere. In this section, the development leading to this flux will be discussed.

By approximating the monochromatic source function as a Planck function, the Eddington flux may be written (see Section 3.10)

---

[2]This section requires the content presented in (the optional) Section 3.10.

$$H_\nu(\tau) = \frac{1}{2}\int_0^\tau B_\nu(t) E_2(\tau - t)\,dt - \frac{1}{2}\int_\tau^\infty B_\nu(t) E_2(t - \tau)\,dt \tag{4.36}$$

Please note that the optical depth has no frequency subscript since a grey atmosphere is considered here.

The Planck function depends on local temperature that is related to the optical depth by Eq. (4.35). It can then be written

$$B_\nu(T) = \frac{2h\nu^3}{c^2} \frac{1}{e^{\frac{h\nu p(\tau)}{kT_{\text{eff}}}} - 1} \tag{4.37}$$

where for conciseness, a function $p(\tau)$ is defined as

$$p(\tau) = \frac{1}{\left[\frac{3}{4}(\tau + q(\tau))\right]^{1/4}} = \frac{T_{\text{eff}}}{T(\tau)} \tag{4.38}$$

By defining a variable

$$\alpha = \frac{h\nu}{kT_{\text{eff}}} \tag{4.39}$$

a Planck function per unit α may be defined. By the fact that the total integrated flux must be the same for any unit used

$$B_\alpha(T)\,d\alpha = B_\nu(T)\,d\nu \tag{4.40}$$

which leads to

$$B_\alpha(T) = B_\nu(T)\frac{d\nu}{d\alpha} = \frac{2k^4 T_{\text{eff}}^4 \alpha^3}{h^3 c^2} \frac{1}{e^{\alpha p(\tau)} - 1} \tag{4.41}$$

The Eddington flux per unit α is given by the following expression

$$H_\alpha(\tau) = \frac{1}{2}\int_0^\tau B_\alpha(t) E_2(\tau - t)\,dt - \frac{1}{2}\int_\tau^\infty B_\alpha(t) E_2(t - \tau)\,dt \tag{4.42}$$

which gives

$$H_\alpha(\tau) = \frac{k^4 T_{\text{eff}}^4 \alpha^3}{h^3 c^2}\left[\int_0^\tau \frac{E_2(\tau - t)}{e^{\alpha p(t)} - 1}\,dt - \int_\tau^\infty \frac{E_2(t - \tau)}{e^{\alpha p(t)} - 1}\,dt\right] \tag{4.43}$$

It is instructive to write the ratio of the Eddington flux per unit α at depth τ to the total integrated flux H

$$\frac{H_\alpha(\tau)}{H} = \frac{4\pi k^4 \alpha^3}{\sigma h^3 c^2}\left[\int_0^\tau \frac{E_2(\tau - t)}{e^{\alpha p(t)} - 1}\,dt - \int_\tau^\infty \frac{E_2(t - \tau)}{e^{\alpha p(t)} - 1}\,dt\right] \tag{4.44}$$

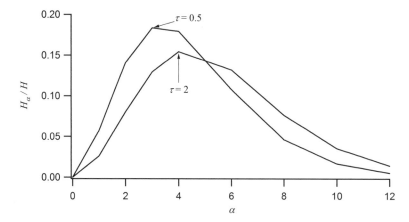

Figure 4.3 The flux from a grey atmosphere at optical depths of $\tau = 0.5$ and 2. The lack of smoothness of the curves is due to the coarseness of the data employed. The data is taken from Chandrasekhar, S., *The Astrophysical Journal*, 101, 328 (1945).

The expression in brackets just above can be tabulated for given values of α and τ. These values were first calculated by S. Chandrasekhar. The monochromatic flux using these results is shown in Figure 4.3. As expected, the maximum of the flux is shifted to higher frequencies (or values of α) as a function of depth (or temperature). This expression may be used to estimate the monochromatic flux at any depth within an atmosphere. However, its validity is limited since it excludes atomic lines and only gives the continuous flux.

4.3 Line Opacities and Broadening

Atomic line opacities play a crucial role in stellar astrophysics. First, they strongly modify the radiative transfer in stars and therefore impact their physical structure. Secondly, the detailed comparison between theoretical line spectra and observed spectra is a powerful diagnostic tool that may be used to better understand stars. It is then fitting to give line opacities appropriate attention. This is the aim of this section. For the sake of briefness, some results shown here emanate from concepts of quantum physics and will be given without demonstration. Such demonstrations are outside the scope of this book.

Quantum theory shows that the opacity of bound–bound atomic transition between a lower atomic energy level i to an upper level j is

$$k_v\rho = \frac{\pi e^2}{m_e c} f_{ij} n_i \varphi_v \qquad (4.45)$$

where f_{ij} is a physical quantity called the oscillator strength of the transition. Its value can vary from 0 to 1. It is proportional to the probability of such a transition taking place. Oscillator strengths can be calculated by quantum-mechanical calculations or obtained experimentally. For example, the oscillator strengths of the first three transitions from the fundamental energy state of the hydrogen atom are: $f_{H\alpha} = 0.6401$, $f_{H\beta} = 0.1193$ and $f_{H\gamma} = 0.04467$. The quantity $\dfrac{\pi e^2}{m_e c} f_{ij}$ is the total integrated cross section of the atomic line. The line profile, which will be discussed in more detail below, is represented by φ_ν. As expected, the opacity of the $i \rightarrow j$ transition is proportional to the number density of potential absorbers n_i (or atoms in level i). In the equation above, the induced (or stimulated) emission of radiation is not included. This topic will be discussed in Section 4.3.4.

As mentioned earlier, atomic lines have a certain width defined by the line profile. The widening of atomic lines is due to three processes: natural, Doppler and pressure (or collisional) broadening. Each one of these three broadening mechanisms will be discussed below.

Spectral studies are very useful and can, for example, lead to an estimate of the abundances of the elements in the atmospheres of stars. This is done by comparing the line intensities in the observed spectra to what is predicted by theoretical calculations. In such calculations, the abundance is adjusted during theoretical computations in order to maximize the quality of the fit of the theoretical atomic lines to the observed lines of the element under consideration. Line widths can also give information concerning the physical conditions in the line-formation regions. For example, two stars with the same spectral type but at different stages of evolution can be distinguished with the help of their observed line widths. A white dwarf, having a much larger density at its surface than a red giant star, has wider atomic lines because of the pressure-broadening mechanism (see Section 4.3.3 for more details). A refined knowledge of spectral line widening mechanisms is essential to extract as much information as possible from observed stellar spectra. The remaining parts of this section are dedicated to describing the physical processes that intervene in atomic line broadening.

For more detailed information than given here about line opacities or more generally about stellar atmospheres, the reader is referred to Mihalas, D., *Stellar Atmospheres*, W.H. Freeman and Company, San Francisco (1970 and 1978).

4.3.1   Natural Broadening

Atomic energy levels can be determined by the resolution of the well-known Schrödinger equation. Such calculations are extremely complicated for all but the simplest configuration: a nucleus with a single electron.[3] These computations determine the energy levels exactly, and the energy difference between two levels defining a given atomic line also gives an exact value. However, the solution found for the energy levels do not take into account Heisenberg's uncertainty principle of quantum mechanics. An electron that finds itself in an energy level has a finite lifetime before transitioning to a lower energy level (with the exception of the fundamental energy level). Because of Heisenberg's uncertainty principle

$$\Delta E \Delta t \geq \frac{h}{4\pi} \tag{4.46}$$

---

[3] The levels of a single-electron atom may also be computed with Bohr's model of the atom.

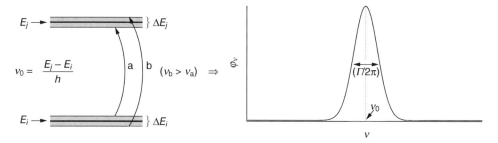

Figure 4.4 Illustration of the effect of the width of atomic energy levels due to the uncertainty principle on the profile of an atomic transition between levels i and j. Here, the energy levels E_i and E_j are those obtained by Schrödinger's equation, while ΔE_i and ΔE_j are the corresponding uncertainties predicted by Heisenberg's uncertainty principle. The value $\Gamma/2\pi$ represents the full width of the profile at half-intensity.

the energy levels have a certain width (ΔE) or uncertainty related to their lifetime (Δt). The effect of the uncertainties of the atomic energy levels on the width of an atomic line is illustrated in Figure 4.4. Schematically, a transition that occurs from let us say the bottom part of level i to the top part of level j has a higher frequency than a transition from the top part of level i to the bottom part of level j. Atoms in level i, can then absorb photons not only at the natural frequency v_0, but also at frequencies surrounding it (see Figures 3.2 and 4.4). Quantum mechanics shows that the line profile caused by this mechanism, called natural broadening, is given by a Lorentz profile

$$\varphi_v = \frac{\dfrac{\Gamma}{4\pi^2}}{(v - v_0)^2 + \left(\dfrac{\Gamma}{4\pi}\right)^2} \tag{4.47}$$

The variable Γ, called the radiative damping constant. The value $\Gamma/2\pi$ represents the full width (as opposed to the half-width) of the profile at half-intensity (see Figure 4.4) of the line profile φ_v.[4] Quantum theory predicts that the value of Γ is equal to the sum of the reciprocal of the mean lifetime of the two atomic energy levels under consideration. It should be noted that the expression for φ_v given above is similar to the amplitude for a classical damped oscillator that is a topic covered in most classical mechanics textbooks. The line profile given above is normalized, i.e.

$$\int_0^\infty \varphi_v dv = 1 \tag{4.48}$$

The line profile may be interpreted in several ways. One way of construing it is that the number density of atoms able to absorb photon of frequency v is equal to $n_i\varphi_v$. The farther

---

[4] Γ is the full width at half-maximum of the profile φ_ω, where $\omega = 2\pi v$.

the frequency of an incoming photon is from the natural frequency of the transition under consideration, the less likely it is for atoms to absorb it.

When the width of a line profile is caused uniquely by natural broadening, a frequency-dependent cross section may then be defined as

$$\alpha(v) = \frac{\pi e^2}{m_e c} f_{ij} \left[\frac{\frac{\Gamma}{4\pi^2}}{(v - v_0)^2 + \left(\frac{\Gamma}{4\pi}\right)^2} \right] \tag{4.49}$$

so that

$$k_v \rho = \alpha(v) n_i \tag{4.50}$$

Typically, the lifetime of excited electrons (i.e. in atomic energy levels other than the fundamental) is on the order of 10^{-8} s (see (advanced optional) Section 4.3.5 for examples). This leads to a line width of $\Delta v \approx 10^8$ Hz or equivalently $\Delta \lambda \approx 10^{-4}$ Å for lines in the visible part of the spectra. This value of line width is much smaller than what is observed in stellar spectra. Therefore, other broadening mechanisms must come into play in stars.

4.3.2   Doppler Broadening

In most stellar plasma, atoms have a velocity distribution that depends on the local temperature: the well-known Maxwell distribution,[5] (see special topic below). Atoms travelling at a certain velocity along the line-of-sight of an observer in the reference frame of the star see a radiation field that is Doppler shifted relative to the field seen by the observer. This causes atoms to have line profiles that are wider than the naturally broadened profile seen above. This line-broadening phenomenon is called Doppler broadening. The aim of this section is to quantify this process and compare it to natural broadening.

Special Topic – Maxwell Distribution

The Maxwell distribution gives the distribution of speeds of particles within an ideal gas. Statistical thermodynamics predicts that particles of mass m inside a gas at temperature T have the probability $f(V)dV$ of having a speed between V and $V + dV$ given by the following expression

$$f(V)dV = 4\pi \left(\frac{m}{2\pi kT} \right)^{\frac{3}{2}} V^2 e^{-\frac{mV^2}{2kT}} dV \tag{4.51}$$

where $V = \sqrt{V_x^2 + V_y^2 + V_z^2}$

---

[5]This is not true for all cases. For example, in very dense stellar plasma such as those found in white dwarf stars, the Maxwell–Boltzmann statistics are not valid there. This topic will be discussed in Chapter 5.

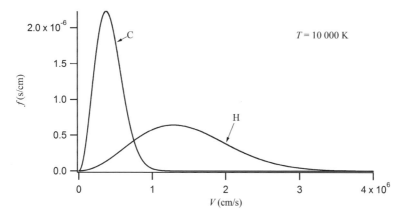

Figure 4.5 The Maxwell distribution for H and C atoms in a gas at 10 000 K. The units of *f* are the reciprocal of those of speed because *f*(*V*)d*V* has no units since it is a probability.

The Maxwell distribution *f* (given in units of s/cm) for two types of atoms in a gas at $T = 10\,000\,\text{K}$ is shown in Figure 4.5. This figure shows that for a given *T*, particles with smaller masses have a larger average speed.

By definition, the mean speed is given by the following expression

$$\bar{V} = \int_0^\infty V f(V) \, dV = \sqrt{\frac{8kT}{\pi m}} \tag{4.52}$$

As mentioned previously, the average speed is larger for less-massive particles. In a gas composed of several types of particles, because of their larger speeds the less-massive particles participate to more collisions per unit time than their more massive counterparts. This explains why, in stellar plasma, free electrons dominate collisions and their related processes such as atomic excitations by collisions for example. It is these collisions that determine the atomic energy-level populations and the ionisation fractions of the species (assuming that radiative processes are negligible).

Meanwhile, the most probable speed V_0, i.e. where $df/dV = 0$, is (see Exercise 4.7)

$$V_0 = \sqrt{\frac{2kT}{m}} \tag{4.53}$$

For a single component of the velocity, V_x for example, the probability distribution is not the same as for the speed *V*. It is given by the following equation

$$f(V_x) \, dV_x = \sqrt{\frac{m}{2\pi kT}} \, e^{-\frac{mV_x^2}{2kT}} \, dV_x \tag{4.54}$$

This last probability distribution will be applied below to evaluate Doppler broadening of atomic lines.

For a given direction (or for the case of interest here, the component along the line-of-sight) the velocity of atoms of mass m in a gas at temperature T is given by the Maxwellian distribution such as the probability of finding an atom with a velocity between V and $V+dV$ is

$$f(V)dV = \sqrt{\frac{m}{2\pi kT}} e^{-\frac{mV^2}{2kT}} dV \tag{4.55}$$

Note here that V defines a single component of velocity and not speed.

By using the Maxwellian distribution for the three-dimensional speed (see above), the most probable speed of an atom is found to be

$$V_0 = \sqrt{V_x^2 + V_y^2 + V_z^2} = \sqrt{\frac{2kT}{m}} \tag{4.56}$$

The one-dimensional Maxwell distribution may then be written as

$$f(V)dV = \frac{1}{\sqrt{\pi}} e^{-\frac{V^2}{V_0^2}} \frac{dV}{V_0} \tag{4.57}$$

Assume an atom that at rest can absorb a photon at frequency v. When travelling at a velocity V relative to the observer, this atom can absorb, according to the Doppler effect, a photon at frequency $v(1-V/c)$ (this equation for the Doppler shift is valid for $V \ll c$, which is almost always the case in stellar plasma). The absorption cross section at frequency v in the observer's reference frame then consists of atoms of various velocities (related to the Maxwell distribution of the line-of-sight velocity) associated to various parts of the naturally broadened profile. A convolution of the naturally broadened cross section with the velocity distribution must then be performed to obtain the true cross section

$$\alpha(v) = \int_{-\infty}^{\infty} \alpha\left(v - \frac{vV}{c}\right) f(V)dV \tag{4.58}$$

By definition of the absorption cross section for a bound–bound transition given in Eq. (4.49), this expression becomes

$$\alpha(v) = \frac{\sqrt{\pi}e^2}{m_e c} f_{ij} \frac{1}{\pi} \int_{-\infty}^{\infty} \left[\frac{\left(\frac{\Gamma}{4\pi}\right) e^{-\frac{V^2}{V_0^2}}}{\left(v - \frac{vV}{c} - v_0\right)^2 + \left(\frac{\Gamma}{4\pi}\right)^2} \right] \frac{dV}{V_0} \tag{4.59}$$

Several variables may be defined to simplify this convolution. First, the Doppler shift Δv at velocity V is

$$\Delta v = \frac{Vv}{c} \tag{4.60}$$

A variable Δv_D, called the Doppler width of the line, is defined by the shift from the line centre for the most probable speed V_0

$$\Delta v_D = \frac{V_0 v_0}{c} \tag{4.61}$$

Its value gives an approximate value of the width of the line due to Doppler broadening.

By defining three other variables v, y and a as follows

$$v = \frac{v - v_0}{\Delta v_D} \tag{4.62}$$

$$y = \frac{\Delta v}{\Delta v_D} \simeq \frac{V}{V_0} \tag{4.63}$$

$$a = \frac{\Gamma}{4\pi\Delta v_D} \tag{4.64}$$

Equation (4.59) becomes

$$\alpha(v) = \frac{\sqrt{\pi}e^2}{m_e c} f_{ij} \frac{H(a, v)}{\Delta v_D} \tag{4.65}$$

where $H(a,v)$ is called the Voigt function. This function is defined as

$$H(a, v) = \frac{a}{\pi} \int_{-\infty}^{\infty} \left[\frac{e^{-y^2}}{(v-y)^2 + a^2} \right] dy \tag{4.66}$$

The integration of this function over all values of v (or in other words over all frequencies) gives $\sqrt{\pi}$. A normalized Voigt function $U(a,v)$ may also be defined (see Example 4.2)

$$U(a, v) = \frac{H(a, v)}{\sqrt{\pi}} \tag{4.67}$$

The Voigt function therefore gives the shape of an atomic line profile with both natural and Doppler broadening. The normalized Voigt profile is then

$$\varphi_v = \frac{U(a, v)}{\Delta v_D} \tag{4.68}$$

Figure 4.6 shows the relative importance of pure Lorentz and Doppler profiles versus a Voigt profile. It shows that in the line core, Doppler broadening dominates. Meanwhile, the wings are determined by the Lorentz profile.

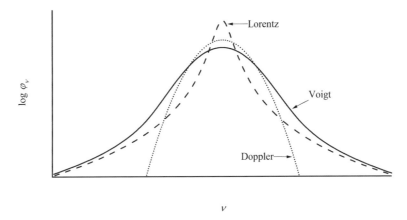

Figure 4.6   Lorentz, Voigt and Doppler profiles of a hypothetical atomic line.

Example 4.2: With the expression found for α_v for a line with both natural and Doppler broadening (Eq. 4.65), prove that, without resorting to an explicit calculation of this integral

$$\int_{c/v_0}^{\infty} H(a, v)\,\mathrm{d}v = \sqrt{\pi}$$

Answer:

The total integrated cross section for an atomic line is

$$\int_{0}^{\infty} \alpha(v)\,\mathrm{d}v = \frac{\pi e^2}{m_e c} f_{ij} \tag{4.69}$$

By using the Voigt profile

$$\int_{0}^{\infty} \alpha(v)\,\mathrm{d}v = \int_{0}^{\infty} \frac{\sqrt{\pi} e^2}{m_e c} f_{ij} \frac{H(a, v)}{\Delta v_D}\,\mathrm{d}v \tag{4.70}$$

and since

$$v = \frac{v - v_0}{\Delta v_D} \tag{4.71}$$

$$\mathrm{d}v = \frac{\mathrm{d}v}{\Delta v_D} \tag{4.72}$$

a change of variable can be performed to give

$$\int_0^\infty \frac{\sqrt{\pi}e^2}{m_e c} f_{ij} \frac{H(a, v)}{\Delta v_D} dv = \int_{c/v_0}^\infty \frac{\sqrt{\pi}e^2}{m_e c} f_{ij} H(a, v) dv \qquad (4.73)$$

Since this integral is equal to $\dfrac{\pi e^2}{m_e c} f_{ij}$, it is trivial to show that

$$\int_{c/v_0}^\infty H(a, v) dv = \sqrt{\pi} \qquad (4.74)$$

Special Topic – Stellar Rotation and Atomic Line Analysis

Stellar rotation also widens the atomic lines in observed spectra. This is caused by the Doppler shifts of the various regions of a star's disc (see Figure 4.7). The spectra obtained are sums of the radiation coming from their entire disc. The radiation coming from the portion of a star's disc moving away from the observer is redshifted, while the radiation coming from the portion of the disc moving towards the observer are blueshifted. The radiation coming from the centre is not Doppler shifted since the velocity is perpendicular to the line-of-sight. The

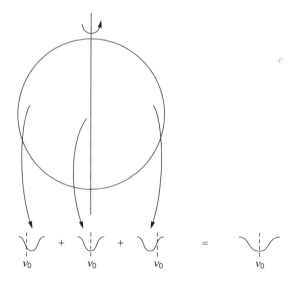

Figure 4.7 Illustration of the effect of a rotating star on an observed atomic line width. The radiation coming from the portion of the star's disc moving away from the observer is redshifted, while the radiation coming from the portion of the disc moving towards the observer are blueshifted. The radiation coming from the centre is not Doppler shifted since the velocity is perpendicular to the line-of-sight. The integrated light signal from the stellar disc thus gives an atomic line that is widened.

no rotation   large rotation

Figure 4.8   Illustration of the blending of two atomic lines. On the left, two atomic lines are shown in a spectrum of a star that does not rotate, while on the right, these two lines are blended together for a star with a large rotational velocity. Note that the blended line is no longer symmetric.

integrated light signal from the stellar disc thus gives atomic lines that are widened. It should be noted that such line broadening is not an intrinsic physical phenomenon, but is an observational artefact that can, however, cause complication for line analysis. Since this process depends on an external (or nonlocal) aspect of the plasma, namely the rotational velocity of stars, it is thus not considered a fundamental line-broadening process such as natural, Doppler or pressure broadening.

For stars with large rotational velocities, some lines may become so broad that they blend with neighbouring lines (with respect to frequency or wavelength) of other species (see Figure 4.8). This makes line identification in such spectra very difficult. It also greatly complicates the determination of the abundance of the elements through line fitting with theoretical models since it is hard to separate the relative importance of each individual line within the blended line.

The effect of line broadening by rotation depends on the angle of the axis of rotation of the star relative to the line-of-sight of the observer. For example, when the axis of rotation points directly towards or away from the line-of-sight, rotational effects are not observable. The reason being, that the bulk velocity of the atoms at the surface of the star is then perpendicular to the line-of-sight and no Doppler shift is observed. The observable rotation velocity at the equator is $V\sin i$, where i is the angle between the line-of-sight and the axis of rotation of the star and V is the rotational velocity at the equator (see Figure 4.9). The value of $V\sin i$ for a given star can be obtained during the fitting of the theoretical spectral lines to those observed.

Figure 4.10 shows the observed spectra of two stars with similar effective temperatures ($T_{eff} \approx 12\,000\,\text{K}$) but very different $V\sin i$. The lines due to various ions are identified in this figure. As expected, the star with the largest $V\sin i$ has much wider lines due to the extra widening effect caused by stellar rotation. Some lines that are clearly defined in the spectrum of the slowly rotating star are blended in the spectra of the fast rotating one.

It should also be noted that turbulent motion of matter in the stellar atmosphere can also broaden atomic lines. Moreover, large movements of matter such as convection may also cause asymmetry in the observed line profiles. All these processes complicate the study of stellar spectra.

Another factor that can come into play when studying stellar spectra is the radial velocity of a star V_r (i.e. the component along the line-of-sight) which can cause Doppler shifts in its atomic lines equal to

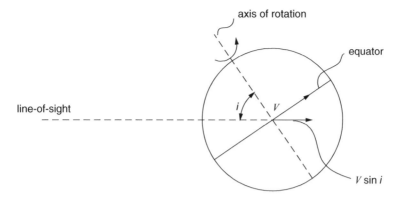

Figure 4.9   Illustration showing that the component of the rotation velocity at the equator (V) along the line-of-sight is $V \sin i$ where i is the angle between the axis of rotation of the star and the line-of-sight.

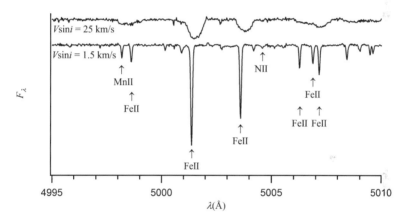

Figure 4.10   Observed flux for two stars of similar effective temperatures ($T_{eff} \approx 12\,000$ K) but with different rotational velocities. The upper curve represents a star that has a relatively large rotational velocity ($V \sin i = 25$ km/s) while the star represented by the lower curve has a small rotational velocity ($V \sin i = 1.5$ km/s). The two spectra are horizontally shifted to distinguish the two curves. The lines from several ions are identified in the figure. These spectra were extracted from the European Space Organization (ESO) Archive, and were processed and kindly obtained from Mouhamadou Thiam.

$$\frac{\Delta \lambda}{\lambda_0} = \frac{\lambda - \lambda_0}{\lambda_0} = \frac{V_r}{c} \qquad (4.75)$$

where λ is the observed wavelength of the centre of a given line and λ_0 is its natural (or theoretical) wavelength (see Exercise 4.5). This fact must be considered when identifying atomic lines in stellar spectra.

4.3.3   Pressure Broadening

A third mode of line broadening is pressure (or collisional) broadening. This process is due to the perturbation of the potential of the atom by neighbouring particles. Fundamentally speaking, when theoretical atomic energy levels are calculated, only the potential due to the nucleus and the bound electrons is considered (i.e. this is the potential that is inserted in Schrödinger's equation). However, in real plasma, other particles may lead to perturbations in this potential and thus modify atomic energy levels. Such perturbations can shift and sometimes split atomic energy levels and they therefore contribute to line broadening. Schematically, at a given time at a given depth in a stellar atmosphere, some atoms are in the process of colliding with other particles while others are between two collisions (and consequently their potential is not strongly affected by other particles in the plasma). Therefore, a portion of the atoms' energy levels is modified by collisions, while other atoms are less affected. Similarly to the process shown in Figure 4.4, these interactions cause line broadening.

One such pressure-broadening process is the Stark effect. This broadening effect is caused by the splitting of degenerate atomic energy levels due to the presence of an external electric field. This effect can also cause a frequency shift for these levels. In stellar plasma, an electric field created by surrounding particles can be felt within the atom under consideration. The lines arising from the individual split levels cannot be observed in stellar spectra because of the relatively small separation between them. They blend into one line that is, however, widened compared to the case where no Stark effect exists.

Another pressure-broadening mechanism is the Van der Waals process. This is related to the perturbation of an atom's potential by neutral atoms. The neutral atoms' electric dipole is at the source of this interaction. Meanwhile, the broadening of a given species due to its interaction with the atoms of the same species is called resonance broadening. These last two types of interaction lead to a Lorentz profile. Therefore, a collisional damping constant Γ_{coll} may be added to the radiative damping constant found in Eq. (4.47) in order to take these last two types of interactions into account.

Figure 4.6 showed that Doppler broadening dominates the central regions of atomic lines. The effect of pressure broadening is mostly seen in the wings of atomic lines. An example of this result is shown in Figure 4.11 for the H_γ line and is discussed in the special topic presented below.

Special Topic – Balmer Lines and Surface Gravity

As discussed in Chapter 1, photometry can be used to estimate the effective temperature of stars. Meanwhile, surface gravity can be determined by spectroscopic means and more specifically by the study of hydrogen Balmer lines. These lines are quite sensitive to pressure broadening (in this case the Stark effect), and their shape depends strongly on the pressure found in the line-formation region and therefore on the surface gravity since the hydrostatic equilibrium

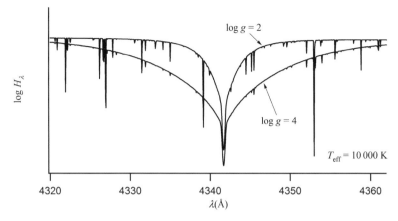

Figure 4.11   The surface flux within the H_γ line at the surface of atmospheres with $T_{\text{eff}} = 10\,000\,\text{K}$ but with different surface gravities typical of main-sequence ($\log g = 4$) and supergiant ($\log g = 2$) stars. Other atomic lines from various metals are also seen within the H_γ line.

equation predicts larger pressures in stellar atmospheres with larger surface gravities.

Figure 4.11 shows the effect of surface gravity on the H_γ line. Its wings are wider in the atmosphere with the larger surface gravity. Since supergiants have lower surface gravities, it is then possible, for example, to distinguish between a main-sequence star and a supergiant with the same effective temperature. The stellar surface gravities obtained in this way are commonly called spectroscopic gravities.

It should also be noted that Figure 4.11 shows that hydrogen lines are much wider than atomic lines of other elements. Also, some atomic lines from metals that are visible within the H_γ line for the $\log g = 2$ model disappear (or are less prevalent) for the $\log g = 4$ model since the H_γ line opacity becomes dominant there. This example illustrates the power of the analysis tool of stellar spectroscopy.

For a given series of hydrogen lines, Stark splitting increases with the principal quantum number n of the upper level. A finite number of Balmer lines is then visible in the spectrum of an A-type star. The reason being that lines near the Balmer jump (i.e. with larger n for the upper level) eventually merge together (see Figure 1.8) because of their larger width. Historically, the number of Balmer lines distinguishable in a stellar spectrum has been used to estimate the electron density in the line-forming region. Nowadays, more detailed comparison between line shapes, depths and widths from theoretical model atmospheres and those observed is used to gauge more precisely the physical structure of the outer regions of stars.

Special Topic – Zeeman Effect

Strong magnetic fields are detected on the surface of some stars. The presence of a magnetic field can cause splitting of degenerate atomic energy levels by lifting this degeneracy. This phenomenon is called Zeeman splitting and also contributes to the widening of the atomic lines in observed spectra. The presence of a magnetic field causes an atomic line to be formed by various components. For example, in the illustration shown Figure 4.12, a line is composed of three components when a magnetic field is present. Generally, since the separation of the Zeeman-splitted atomic levels is relatively small (except for very large magnetic fields), the individual components of an atomic line cannot be separated in stellar spectra but instead lead to a single line that is wider than when no magnetic field is present. It should be noted that such line broadening is not considered a fundamental broadening process such as natural, Doppler or pressure broadening, since it depends on an aspect external (or nonlocal) to the plasma, namely the presence of a magnetic field.

The study of polarized line transfer can give additional information about the atomic lines affected by the Zeeman effect and is a powerful tool in the study of stars with magnetic fields such as Ap-type stars (see (optional) Section 7.2.2 for more details about these magnetic stars). Detailed observations of stellar spectra using a spectropolarimeter are compared to theoretical calculations. This observing method can be used, for example, not only to evaluate the intensity of the magnetic field in the atmospheres of stars but also the configuration of this field on the stellar surface.

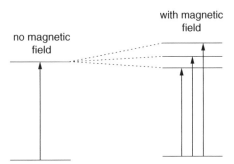

Figure 4.12 Illustration of Zeeman splitting of an atomic energy level and its effect on the atomic transitions.

4.3.4 Stimulated Emission and Masers

There are three modes by which photons can be absorbed or emitted via transitions between atomic energy levels. Absorption and spontaneous emission were already discussed in Chapter 3. Absorption contributes to the line opacity, while spontaneous

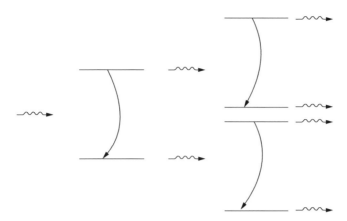

Figure 4.13 Illustration of stimulated (or induced) emission of radiation from a bound–bound atomic transition. First, a photon induces an emission from an excited atom that cascades to a lower level. Thereafter, the two photons can in turn induce two other excited atoms to emit two additional photons. If there exists a sufficient number of exited atoms, the number of photons in the beam of radiation may grow exponentially.

emission is counted as emissivity. A third mode, namely stimulated (or induced) emission can also come into play. Induced emission, is a de-excitation process due to the presence, near the atom, of a passing photon with same energy than the atomic transition under consideration. Theory predicts that the photon that is emitted following the de-excitation of the electron in this manner travels in the same direction than the photon that induced the transition (see Figure 4.13). The intensity of induced emission therefore depends on the radiation field (or I_v). This process is thus not equivalent to a spontaneous emission that is isotropic with respect to direction and is independent of I_v. Stimulated emission is rather treated like a negative opacity because it adds photons to I_v instead of taking some away as for the case of opacity. Quantum theory shows that the opacity due to an atomic transition between a lower atomic level i and an upper level j while including stimulated emission is given by the following expression

$$k_v\rho = \frac{\pi e^2}{m_e c} f_{ij}\varphi_v n_i \left(1 - \frac{n_j g_i}{n_i g_j}\right) \tag{4.76}$$

as opposed to the line opacity given in Eq. (4.45). The portion of Eq. (4.76) in parenthesis is related to stimulated emission. In LTE (i.e. where the Boltzmann equation applies), $\frac{n_j g_i}{n_i g_j} = e^{\frac{-hv_0}{kT}}$ and the bound–bound opacity may then be written

$$k_v\rho = \frac{\pi e^2}{m_e c} f_{ij}\varphi_v n_i \left(1 - e^{\frac{-hv_0}{kT}}\right) \tag{4.77}$$

The correction factor due to stimulated emission $\left(1 - e^{\frac{-hv_0}{kT}}\right)$ decreases the opacity.

However, this correction factor is only valid for LTE. When nonthermal processes are present, they can invert the population (i.e. increase the population of the upper level of the transition relative to the lower level as compared to LTE populations) and in some cases lead to $\dfrac{n_j g_i}{n_i g_j} > 1$, and this therefore gives a negative total opacity. In this case, the specific intensity is exponentially amplified when traversing the plasma instead of being attenuated (such as in Example 3.3 for instance). This physical phenomenon is the source of the laser (light amplification by stimulated emission of radiation).

Lasers consist of a gas whose atoms are 'pumped' to an excited atomic state by a nonthermal process such as electricity (or collisions). To be efficient, the upper level needs to have a sufficiently long half-life so that a large number of excited atoms may exist at a given time. Such atomic energy levels are commonly called metastable levels. When photons at the natural frequency of the atomic transition that intervenes traverse the laser, the signal is amplified (see Figure 4.13). They induce stimulated emission via de-excitations of electrons from the metastable level to the proper lower atomic level.

Astronomical masers (microwave amplification by stimulated emission of radiation) naturally occur in the universe. These microwave sources are due to molecules (such as OH, H_2O, SiO, etc.) found in interstellar or circumstellar gas. They are observed as emission lines. The agent pumping the molecules to the proper excited state is often due to collisions in the plasma or sometimes caused by radiation coming from nearby stars for masers in interstellar clouds or the central star for circumstellar masers. For example, H_2O has a maser transition at a frequency of 22.23 GHz, while SiO has such transitions at 43.12 and 86.24 GHz.

Stimulated emission discussed here was related to bound–bound transitions. However, the opacity due to bound–free and free–free opacities also needs to be corrected for induced emission.

4.3.5   Einstein Coefficients[††]

As discussed previously, there are three types of transitions possible for bound–bound transitions: spontaneous de-excitation, radiative excitation and stimulated emission. The so-called Einstein coefficients define the transition probabilities between two atomic levels for each of these three types of transitions. The coefficients A_{ji}, B_{ij} and B_{ji} measure the respective transition probability for spontaneous de-excitations, radiative excitation and stimulated emission. The rate of transitions in stellar plasma for each of these three types of transitions depend on the density of atoms in the proper level (i.e. in level i for radiative excitations, and in level j for spontaneous and stimulated emissions). The rate of transitions for both radiative excitations and stimulated emissions also depend on the strength of the radiative field.

By making the approximation that at each frequency the same number of photons is emitted than the quantity that is absorbed (this approximation is commonly called detailed balancing), the following equation may be written

$$n_i B_{ij} I_v = n_j A_{ji} + n_j B_{ji} I_v \tag{4.78}$$

The left-hand side of this equation is proportional to the rate of absorption, while the two terms on the right-hand side, respectively, describe the rates of spontaneous and stimulated emissions. Specific intensity appears for radiative absorption and stimulated emission because these processes depend on the radiation field. Relations between the Einstein coefficients can be obtained by simple algebraic manipulations and logic. First, the specific intensity may be isolated in the equation above

$$I_v = \frac{n_j A_{ji}}{n_i B_{ij} - n_j B_{ji}} = \frac{A_{ji}}{B_{ji}} \frac{1}{\dfrac{n_i B_{ij}}{n_j B_{ji}} - 1} \tag{4.79}$$

Assuming that the atomic populations respect the Boltzmann equation

$$\frac{n_j}{n_i} = \frac{g_j}{g_i} e^{-\frac{h v_0}{kT}} \tag{4.80}$$

Eq. (4.79) becomes

$$I_v = \frac{n_j A_{ji}}{n_i B_{ij} - n_j B_{ji}} = \frac{A_{ji}}{B_{ji}} \frac{1}{\dfrac{g_i B_{ij}}{g_j B_{ji}} e^{\frac{h v_0}{kT}} - 1} \tag{4.81}$$

The relation between the Einstein coefficients can be found by assuming that the specific intensity is equal to the Planck function (i.e. assuming strict LTE) and since at the line centre

$$B_{v_0} = \frac{2h v_0^3}{c^2} \frac{1}{e^{\frac{h v_0}{kT}} - 1} \tag{4.82}$$

a comparison of the two equations above gives rise to the following relations between the Einstein coefficients

$$\frac{A_{ji}}{B_{ji}} = \frac{2h v_0^3}{c^2} \tag{4.83}$$

$$\frac{g_i B_{ij}}{g_j B_{ji}} = 1 \tag{4.84}$$

The results were obtained by assuming Boltzmann statistics for the atomic populations and strict LTE ($I_v = B_v$). However, since the Einstein coefficients are physical constants, the relations found for these coefficients are universal because their value does not depend on the physical conditions imposed. It should also be mentioned that for a given transition, only one Einstein coefficient is needed to calculate the other two.

Table 4.1   Examples of Einstein coefficients and oscillator strengths.

Transition	A_{ji}	f_{ij}
L_α	$4.699 \times 10^8 \, \text{s}^{-1}$	0.4162
L_β	$5.575 \times 10^7 \, \text{s}^{-1}$	7.910×10^{-2}
L_γ	$1.278 \times 10^7 \, \text{s}^{-1}$	2.899×10^{-2}
H_α	$4.410 \times 10^7 \, \text{s}^{-1}$	0.6407
H_β	$8.419 \times 10^6 \, \text{s}^{-1}$	0.1193
H_γ	$2.530 \times 10^6 \, \text{s}^{-1}$	4.467×10^{-2}

Quantum theory shows that there also exists a relation between the Einstein coefficients and the oscillator strength that is given by the following expression

$$B_{ij} \frac{h\nu_0}{4\pi} = \frac{\pi e^2}{m_e c} f_{ij} \tag{4.85}$$

and the opacity of the atomic line $i \rightarrow j$ can thus be written

$$k_\nu \rho = B_{ij} \frac{h\nu_0}{4\pi} \varphi_\nu n_i \left(1 - e^{\frac{-h\nu_0}{kT}} \right) \tag{4.86}$$

The Einstein coefficients are obtained with complex quantum-mechanical calculations. These quantities (or equivalent data like the oscillator strength) are given in several atomic data bases commonly used in astrophysics. Here are a few examples of such databases: National Institute of Standards and Technology (NIST) atomic spectra database (http://physics.nist.gov/PhysRefData/ASD/), Robert L. Kurucz's database (http://www.pmp.uni-hannover.de/cgi-bin/ssi/test/kurucz/sekur.html) and The Opacity Project atomic database or TOPBase (http://cdsweb.u-strasbg.fr/topbase/topbase.html). Table 4.1 gives examples of such coefficients for the first three lines of the Lyman and Balmer series of hydrogen.

The radiative damping constant Γ_{ij}, which defines the width of a naturally broadened line, is related to the spontaneous de-excitation Einstein coefficients. The radiative damping constant of a transition between levels i and j is

$$\Gamma_{ij} = \sum_{k<j} A_{jk} + \sum_{k<i} A_{ik} \tag{4.87}$$

The two terms on the right-hand side of this equation are, respectively, the reciprocal of the mean lifetime of levels j and i. These sums give the total spontaneous de-excitation rates from each level. They include all of the channels by which an electron on a given level j or i may cascade to a lower level. For example, the radiative-damping constant of the L_α atomic line is simply A_{21} since the spontaneous de-excitation rate is nil for the fundamental state (or in other words, its mean lifetime is infinite) and only one channel of decay exists for the level under consideration.

4.4   Equivalent Width and Formation of Atomic Lines

As discussed above, the detailed study of the shape of an atomic line in a stellar spectrum can furnish precious information regarding the physical conditions in the line-forming region of the atmosphere. However, when the spectral resolution of the spectrometer is low or/and when the observed signal is weak, the stellar spectrum is sometimes not of sufficient quality for a detail comparison between these observations and theoretical spectra. Also, since some parameters related to line-profile calculations such as those related to pressure broadening are often not well known, detailed spectroscopic studies are not always valuable due to lack of accuracy. In such cases, a more general way of interpreting and studying atomic lines without having to measure or calculate the detailed shape of the atomic lines is warranted. This is one of the reasons for defining an equivalent width for atomic lines. Following a discussion dedicated to the definition of the equivalent width, its application to weak atomic lines and general results for lines of any intensity will be discussed.

4.4.1   Equivalent Width

The equivalent width W_λ of an atomic line is defined as the width of a hypothetical atomic line of rectangular shape that absorbs all of the radiation within it and the same total amount of energy as the atomic line associated to it. Figure 4.14 shows graphically the meaning of the equivalent width where F_c is the continuum flux (i.e. outside the atomic line) and F_λ is the monochromatic flux. Therefore, since the real atomic line and the fictitious rectangular line absorb the same amount of radiative energy

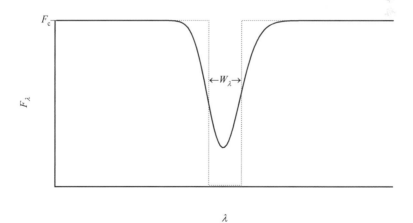

Figure 4.14   Schematic definition of the equivalent width (W_λ) of an atomic line. The fictitious rectangular line which absorbs all photons within it has a width such that it absorbs the same quantity of energy as the atomic line to which it is associated. The quantity F_c is the flux of the continuum.

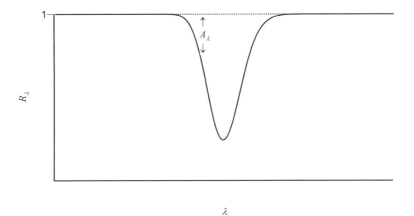

Figure 4.15   Illustration of the residual intensity (R_λ) and the line (or absorption) depth (A_λ).

$$F_c W_\lambda = \int [F_c - F_\lambda] d\lambda \tag{4.88}$$

or

$$W_\lambda = \int \left[1 - \frac{F_\lambda}{F_c} \right] d\lambda = \int [1 - R_\lambda] d\lambda = \int A_\lambda d\lambda \tag{4.89}$$

where the integration is done over the entire spectrum. In reality, since the absorption of an atomic line falls off quickly at wavelengths far from its natural wavelength, this integration can be limited to a range of several times the value of the equivalent width on each side of the line centre. The quantity R_λ (see Figure 4.15) is called the residual intensity of the atomic line and is given by the ratio of the flux within a line to the value of the flux in the continuum

$$R_\lambda = \frac{F_\lambda}{F_c} \tag{4.90}$$

while A_λ is called the line (or absorption) depth

$$A_\lambda = 1 - R_\lambda = 1 - \frac{F_\lambda}{F_c} \tag{4.91}$$

The equivalent width is a measure of the strength of an atomic line and is normally given in units of Å. It is clear that this quantity is a global property of a line, since lines with very different profiles can have the same equivalent width or in other words, absorb the same amount of energy. It will be shown below (in (optional) Section 4.4.3) that the equivalent width can be used to estimate the abundance of the element responsible for the atomic line under consideration.

Example 4.3: Calculate the equivalent width of a rectangular line with a 4-Å width and with a flux in its interior that is 2/3 of that of its value in the continuum.

Answer:

The line depth for the line under consideration is

$$A_\lambda = 1 - \frac{F_\lambda}{F_c} = \frac{1}{3} \tag{4.92}$$

To calculate the equivalent width, Eq. (4.89) must, in theory, be integrated over the whole spectrum. But since the line depth under consideration is nil outside the rectangular width, the integration may be done from x to $x + 4\,\text{Å}$, where x is the wavelength where the rectangular line begins

$$W_\lambda = \int A_\lambda d\lambda = \int_x^{x+4} \frac{d\lambda}{3} = \frac{4}{3}\,\text{Å} \tag{4.93}$$

4.4.2 Formation of Weak Atomic Lines

One of the principal aims of studying stellar atmospheres is to calculate theoretical stellar spectra. Such detailed calculations are complex and are outside the reach of analytical analysis and require numerical modelling. However, by making simple approximations, an analytical expression for the line depth of weak atomic lines may be found. This is the aim of this section. Weak atomic lines are lines for which their opacity is much smaller than the opacity of the continuum. Such lines are commonly called optically thin lines because of their relatively weak opacity.

The intensity of the flux in the core of absorption lines is weaker than in the continuum because the radiation near their natural wavelength λ_0 comes from shallower regions than for the continuum. This is due to the fact that the opacity is larger in the core than in the wings. As discussed in Section 4.3, the atmospheric layers responsible for most of the flux at a given wavelength λ come from optical depths $\tau_\lambda \approx 2/3$. The value of $\tau_\lambda \approx 2/3$ is situated at geometrically deeper (and thus hotter) regions in the wings as compared to wavelengths in the core. The radiation from different parts of an atomic line comes from various geometrical depths (of varying temperature) of the atmosphere. The intensity of the radiation in the line centre is weaker than in the wings thus explaining the shape of absorption lines (more quantitative details are given below).

By approximating the monochromatic flux at the surface of stars by the flux of a blackbody $F_\lambda = \pi B_\lambda$ (see Example 3.1) and by supposing that all of the radiation comes from a single layer situated at an optical depth $\tau_\lambda = 2/3$, the line depth may be approximated by the following expression

$$A_\lambda = \frac{F_c - F_\lambda}{F_c} \approx \frac{B_\lambda(\tau_c = 2/3) - B_\lambda(\tau_\lambda = 2/3)}{B_\lambda(\tau_c = 2/3)} \tag{4.94}$$

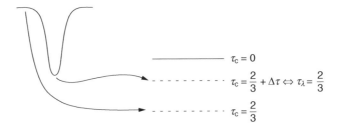

Figure 4.16   Illustration showing the depth at which the continuum is formed ($\tau_c = 2/3$) and where the atomic line under consideration is formed ($\tau_c = 2/3 + \Delta\tau$) where $\Delta\tau < 0$. The depth $\tau_c = 2/3 + \Delta\tau$ corresponds to the depth $\tau_\lambda = 2/3$.

where

$$d\tau_\lambda = -k_\lambda \rho dz \qquad (4.95)$$

$$d\tau_c = -k_c \rho dz \qquad (4.96)$$

$$k_\lambda = k_l + k_c \qquad (4.97)$$

The variable k_l is the monochromatic opacity due to the atomic line under consideration, k_c is the opacity due to continuous sources of opacities (such as electron scattering for example) and k_λ is the total monochromatic opacity. Since the total opacity in the continuum (i.e. far from any atomic line centre) is due mostly to k_c, the total optical depth there is such as $\tau_\lambda = \tau_c$ and therefore the flux in the continuum is proportional to the Planck function found at $\tau_c = 2/3$ (i.e. $F_c \approx \pi B_\lambda(\tau_c = 2/3)$). Meanwhile, the flux inside the line can be approximated by $F_\lambda \approx \pi B_\lambda(\tau_\lambda = 2/3)$. Typically, within the line widths, the continuum opacity can usually be considered constant (or independent of wavelength), except for very wide lines such as hydrogen lines.

The geometrical depth where $\tau_\lambda = 2/3$ can be related to an optical depth in the continuum $\tau_c = 2/3 + \Delta\tau$ (see Figure 4.16) where $\Delta\tau < 0$. The line depth may then be written

$$A_\lambda \approx \frac{B_\lambda(\tau_c = 2/3) - B_\lambda(\tau_c = 2/3 + \Delta\tau)}{B_\lambda(\tau_c = 2/3)} \qquad (4.98)$$

For weak lines, the core is formed at layers not too far from those where the continuum radiation emanates ($|\Delta\tau| \ll 1$) and the Planck function in the line may be approximated by the following Taylor expansion near $\tau_c = 2/3$

$$B_\lambda(\tau_\lambda = 2/3) = B_\lambda(\tau_c = 2/3 + \Delta\tau) \approx B_\lambda(\tau_c = 2/3) + \frac{dB_\lambda}{d\tau_c}\bigg|_{\tau_c=2/3} \Delta\tau \qquad (4.99)$$

Since $\Delta\tau < 0$ and $\dfrac{dB_\lambda}{d\tau_c} > 0$ (assuming that the temperature increases with depth[6]), within the line $B_\lambda(\tau_\lambda = 2/3) < B_\lambda(\tau_c = 2/3)$, and as explained previously, the radiative intensity in

---

[6]Generally speaking, the temperature increases with optical depth in stellar atmospheres. However, non-LTE effects can in some instances cause a temperature rise in parts of the outer atmosphere. Such effects are not considered here.

the core is smaller than in the continuum. This gives rise to an absorption profile for atomic lines in the stellar flux.

Therefore, for weak atomic lines, the line depth (Eq. 4.94) may be approximated by

$$A_\lambda \approx -\frac{1}{B_\lambda}\frac{dB_\lambda}{d\tau_c}\bigg|_{\tau_c=2/3} \Delta\tau = -\frac{d\ln B_\lambda}{d\tau_c}\bigg|_{\tau_c=2/3} \Delta\tau \tag{4.100}$$

where $\Delta\tau = \tau_c - 2/3$ and τ_c is the optical depth in the continuum associated to the geometrical depth where the line forms (see Figure 4.16). From the definition of the two optical depth scales τ_c and τ_λ, the following relation exists between the optical depth scales and the opacities

$$\frac{\tau_c}{\tau_\lambda} = \frac{k_c}{k_\lambda} = \frac{k_c}{k_l + k_c} \tag{4.101}$$

At the depth τ_c linked to the geometrical depth where the line forms (or where $\tau_\lambda = 2/3$), the equation above leads to the expression

$$\tau_c = \frac{2}{3}\left[\frac{k_c}{k_l + k_c}\right] \tag{4.102}$$

Since for weak atomic lines, the line opacity is much smaller than that of the continuum ($k_l \ll k_c$), the portion of the equation above found in brackets may be approximated by the following expression

$$\frac{k_c}{k_l + k_c} \approx 1 - \frac{k_l}{k_c} \tag{4.103}$$

Also, since $\Delta\tau = \tau_c - 2/3$, it is trivial to show that

$$\Delta\tau = -\frac{2}{3}\frac{k_l}{k_c} \tag{4.104}$$

The line depth can then be written

$$A_\lambda \approx \frac{2}{3}\frac{k_l}{k_c}\frac{d\ln B_\lambda}{d\tau_c}\bigg|_{\tau_c=2/3} \tag{4.105}$$

It was seen in Section 4.2 that the line opacity k_l is proportional to the abundance of the element responsible for the atomic line. Generally speaking, the opacity of the continuum k_c is more or less independent of the abundance of a given element, except when an element's abundance is relatively large such as for hydrogen for instance. Therefore, A_λ and also W_λ are proportional to the elemental abundance for the case of optically thin lines treated here. The measurement of the equivalent width of atomic lines can then be a useful tool to estimate the abundance of the elements present in the atmospheres of stars. This topic will be discussed further in (optional) Section 4.4.3.

With the assumptions made above, for the case of an isothermal plasma $\dfrac{d \ln B_\lambda}{d\tau_c} = 0$, therefore $W_\lambda = 0$. Under such conditions, no atomic lines are visible in the spectrum. The presence of atomic lines in stellar spectra is then intrinsically linked to the variation of the physical conditions of the atmosphere as a function of depth.

4.4.3   Curve of Growth[†]

In the previous section, it was shown that for weak atomic lines, the equivalent width increases linearly with the abundance of the element. In this case, the quantity of energy absorbed is proportional to the number of atoms of that element that is present in the line-forming region. Such a line is said to be unsaturated. However, if the abundance could be increased in the star's atmosphere, the flux in the line's core would eventually reach a minimum set at $\pi B_\lambda(\tau = 0)$ (or the flux emitted by a blackbody of temperature $T(\tau = 0)$, see Figure 4.17). The line is then said to be saturated, and the amount of energy absorbed by it is no longer linearly proportional to the abundance but rather it is found that $W_\lambda \propto \sqrt{\ln N}$, where N represents the abundance of the element. At even larger abundances, the absorption in the wings of the lines increases due to pressure broadening and the equivalent width then varies as $W_\lambda \propto \sqrt{N}$ (the abundance may be given for instance as the fraction of the total number of atoms that is of that element, see Appendix E). The dependence of the equivalent width as a function of abundance is called the curve of growth and is represented schematically in Figure 4.18. This curve can be used to estimate the abundance of an element found in a given star by measuring the equivalent width of its atomic lines.

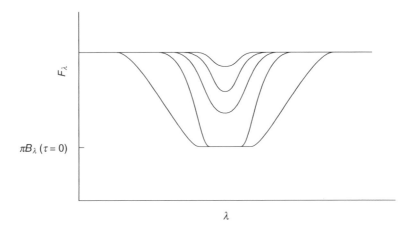

Figure 4.17   Illustration of the varying shape of an atomic line as the abundance increases. It goes from an unsaturated to a saturated condition. As the line deepens, the flux eventually attains a minimum, which when assuming LTE is equal to $\pi B_\lambda(\tau = 0)$.

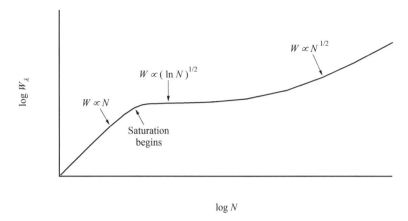

Figure 4.18   Illustration of the equivalent width (W_λ) as a function of the abundance (N) of the species for a given atomic line (commonly called the curve of growth). The dependence of the equivalent width with respect to abundance for the various parts of the curve is given in the figure. The approximate position where the line begins to be saturated is also shown in this figure.

4.5   Atmospheric Modelling

Model atmospheres are numerical models that attempt to properly estimate the structure of the outer regions of stars, the structure being defined as the various physical quantities such as T, ρ, r, etc. as a function of a chosen optical depth. Since the knowledge of the physical structure of atmospheres is critical for calculating precise theoretical spectra, the field of stellar atmospheres is very important for better understanding stars. Such theoretical spectra serve as a critical link between theory and observational stellar astronomy.

In this section, the various aspects surrounding the numerical computation of stellar atmosphere models will be discussed. To render the stellar atmosphere problem numerically feasible, a number of approximations are normally assumed. Such approximations, along with the data necessary for atmospheric modelling will be presented. To calculate a stellar atmosphere model, a series of physical equations must be solved in a deliberate way. The algorithm typically used to achieve the goal of computing the atmospheric structure will be described in (advanced optional) Section 4.5.2. Examples of stellar atmosphere models will then be presented. Finally, an advanced and optional section depicting a method for correcting the temperature in a stellar atmosphere calculation will be presented. As will be seen below, proper temperature correction during atmospheric modelling is a critical part of the stellar atmosphere algorithm.

4.5.1   Input Data and Approximations

As previously mentioned in Chapter 1, global properties of stars can be defined by three fundamental parameters: mass, radius and luminosity. However, other equivalent physical quantities can also be used to define stars. For example, with the knowledge of the radius and the luminosity, the effective temperature can be obtained (see Eq. 1.10). Any two of the parameters: luminosity, radius or effective temperature, plus the mass are equivalent to the three fundamental parameters listed above. Also, since $g = GM_*/R_*^2$, the surface gravity can also be calculated with the knowledge of the mass and the radius. Again, a set composed of any two of the parameters: mass, radius or surface gravity, plus either the luminosity or effective temperature completely defines the global properties of a star.

For stellar atmosphere modelling, only the outer regions are of interest. We may then suppose that the gravitational acceleration is constant since r does not vary much from R_* within the atmosphere. The geometrical depth of atmospheres is typically on the order of 1 % of the stellar radius. This depth is determined by the optical depth chosen for the deepest layer of the atmosphere that is typically set at $\tau \approx 10^3$ (see discussion below). The geometrical depth of atmospheres of supergiant stars are relatively larger (up to several percentage points of their radius) than the depth of atmospheres of main-sequence stars because of the lower density of their plasma. The density at the surface of supergiant stars is smaller due to their relatively weak surface gravity. The three main input parameters that must be furnished to the model can then be the radius, surface gravity and effective temperature. However, since the thickness of stellar atmosphere models is much smaller than the stellar radii, they can be modelled by plane-parallel layers. In the plane-parallel approximation for stellar atmospheres, the radius is of no importance and only two parameters remain: gravity and effective temperature. Gravity determines the pressure profile inside the atmosphere via the hydrostatic equilibrium equation, assuming a star at equilibrium or not in a quickly evolving evolutionary stage. Most atmospheric models are only valid for stars in a static phase; however, some more modern codes can take global movements of the medium due to evolutionary effects into account. Meanwhile, the effective temperature defines the integrated flux at the stellar surface.

Typically, stellar atmosphere models are divided into 50 to 100 layers, for each of which the physical properties such as local temperature, pressure, density, etc. are desired. These layers are chosen in a way to properly sample optical depths from approximately 10^{-8} to 10^3. The optical depth is normally chosen at a wavelength in the visible region of the spectrum.

The resolution of the radiative-transfer equation requires the knowledge of monochromatic radiative opacities. The abundances of the elements present must then be given as input along with the pertinent atomic data for all of the ions encountered within the atmosphere. Since only the exterior parts of the star are modelled, only the first few ions of each element present are of importance. The following atomic data are needed for opacity computations: the atomic energy levels and their degeneracy, the ionisation energies of the various ions, the photoionisation cross sections from the various atomic levels and the *gf* values of the atomic transitions. The *gf* value for an atomic transition $i \rightarrow j$ is equal to $g_i f_{ij}$ and is sometimes called the weighted oscillator strength. The knowledge of the product of g_i and f_{ij} is sufficient for line-opacity calculations because line opacity is proportional to this value (see Exercise 4.2). For cool stars, the relevant data for the molecules present there must also be given since they can contribute significantly to the radiative opacity.

The radiative-transfer equation is solved at a given number (typically around 10) of directions (or values of u) on a chosen frequency grid. This grid is typically composed by on the order of 10^4 points. The frequency points are chosen so that they properly sample the spectral regions where the radiation field is strong. A detailed knowledge of the radiative field is not critical for stellar atmosphere models because their structural properties depend on global aspects of the radiation field. Once an atmospheric model is obtained it can then be used to calculate the flux on a finer frequency grid that may be compared to observed spectra.

In stellar atmospheres, the equation of state chosen is normally that of an ideal gas. This is a good approximation except, for example, for white dwarfs, where the densities typically found in their atmosphere can necessitate a more elaborate equation of state. This topic will be discussed in Chapter 5.

In cooler stars, where convection can occur in the atmosphere, convection must be included in the modelling since it participates in energy transport and can therefore modify the atmospheric structure. A description of convection and the physical data that intervene for this physical process will be discussed in more detail in Chapter 5.

4.5.2   Algorithm for Atmospheric Modelling[††]

Detailed numerical calculation of a stellar atmosphere necessitates a number of successive and purposeful steps. It is an iterative process in which, for plane-parallel atmospheres for example, a structure is sought such that it leads to a constant integrated flux $F = \sigma T_{eff}^4$ throughout the atmosphere or in other words in each layer of the numerical model. Such numerical modelling requires powerful computing resources since a variety of equations must be solved and a large quantity of data is needed.

The algorithm begins with loading fundamental physical quantities defining the atmosphere: T_{eff}, log g and the abundances of the elements present. The various atomic and molecular data required must also be loaded by the computer code. For spherical model atmospheres, the radius of the star is also needed. Figure 4.19 illustrates the algorithm discussed in this section.

The atmosphere is divided into a finite number of layers (typically 50 to 100). The layers of the model atmosphere are chosen in such a way that they properly sample the optical depths from approximately log $\tau = -8$ to +3. The layers are usually divided using a constant increment $\Delta(\log \tau)$, where τ is an optical depth scale at a predetermined wavelength. The hydrostatic equilibrium equation may be written (see Exercise 4.11)

$$\frac{d \ln P}{d \ln \tau} = \frac{g \tau}{kP} = \frac{d \log P}{d \log \tau} \qquad (4.106)$$

where k is the opacity associated to the chosen optical depth scale. This form of the hydrostatic equilibrium equation is numerically useful since it incorporates the variable log τ, which is the quantity with which the atmosphere is divided. The first step for modelling the atmosphere is to solve the hydrostatic equilibrium equation. However, the opacity is not yet known. A crude estimate of the opacity profile (i.e. as a function of depth) can be used during the first iteration. In future iterations, the opacity is calculated more precisely and gives a more realistic solution. Also, to solve the hydrostatic equilibrium equation, the pressure at a given point in the atmosphere must be known. Usually, a more or less arbitrary value for the pressure in the first layer (at the surface) is given in input. The

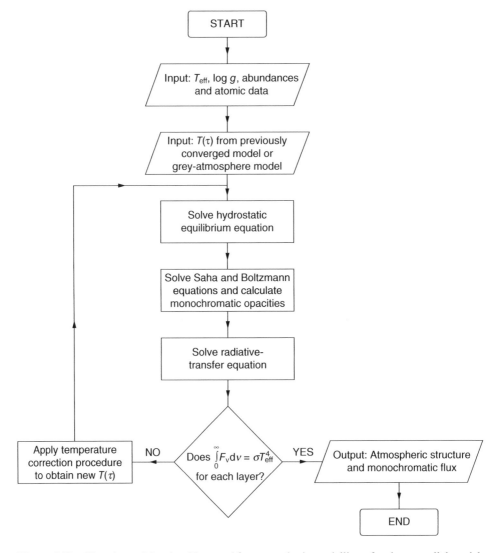

Figure 4.19   Flowchart of the algorithm used for atmospheric modelling of a plane-parallel model. Once the radiative-transfer equation is solved, radiative pressure that is not mentioned in this figure or in the text may also be included in the hydrostatic equilibrium equation for subsequent iterations (see optional Section 3.12). Figure reproduced and adapted with permission from François Wesemaël.

value chosen affects the structure of the first few layers, but does not appreciably affect the deeper layers as long as the pressure chosen for the first layer is negligible compared to the pressure in these deeper layers.

Once the pressure stratification is known and assuming a given equation of state (for most cases the ideal-gas approximation is sufficient), the total number density of particles

Figure 2.3 Molecular clouds in the M16 nebula where star formation is present (NASA/courtesy of nasaimages.org).

Figure 3.12 The solar corona as seen during a total solar eclipse (NASA/courtesy of nasaimages. org).

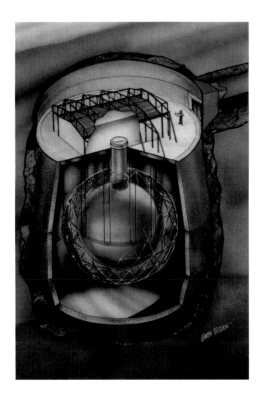

Figure 6.8 An artist's conception of the Sudbury Neutrino Observatory (photo courtesy of SNO).

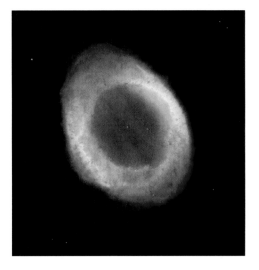

Figure 6.15 M57 (also called the Ring Nebula) is a well-known planetary nebula. The central white dwarf is the pinpoint body visible in the middle of the expanding shell of matter. This object has a diameter of approximately 2.4 ly and is situated 2300 ly from Earth. (NASA/courtesy of nasaimages.org).

Figure 6.17 The Crab nebula (M1), which is a supernova remnant. (See text for full caption.)

(a)

(b)

(c)

(d)

Figure 6.18 Examples of the various types of normal galaxies. (See text for full caption.)

Figure 6.20 The open cluster M45 commonly called the Pleiades. It is found at a distance of approximately 440ly from Earth. (NASA/courtesy of nasaimages.org).

Figure 6.21 The globular cluster M4. It is found in the Scorpio constellation at a distance of approximately 7200ly from Earth. (NASA/courtesy of nasaimages.org).

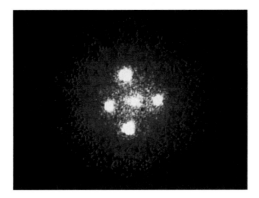

Figure 6.30 An image of Einstein's cross. This system is made up of a distant quasar that is found behind another galaxy that is seen in the centre of this image. This central galaxy serves as a gravitational lens that bends the light coming from the quasar and giving four distinct images of the distant quasar. (NASA/courtesy of nasaimages.org).

present in each layer may be calculated providing the local temperature is a known quantity. Since at this point of the algorithm the temperature is unknown, an initial temperature profile with respect to depth must be employed. This initial temperature used can be, for instance, the one for a grey atmosphere (Eq. 4.21). A previously calculated model (even one with different fundamental parameters) can also be used during the first iteration. The total number density of particles and the temperature may be used to solve the Saha equations that give the ionisation fractions of the species present (see the special topic below for more details). The Boltzmann equations are then solved to obtain the populations of the various atomic energy levels that may then be used to calculate the monochromatic opacity spectrum. These calculations are done with the aid of the atomic data loaded at the onset of the computations.

Special Topic – Saha Equations: A More Realistic View

In Chapter 1, several examples were shown where ionisation fractions were calculated for relatively simple cases. Here, an example illustrating the complexity in obtaining the ionisation fractions in a more realistic case is presented. A star composed of hydrogen and helium is considered here. For the calculation of a stellar atmosphere model, the number densities of the five ions present: n_{HI}, n_{HII}, n_{HeI}, n_{HeII} and n_{HeIII} as well as n_e must be obtained to enable the calculation of the radiative opacities. To obtain these six unknown quantities, six independent equations are required. At the point where the opacities are needed in atmospheric modelling, both T and P are known. In addition to the following three Saha equations (assuming of course that the proper atomic data are known): n_{HII}/n_{HI}, n_{HeII}/n_{HeI} and n_{HeIII}/n_{HeII}, the following equations come into play

$$P = n_{tot}kT = (n_{HI} + n_{HII} + n_{HeI} + n_{HeII} + n_{HeIII} + n_e)kT \qquad (4.107)$$

$$n_e = n_{HII} + n_{HeII} + 2n_{HeIII} \qquad (4.108)$$

The first of the two equations above assumes the equation of state of an ideal gas. The second equation relates that the free electrons in the plasma come from the ionisation of H and He.

An additional equation is therefore needed to obtain the various populations listed above. This last equation is obtained from the abundances of the elements present in the star. Assuming that the fraction of hydrogen atoms is A_H and that of helium atoms is A_{He}, and since only H and He is present in the star under consideration, $A_H + A_{He} = 1$. Two equations defining these abundances can then be written

$$A_H = \frac{n_{HI} + n_{HII}}{n_{HI} + n_{HII} + n_{HeI} + n_{HeII} + n_{HeIII}} \qquad (4.109)$$

$$A_{He} = \frac{n_{HeI} + n_{HeII} + n_{HeIII}}{n_{HI} + n_{HII} + n_{HeI} + n_{HeII} + n_{HeIII}} \qquad (4.110)$$

However, since a relation exists between A_H and A_{He}, these two equations are not independent and in reality can be simplified to only one equation with the knowledge that for the case under consideration $A_H + A_{He} = 1$. This gives the sixth and final equation needed to obtain the six unknown quantities enumerated above.

For the example discussed here, only two elements were considered. In typical atmospheric model, dozens of elements are often included. A large number of equations must then be solved, leading to another reason why considerable computation resources are needed for atmospheric modelling.

The opacity spectrum is needed to solve the radiative-transfer equation and it may be calculated with the populations obtained by the procedure described above (supposing that the proper atomic data are given as input). The radiative-transfer equation can then be solved for each value of u and v of the respective grids (specified by the user of the atmospheric code) thereby giving the specific intensity in each atmospheric layer. Since the frequency grid typically comprises of tens of thousands of points so that it properly samples the radiation field, large numerical resources are needed for such calculations. The number of directions (or values of u) chosen for which the transfer equation is solved is normally on the order of 10. Once the specific intensity is known, the monochromatic and integrated fluxes may be computed. If the integrated flux is equal to σT_{eff}^4 (within a certain tolerance) for each layer (which is of course never the case for the first iteration!) the structure of the atmosphere is then considered to be sufficiently precise and may be employed for various astrophysical applications. When the value of integrated flux is not within this tolerance, the temperature profile is modified via a temperature-correction procedure (see (advanced optional) Section 4.5.4) that is designed to lead to a flux nearer to the desired value. This new temperature profile is then used to begin a new iteration (see Figure 4.19). The succession of steps described above is repeated until an atmospheric structure with a temperature profile that converges towards the desired value for the integrated flux is obtained. The atmospheric code then gives as output the structure of the atmosphere and its monochromatic flux.

4.5.3   Example of a Stellar Atmosphere Model

In this section, some results from a model atmosphere of a star with $T_{eff} = 10\,000\,K$, log g = 4.0 and solar abundances are presented. The aim here is to initiate the reader to the typical order of magnitudes of the various physical quantities as a function of optical depth in a typical stellar atmosphere. The atmospheric model shown in this section was calculated with the Phoenix stellar atmosphere code (Hauschildt, Allard and Baron, *The Astrophysical Journal*, 512, 377 (1999)). This is a multi-purpose stellar atmosphere code that can be used to calculate atmospheres for various types of stars.

The temperature profile of the model atmosphere under consideration is shown in Figure 4.1. This figure shows that the temperature at the bottom of the atmosphere is equal to several times the value of T_{eff}. This result is important because it permits an appropriate selection of the number of ions for each element that is required for proper modelling (see Exercise 4.13).

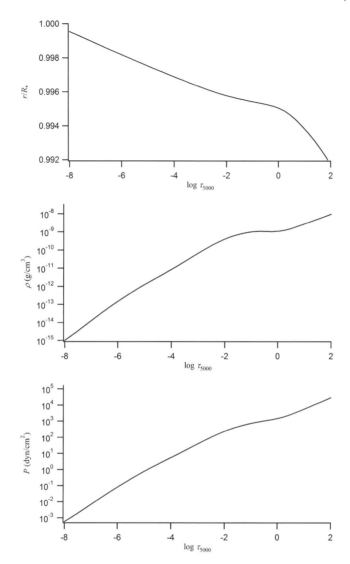

Figure 4.20   The ratio of the local radius to that of the stellar radius, density and pressure as a function of optical depth calculated at 5000 Å for a plane-parallel model atmosphere with $T_{\mathrm{eff}} = 10\,000\,\mathrm{K}$, $\log g = 4.0$ and solar abundances. The temperature profile for this stellar atmosphere model is shown in Figure 4.1.

Figure 4.20 shows the radius, density and pressure as a function of optical depth at 5000 Å in the atmosphere under consideration. This figure spans the depths $-8 \leq \log \tau_{5000} \leq 2$, which is a typical extent used for atmospheric modelling. As mentioned previously, the geometrical extension of atmosphere is on the order of one per cent of the stellar radius (see top part of Figure 4.20). The density of air at sea level on Earth is approximately $1.3 \times 10^{-3}\,\mathrm{g/cm^3}$. The density in stellar atmospheres is typically many orders of magnitudes smaller than this value (see middle part of Figure 4.20). A stellar atmosphere is thus composed of very tenuous gases. Meanwhile, the pressure within the atmosphere increases by several orders of magnitude and remains much smaller than the atmospheric pressure at sea level on Earth, which is equal to 1.01×10^6 dyn/cm^2 (see bottom part of Figure 4.20) throughout most of the stellar atmosphere.

In Figure 4.20, the shapes of the various curves change relatively suddenly near $\log \tau_{5000} = 0$. This is due to the ionisation of hydrogen which occurs in this region for the stellar atmosphere shown here. The excitation and ionisation of the most abundant element present (i.e. hydrogen) causes a considerable increase of the opacity and leads to a fast increase of the geometrical depth as a function of optical depth. This causes the abrupt slope changes seen in Figure 4.20.

4.5.4   Temperature-Correction Procedure[††]

In Section 4.5.2, it was shown that a temperature correction scheme leading to the conservation of the flux throughout the atmosphere is a critical component of the algorithm used to calculate stellar atmospheres. In this section, a method used for modifying the temperature profile during atmospheric modelling is developed. The temperature-correction method shown here is called the lambda-iteration procedure. Its name stems from the fact that the average intensity intervenes and it is calculated with the Λ operator (see (advanced optional) Section 3.10). Since more efficient methods exist and are usually employed in modern model atmosphere codes, this procedure is mainly presented for illustrative and instructive purposes.

In Section 3.7, when an atmosphere is in radiative equilibrium and assuming $S_\nu = B_\nu$, it was shown that

$$\int_0^\infty \kappa_\nu B_\nu d\nu = \int_0^\infty \kappa_\nu J_\nu d\nu \tag{4.111}$$

However, during a model atmosphere calculation and before the atmospheric structure is properly converged, each layer i does not respect the equation given above. In other words, the following inequality ensues

$$\int_0^\infty \kappa_\nu(\tau_i) B_\nu(T_i) d\nu \neq \int_0^\infty \kappa_\nu(\tau_i) J_\nu(\tau_i) d\nu \tag{4.112}$$

The temperature T_i of a given layer must then be modified by a value ΔT_i in the aim of attaining radiative equilibrium

$$\int_0^\infty \kappa_\nu(\tau_i) B_\nu(T_i + \Delta T_i) d\nu = \int_0^\infty \kappa_\nu(\tau_i) J_\nu(\tau_i) d\nu \tag{4.113}$$

The Planck function at temperature $T_i + \Delta T_i$ may be approximated by the following Taylor expansion

$$B_v(T_i + \Delta T_i) \approx B_v(T_i) + \Delta T_i \frac{dB_v}{dT}\bigg|_{T_i} \tag{4.114}$$

and therefore, the radiative equilibrium equation becomes

$$\int_0^\infty \kappa_v(\tau_i)\left[B_v(T_i) + \Delta T_i \frac{dB_v}{dT}\bigg|_{T_i}\right]dv \approx \int_0^\infty \kappa_v(\tau_i)J_v(\tau_i)dv \tag{4.115}$$

The temperature correction ΔT_i may then be isolated to give

$$\Delta T_i \approx \frac{\int_0^\infty \kappa_v(\tau_i)[J_v(\tau_i) - B_v(T_i)]dv}{\int_0^\infty \kappa_v(\tau_i)\frac{dB_v}{dT}\bigg|_{T_i} dv} \tag{4.116}$$

Such temperature corrections may be applied (at each iteration) to the temperature profile of the atmosphere that eventually leads to a model that converges to the desired solution (within a certain tolerance). This equation shows that when T_i is such that the related Planck function is smaller that J_v, the temperature must be increased (i.e. $\Delta T_i > 0$). Inversely, when the local Planck function is larger than J_v, the local temperature must be decreased.

Unfortunately, the lambda-iteration procedure has several drawbacks. For example (and as discussed in Section 3.8), the average intensity at large optical depths is approximately equal to the local Planck function. Therefore, at large depths ΔT_i is very small and this procedure converges very slowly. As previously mentioned, more efficient temperature-correction procedures have been developed and are employed in modern stellar-atmosphere codes. These temperature-correction methods also have the capability of taking into account energy transport due to convection that was neglected in the theoretical development shown above. This feature is critical since convection is present in the atmospheres of late-type (or cooler) stars.

4.6 Summary

Grey-atmosphere temperature profile: $T(\tau) = T_{\text{eff}}\left[\frac{3}{4}\left(\tau + \frac{2}{3}\right)\right]^{1/4}$ $\qquad$ (4.117)

Opacity for an atomic line: $k_v\rho = \frac{\pi e^2}{m_e c}f_{ij}\varphi_v n_i\left(1 - e^{\frac{-hv_0}{kT}}\right)$ $\qquad$ (4.118)

Line-broadening mechanisms: natural, Doppler and pressure

Lorentz profile: $\varphi_v = \dfrac{\dfrac{\Gamma}{4\pi^2}}{(v - v_0)^2 + \left(\dfrac{\Gamma}{4\pi}\right)^2}$ $\qquad$ (4.119)

Voigt profile (natural and Doppler broadening): $\varphi_v = \dfrac{U(a, v)}{\Delta v_D}$ (4.120)

Line depth: $A_\lambda = 1 - \dfrac{F_\lambda}{F_c}$ (4.121)

Equivalent width: $W_\lambda = \int A_\lambda d\lambda$ (4.122)

Line depth for weak lines: $A_\lambda \approx \dfrac{2}{3} \dfrac{k_1}{k_c} \dfrac{d \ln B_\lambda}{d\tau_c}\bigg|_{\tau_c = 2/3}$ (4.123)

Input for plane-parallel atmosphere modelling: T_{eff}, log g, abundances and atomic data

4.7 Exercises

4.1   Show that Eddington's approximation ($J_v(\tau) = 3K_v(\tau)$) is valid when the specific intensity is given by $I_v(\tau, u) = a_v(\tau) + b_v(\tau)u$.

4.2   Show that in a grey plane-parallel atmosphere in radiative equilibrium, the integrated flux is $H = \dfrac{S(\tau = 2/3)}{4}$.

4.3   Show that the full width at half-maximum for a Lorentzian line profile (Eq. 4.47) is equal to $\Gamma/2\pi$.

4.4   At what interval (in units of $\Gamma/4\pi$, or in other words the half-width at half-maximum) from the natural frequency does the Lorentz profile have a value of 1 % of its central intensity? Calculate this interval for 0.1 % of the central intensity.

4.5   The atomic lines for a star are observed to be shifted relative to their normal positions. This is due to a radial velocity of the star (i.e. the component of the star's velocity along the line-of-sight). If the shift of the H_β line is $\Delta\lambda = +0.4\,\text{Å}$, what is the value and the direction of the radial velocity of the star?

4.6   Show that the opacity of an atomic line (Eq. 4.86) is proportional to gf_{ij} (this quantity is commonly called the gf value and is often given in atomic databases and used in astrophysical applications).

4.7   Demonstrate that the most probable speed V_0 in a gas of particles with mass m and at temperature T that possesses a Maxwell speed distribution is

$$V_0 = \sqrt{\dfrac{2kT}{m}}$$ (4.124)

4.8 Calculate the percentage of free electrons that have sufficient kinetic energy in a gas at $T = 10\,000\,\text{K}$ to ionise hydrogen atoms found in the first exited state. (Numerical problem)

4.9 Calculate the equivalent width of an absorption atomic line of triangular shape where the flux in the centre is ¼ of the corresponding value in the continuum and the width of the base of the triangle is $6\,\text{Å}$.

4.10 Assume a beam of radiation enters an interstellar cloud with a thickness of $10\,\text{AU}$. At a given frequency, a transition can cause stimulated emission in this gas. If the intensity when the beam exits the cloud is 20 times its initial intensity, calculate $k_\nu \rho$ in this cloud at this frequency. Assume that the emissivity is negligible at this frequency and that the opacity and density is constant throughout the cloud.

4.11 Show that the hydrostatic equilibrium equation may be written

$$\frac{d \ln P}{d \ln \tau} = \frac{g\tau}{kP} = \frac{d \log P}{d \log \tau} \tag{4.125}$$

4.12 In Figure 4.10, for the star where $V \sin i = 1.5\,\text{km/s}$, the line widths are on the order of $0.2\,\text{Å}$ in the part of the electromagnetic spectrum shown. In this case, the rotational broadening can be considered negligible as compared to the observed line width. For approximately what value of $V \sin i$ does the widening due to rotation begin to dominate line broadening there?

4.13 In order to diminish the computing time during the modelling of stellar atmospheres, only the atomic data for the ions that are present there are taken into account. By knowing the temperature at which certain elements ionise (see Chapter 1 and Appendix D) and by using the results for the grey atmosphere estimate the number of magnesium ions that need to be considered when modelling the atmosphere of a star with $T_{\text{eff}} = 8000\,\text{K}$ up to a depth of $\tau = 500$.

4.14 In a star where the temperature of the atmosphere decreases with depth in the line-formation region, what particularity is observed concerning the atomic lines in the spectra formed in these depths? Explain why. According to the definition of the equivalent width, what is particular in this instance? Is this consistent with Eq. (4.105)?

5

Stellar Interiors

5.1 Introduction

The previous chapter was dedicated to modelling the outer layers of stars or its atmosphere. In stellar atmosphere modelling, the basic parameters (T_{eff}, $\log g$ and R_*) of the star are needed. Physically, these parameters are determined by the overall structure of the star. In the present chapter, the equations needed to properly understand the general structure of stars will be seen. This field of stellar astrophysics is called stellar structure or sometimes stellar interiors. The precise determination of the structure of stellar interiors is important for several reasons. For instance, a good knowledge of the interior structure of stars is important to properly estimate the nuclear reaction rates in the core of stars that is critical for the study of their evolution. The stellar structure is also necessary for theoretically predicting of the oscillation frequencies of pulsating stars. On the other hand, the observational study of stellar pulsations can give information about the interior of stars.

The chapter will begin by a review of the four basic equations of stellar structure: hydrostatic equilibrium, mass conservation, energy transport and energy conservation. Up to now, it was often assumed that all of the energy in stars was transported by radiation. The other two energy-transportation modes sometimes present in stars, namely convection and conduction, will be described. Convection, which is the more prevalent of these two modes in stars, will be discussed in more detail. The physical conditions required for convective-energy transport will be reviewed. A simple theoretical framework for convection, namely the mixing-length theory, will also be presented.

The resolution of the equations of stellar structure requires the knowledge of the equation of state of the stellar plasma. It will be seen that a type of equation of state called a polytrope leads to a relatively simple solution for the stellar structure and is very instructive. In the previous chapters, the ideal-gas equation of state was assumed to be valid in stars. However, under certain conditions, like for example when the stellar plasma is very dense, the ideal-gas approximation is not suitable. This is the case for white dwarfs for

An Introduction to Stellar Astrophysics Francis LeBlanc
© 2010 John Wiley & Sons, Ltd

instance, where the gas is said to be degenerate. The equation of state there is quite different from an ideal gas and it strongly affects the physical structure of stars. A section describing the complexity of the equation of state in stars will therefore be presented.

Sections pertaining to the structure of the Sun and to variable stars and asteroseismology will also be presented in this chapter. Several types of variable stars will be discussed along with some of the physics behind the pulsation of stars. The importance of asteroseismology as a tool for better understanding stars when used in conjunction with observations will be outlined. The utility of using variable stars to gauge distances will also be discussed.

5.2   Equations of Stellar Structure

The structure of a star can be calculated by solving four basic equations commonly called the equations of stellar structure. In this section, these four equations along with other ingredients necessary to properly define the structure of a star will be discussed. Please note that spherical symmetry will be assumed throughout this chapter.

5.2.1   Hydrostatic Equilibrium Equation

As seen in Chapter 2, when a star is in hydrostatic equilibrium, a differential equation that relates the gas pressure[1] P caused by the weight of the matter above point r to the pertinent physical quantities was shown to be

$$\frac{dP(r)}{dr} = -\frac{\rho(r)GM(r)}{r^2} \tag{5.1}$$

This is the first of the equations of stellar structure.

However, when a star is in a rapid evolutionary phase such as the supernova stage or if a star is pulsating, it cannot be considered in hydrostatic equilibrium and therefore the following equation of motion must then be taken into account

$$\rho\frac{d^2r}{dt^2} = -\frac{\rho GM}{r^2} - \frac{dP}{dr} \tag{5.2}$$

In such instances, the hydrodynamics of the medium must be considered. For most cases studied in this chapter, stars are assumed to be in hydrostatic equilibrium.

5.2.2   Equation of Mass Conservation

The variable $M(r)$, which is defined as the mass inside the radius r, must be known to solve the hydrostatic equilibrium equation. Figure 5.1 shows that the mass $dM(r)$ of a spherical shell found between the radii r and $r + dr$ is given by the equation

$$dM(r) = 4\pi r^2 \rho(r)dr \tag{5.3}$$

---

[1] Here, only gas pressure is included. However, radiation pressure can come into play in certain circumstances.

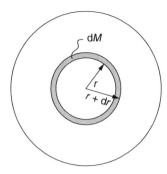

Figure 5.1   Illustration of the mass dM found inside the spherical shell found between the radii r and $r + dr$ (shaded area).

where $4\pi r^2 dr$ is the volume of the spherical shell and $\rho(r)$ is the local density. The equation above may be written

$$\frac{dM(r)}{dr} = 4\pi r^2 \rho(r) \tag{5.4}$$

This is called the equation of mass conservation. The amount of mass $M(r)$ inside radius r is obtained by integrating this equation from the centre to radius r

$$M(r) = \int_0^r dM(r) = \int_0^r 4\pi r^2 \rho(r) dr \tag{5.5}$$

It is self-evident that the integration of this equation over the whole star must give the mass M_* of the star under consideration.

Example 5.1: Find the pressure stratification $P(r)$ inside a star with mass M_* and radius R_* in which the density decreases linearly with r via the expression.

$$\rho(r) = \rho_c \left(1 - \frac{r}{R_*} \right) \tag{5.6}$$

where ρ_c is the central density.

Answer:

The pressure may be calculated by integrating the hydrostatic equilibrium equation

$$\frac{dP(r)}{dr} = -\frac{\rho(r)GM(r)}{r^2} \tag{5.7}$$

By Eq. (5.5), $M(r)$ is

$$M(r) = \int_0^r 4\pi x^2 \rho(x)\,dx = \int_0^r 4\pi x^2 \rho_c \left(1 - \frac{x}{R_*}\right)dx = \frac{4\pi}{3}\rho_c r^3 \left(1 - \frac{3r}{4R_*}\right) \qquad (5.8)$$

Here, the mute variable x was used instead of r in order to distinguish it from the integration boundary. The central density may be evaluated with the knowledge that $M(R_*) = M_*$, which leads to

$$\rho_c = \frac{3M_*}{\pi R_*^3} \qquad (5.9)$$

With these results, the pressure gradient can be obtained

$$\frac{dP(r)}{dr} = -\frac{4\pi G}{3}\rho_c^2 r \left(1 - \frac{3r}{4R_*}\right)\left(1 - \frac{r}{R_*}\right) \qquad (5.10)$$

The pressure at radius r can then be calculated by integrating this equation from the centre of the star to radius r

$$\int_{P_c}^{P(r)} dP = -\frac{4\pi G}{3}\rho_c^2 \int_0^r x \left(1 - \frac{3x}{4R_*}\right)\left(1 - \frac{x}{R_*}\right)dx = -\frac{4\pi G}{6}\rho_c^2 r^2 \left(1 - \frac{7r}{6R_*} + \frac{3r^2}{8R_*^2}\right) \qquad (5.11)$$

$$P(r) = P_c - \frac{4\pi G}{6}\rho_c^2 r^2 \left(1 - \frac{7r}{6R_*} + \frac{3r^2}{8R_*^2}\right) \qquad (5.12)$$

By using the value for the central density found above (Eq. 5.9), the pressure as a function of radius is

$$P(r) = P_c - \frac{6G}{\pi}\frac{M_*^2}{R_*^6}r^2 \left(1 - \frac{7r}{6R_*} + \frac{3r^2}{8R_*^2}\right) \qquad (5.13)$$

By assuming that the pressure is nil at the surface of the star, the central pressure is found to be equal to

$$P_c = \frac{5G}{4\pi}\frac{M_*^2}{R_*^4} \qquad (5.14)$$

Therefore, with these last two equations, an expression for $P(r)$ as a function of the fundamental parameters of the star may be written. The result for the central pressure found here is similar to the value found in Example 2.4. The values for the central density and pressure found here, when applied to the Sun, are more than an order of magnitude smaller than the corresponding values obtained by detailed numerical solar models. This is not surprising since the density stratification inside the Sun is not given by Eq. (5.6).

5.2.3   Energy-Transport Equation

As discussed earlier, there are three modes of energy transport in stars: radiation, conduction and convection. In this section, a relation between the temperature gradient and the luminosity of the star will be found. For simplicity, all of the energy will be assumed to be transported by radiation. The other two modes of energy transport will be discussed later in the chapter.

The theoretical developments seen in Sections 3.8 and 3.9 lead to the integrated Eddington flux at large optical depths given by

$$H(r) = -\frac{1}{3k_R\rho}\frac{dB}{dT}\frac{dT}{dr} \tag{5.15}$$

The variable r is used in the equation above instead of z (Eq. 3.74) since spherical symmetry is assumed here. The integrated Eddington flux may also be written as a function of luminosity

$$H(r) = \frac{1}{4\pi}\left(\frac{L(r)}{4\pi r^2}\right) \tag{5.16}$$

Contrarily to stellar atmospheres, the integrated flux in stellar interiors is not constant because of the following two reasons. First, the surface on which the luminosity is distributed depends on r, while for the plane-parallel stellar atmospheres discussed in Chapter 4, the surface of the atmosphere is independent of depth. Secondly, in the stellar core where nuclear reactions take place, the luminosity in not constant and decreases as the stellar centre is approached (see Section 5.2.4 for more details). This therefore modifies the flux.

Since the integrated Planck function is

$$B(r) = \frac{\sigma T^4}{\pi} \tag{5.17}$$

the temperature gradient as a function of r using the three equations above lead to the following result

$$\frac{dT(r)}{dr} = -\frac{3k_R\rho}{64\pi r^2\sigma T^3}L(r) \tag{5.18}$$

This equation is commonly called the energy-transport equation (for radiation only). To simplify this expression, the depth dependence of T, ρ and k_R is not written explicitly. The following physical interpretation of this equation is useful to better understand the transfer of radiation in stars. To have energy transfer, a temperature gradient is needed. This equation shows that the temperature gradient is proportional to luminosity. Assuming all else equal, if the luminosity increases, the temperature gradient increases to enable the transfer of the extra amount of energy. Also, when all else is equal, if the opacity increases, once again the temperature gradient increases in order to compensate for the higher difficulty for the photons to cross the medium.

A variable often used in stellar astrophysics ∇, commonly called the temperature gradient, is defined as

$$\nabla = \frac{d\ln T}{d\ln P} \tag{5.19}$$

Such a temperature gradient is used when discussing convection. This topic will be presented later in this chapter. For the case of pure radiative transport and assuming hydrostatic equilibrium, this variable can be shown to be equal to (see Exercise 5.2)

$$\nabla_{rad} = \frac{d\ln T}{d\ln P} = \frac{3k_R}{64\pi r^2 g} \frac{P}{\sigma T^4} L(r) \tag{5.20}$$

Where the subscript rad is used to underline that the medium is in radiative equilibrium. Once again, the explicit dependence of g, k_R, T and P on r is not written.

The equation above shows that when the opacity increases, so does ∇_{rad}, and it will be shown that under certain conditions pertaining to this physical quantity convective energy transport can become efficient (see Section 5.3.3). The temperature gradient increases when radiative opacity is large and it becomes increasingly difficult to transport energy via radiation. This can lead to convective transport of energy that takes over from radiative transport.

The energy-transport equation found above depends on the integrated flux (i.e. $\frac{L(r)}{4\pi r^2}$).

The structure of stars therefore depends on the global properties of the radiation field. Even though it is not of critical importance for the global properties of stars, the detailed monochromatic flux in stellar interiors will be discussed below in Section 5.3.1. When taking into account more physical processes such as atomic diffusion, the monochromatic flux can in an indirect way affect the structure of stars (see Chapter 7) and can therefore be of interest.

5.2.4   Equation of Energy Conservation

The final equation of stellar structure is related to the local luminosity $L(r)$ due to all sources of energy within the radius r. When a star is gravitationally stable (not in a phase of contraction), its source of energy is solely thermonuclear fusion. The luminosity therefore depends on the thermonuclear energy production rate in the stellar core. In the equations shown here, the nuclear production rate per unit mass (i.e. the total quantity of nuclear energy produced per gram of matter per second) defined by the quantity $\varepsilon(r)$ will be employed. The evaluation of this quantity is quite complicated since the reaction rates of all important fusion reactions must be known for its calculation. This physical quantity depends on the local density and temperature as well as on the abundances of the atomic species present. More details surrounding the various nuclear reactions present in stellar cores will be given in Chapter 6.

Similarly to the mathematical development seen above regarding the conservation of mass, the quantity of energy per second (or luminosity) generated inside a spherical shell found between the radii r and $r + dr$ (see Figure 5.2) is given by the equation

$$dL(r) = 4\pi r^2 \rho(r) \varepsilon(r) dr \tag{5.21}$$

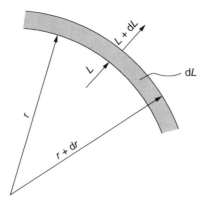

Figure 5.2 Illustration of the energy generated per unit time dL inside the spherical shell found between the radii r and $r + $ dr (shaded area).

where $\rho(r)$ is the local density and $\rho(r)\varepsilon(r)$ represents the energy per unit time per unit volume generated at radius r. The equation above may be written as follows

$$\frac{dL(r)}{dr} = 4\pi r^2 \rho(r)\varepsilon(r) \tag{5.22}$$

This equation is called the energy-conservation equation or sometimes simply the energy equation. Its integration (assuming the knowledge of $\rho(r)$ and $\varepsilon(r)$) leads to the luminosity at a given radius

$$L(r) = \int_0^r dL(r) = \int_0^r 4\pi r^2 \rho(r)\varepsilon(r)dr \tag{5.23}$$

Since all of the variables on the right-hand side of Eq. (5.22) are positive (or nil), $\frac{dL}{dr} \geq 0$ and therefore the local luminosity increases with r. Not surprisingly, this equation shows that if the nuclear energy production rate increases (as it does during stellar evolution, see Chapter 6 for more details), so does the luminosity. As mentioned previously, nuclear fusion only takes place in the central regions of stars. Consequently, $\varepsilon = 0$ in the exterior regions, and thus $\frac{dL}{dr} = 0$ there. In these outer regions of a star, the luminosity is therefore constant and equals L_*, the total luminosity of the star. Figure 5.3 shows the luminosity as a function of radius for the Sun. From this figure, it may be inferred that the rate of nuclear energy generation goes from a maximum value in the centre to zero near $r = 0.25R_\odot$.

5.2.5 Other Ingredients Needed

In addition to the four equations of stellar structure, other ingredients must be included to calculate a stellar model. The equation of state $P(\rho,T,X_i)$ must be known, where X_i are the mass fractions of the various elements present (see Eq. 5.109). The Rosseland

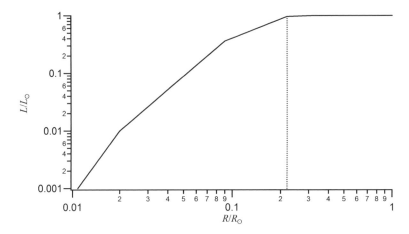

Figure 5.3   Luminosity as a function of radius inside the Sun. The data used here are those found in Table 5.1 (see Section 5.5). The dotted line shows approximately the radius where $\varepsilon \to 0$ or $L \to L_\odot$.

opacity $k_R(\rho,T,X_i)$ and the rate of thermonuclear energy production per unit mass $\varepsilon(\rho,T,X_i)$ must also be known quantities. When relevant, the importance of conduction and convection relative to radiation transport of energy must also be taken into account. Similarly to the case of stellar atmospheres seen in Chapter 4, the calculation of a detailed stellar interior model necessitates considerable computing resources along with the knowledge of nuclear and atomic data. Instead of loading all of the atomic data needed for opacity calculations, the computer codes that calculate stellar interior models often use pretabulated opacity tables. This diminishes the computing time necessary for modelling stars.

As mentioned previously, when calculating a stellar model, the star is often assumed to be static. However, on longer timescales, the abundances of the elements in the stellar core change as a function of time due to the nuclear reactions present there. This modifies the structure of the star. Computer codes specifically developed for studying stellar evolution take these abundance changes into account and calculate a series of stellar models at various times during a star's life.

Other factors such as stellar rotation, the presence of a magnetic fields or the accretion of matter at the stellar surface from the surrounding medium can also affect the structure of a star. The relative diffusion of the elements within the star (see Chapter 7) can cause abundance stratification and therefore also modifies the stellar structure. All of these factors are usually neglected in ordinary stellar models. However, specialized stellar models do exist that take these physical phenomena into account.

To solve the four differential equations defining the structure of stars, proper boundary conditions must be imposed. At the star's centre, the following conditions must be respected

$$L(r \to 0) = 0 \quad \text{and} \quad M(r \to 0) = 0 \tag{5.24}$$

Meanwhile, at the surface of the star the following conditions may be used

$$P(r \rightarrow R_*) = 0, \rho(r \rightarrow R_*) = 0 \quad \text{and} \quad T(r \rightarrow R_*) = T_{\text{eff}} \tag{5.25}$$

These last three boundary conditions are theoretical because the stellar surface is difficult to define. However, since the density at the surface is very small, this difficulty in defining a true surface does not significantly affect the model obtained for the interior. Alternatively, the values for these physical quantities at the surface used in stellar structure modelling can be taken from a predetermined stellar atmosphere model (by using their value at a given small optical depth). Evidently, to obtain a model for a star, its mass, radius, luminosity and composition (i.e. X_i) must be given as input to the computer code.

Special Topic – Equations of Stellar Structure as a Function of $M(r)$

Similarly to stellar atmospheres, the numerical process of modelling stellar interiors first consists of dividing the star into a number of layers (or concentric shells). For atmospheric modelling, it was shown that it is preferable to divide the layers as a function of a standard optical depth instead of geometrical depth. For stellar interiors, instead of dividing the layers used in the numerical model as a function of r, it is often more useful to divide them relative to $M(r)$. The various physical quantities defining stellar structure can then be written as a function of $M(r)$: $r = r(M)$, $P = P(M)$, $L = L(M)$, $T = T(M)$, $\rho = \rho(M)$ and $\varepsilon = \varepsilon(M)$. It can be shown (see Exercise 5.4) that the equations of stellar structure when written as a function of $M(r)$ are

$$\frac{dP(M)}{dM} = -\frac{GM}{4\pi r^4(M)} \tag{5.26}$$

$$\frac{dr(M)}{dM} = \frac{1}{4\pi r^2 \rho(M)} \tag{5.27}$$

$$\frac{dL(M)}{dM} = \varepsilon(M) \tag{5.28}$$

$$\frac{dT(M)}{dM} = -\frac{3k_R}{256\pi^2 r^4 \sigma T^3} L(M) \tag{5.29}$$

5.3 Energy Transport in Stars

Energy transport in stars is an extremely important physical process that is critical to properly understand the functioning of stars. It has been previously mentioned that three energy transport modes exist in stars: radiation, conduction and convection. There is also a fourth manner by which energy can travel from the central regions of stars to interstellar space: neutrinos. Neutrinos are particles that are emitted by certain fusion reactions.

However, since neutrinos interact with matter only via the weak nuclear force, except of course for gravity, almost all of these particles simply traverse the star without interacting with the stellar plasma. Therefore, the energy transported by neutrinos is directly lost to the interstellar space and this energy is usually simply subtracted from the thermonuclear energy production rate (see Chapter 6 for more details) and is thus simply considered as an energy sink and not as a mode of energy transport.

Up until now, a lot of attention has been given to the radiative mode of energy transport. The aim of this section is to expand the discussion to both conduction and convection. After a general description of both these transport processes, a more detailed discussion of convection will be undertaken since it is the second most important transport process after radiation. Convection occurs in a large portion of certain stars, our Sun being an example. Convection zones may also appear (or disappear) in a star during its evolution due to the changes in its internal structure. Conduction is much less prevalent in stars, occurring only in certain stars such as white dwarfs. However, before delving into these last two transport modes, a section describing the monochromatic radiative flux in stellar interiors is presented.

5.3.1   Monochromatic Radiative Flux in Stellar Interiors

For the equations of stellar structure seen earlier in this chapter, only the integrated radiative flux intervened. The reason being, that the global structure of a star does not depend on the detailed monochromatic flux (or luminosity) but rather on its integrated value. However, certain astrophysical applications such as diffusion of the elements in stars (see Chapter 7) strongly depend on the detailed monochromatic flux inside stars. It is then essential to be able to properly evaluate the monochromatic flux in the interior regions of stars. In this section, an approximate formula for this physical quantity will be found.

From the energy-transport equation (Eq. 5.18) for which it is assumed that all the energy is transported by photons, the temperature gradient (relative to r) in the stellar regions outside where nuclear reactions take place (or in other words where the luminosity is constant and equal to $L(r) = 4\pi R^2 \sigma T_{\text{eff}}^4$) is

$$\frac{dT}{dr} = -\frac{3k_R \rho R^2 T_{\text{eff}}^4}{16r^2 T^3} \tag{5.30}$$

Inserting this result into the equation of the monochromatic flux valid at large optical depths found in Eq. (3.73) gives

$$H_\nu = \frac{1}{16} \frac{k_R}{k_\nu} \frac{T_{\text{eff}}^4}{T^3} \left(\frac{R}{r}\right)^2 \frac{dB_\nu}{dT} \tag{5.31}$$

It can be shown (see Exercise 5.5) that

$$\frac{dB_\nu}{dT} = \frac{2k^3 T^2}{h^2 c^2} \left[\frac{u^4 e^u}{(e^u - 1)^2}\right] \tag{5.32}$$

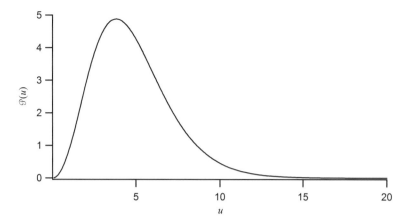

Figure 5.4 The dependence of $\mathcal{P}(u)$ as a function of u. Its maximum is found at $u \approx 3.8$.

where a new variable[2] $u = \dfrac{h\nu}{kT}$ is introduced. The monochromatic Eddington flux outside the thermonuclear core may then be expressed as

$$H_\nu = \frac{k^3}{8h^2c^2} \frac{k_R}{k_\nu} \frac{T_{\text{eff}}^4}{T} \left(\frac{R}{r}\right)^2 \mathcal{P}(u) \tag{5.33}$$

where

$$\mathcal{P}(u) = \left[\frac{u^4 e^u}{\left(e^u - 1\right)^2}\right] \tag{5.34}$$

This function gives the global shape of the flux and is shown in Figure 5.4. Its maximum is found at $u \approx 3.83$ (see Exercise 5.6). The introduction of the variable u is useful since the maximum of the flux becomes independent of both T and ν when using it. The detailed frequency dependence of the flux is determined by the monochromatic opacity that appears in the denominator of Eq. (5.33). Also appearing in this equation is the term $\left(\dfrac{R}{r}\right)^2$ that represents the increase of the flux as r decreases. This increase of the flux is due to the fact that the luminosity is distributed on a smaller surface of the imaginary sphere with radius r. Also, and not surprisingly, the flux is proportional to T_{eff}^4.

The monochromatic flux given by Eq. (5.33) is quite useful. In a practical sense, it can be used to estimate the flux in a predetermined stellar model. Most stellar models give not only the various variables such as T, ρ, etc. as a function of r but also k_R. Therefore, a calculation of the monochromatic opacity spectrum leads to the knowledge of H_ν.

---

[2]Beware, do not confuse the angular dependence of the specific intensity $u = \cos\theta$ used in previous chapters with this new variable.

Example 5.2: Find the frequency (and the related wavelength) at the maximum of the function $\mathcal{R}(u)$ for $T = 10^4$, 10^5 and 10^7 K. In which part of the electromagnetic spectrum are each of these wavelengths found?

Answer:

The maximum of the function $\mathcal{R}(u)$ is found at $u \approx 3.83$ (see Exercise 5.6). Since

$$u = \frac{h\nu}{kT} \tag{5.35}$$

The values for the frequency and wavelength for each temperature are then easily found to be

$$\nu \approx 7.9 \times 10^{14} \text{ Hz or } \lambda \approx 3800 \text{ Å for } T = 10^4 \text{ K} \tag{5.36}$$

$$\nu \approx 7.9 \times 10^{15} \text{ Hz or } \lambda \approx 380 \text{ Å for } T = 10^5 \text{ K} \tag{5.37}$$

$$\nu \approx 7.9 \times 10^{17} \text{ Hz or } \lambda \approx 3.8 \text{ Å for } T = 10^7 \text{ K} \tag{5.38}$$

The first two maxima found above are in the ultraviolet (the one for $T = 10^4$ K is very close to the visible part of the spectrum) and the third maximum is in the X-ray region of the electromagnetic spectrum. The largest of these three temperatures is on the order of the central temperature found in main-sequence stars. Therefore, the radiation flux in central regions of main-sequence stars is composed mostly of X-ray photons.

5.3.2   Conduction

Conduction is a mode of energy transport that is important under specific physical conditions where particles of matter may carry energy from hotter (or deeper) to cooler (or shallower) regions within a star. Such physical conditions are found inside white dwarfs where the free electron gas is degenerate, which leads to an equation of state that is much different from an ideal gas (the concept of degeneracy for a gas will be discussed in Section 5.6.3). In stellar plasma, the particles participating in conduction are these degenerate free electrons. For conduction to be important, the mean free path of the electrons must be of the same geometrical scale (or larger) on which the local temperature changes. Electrons from hotter (deeper) layers can therefore travel to cooler (shallower) layers without being impeded. The kinetic energy of the electrons from the hotter layers can then be transferred to cooler layers. The extra energy carried by the conductive electrons arriving in these cooler regions is conveyed to the local plasma following collisions where these high-energy electrons are eventually thermalized to the local temperature.

As for the case of radiative transport, the flux of energy due to conduction is proportional to the temperature gradient. An opacity related to the conductive flow of energy k_{cond} may then be defined such that the Eddington flux of energy due to conduction H_{cond} is

$$H_{cond} = -\frac{4\sigma T^3}{3\pi k_{cond}\rho}\frac{dT}{dr} \tag{5.39}$$

While the Eddington flux due to radiation H_{rad} is

$$H_{rad} = -\frac{4\sigma T^3}{3\pi k_{rad}\rho}\frac{dT}{dr} \tag{5.40}$$

where k_{rad} is the opacity due to radiation that is simply equal to the Rosseland mean opacity. The expression k_{rad} is used instead of k_R for clarity when comparing radiative and conductive transport. In the absence of convection, the total flux of energy H_{tot} is

$$H_{tot} = H_{rad} + H_{cond} = -\frac{4\sigma T^3}{3\pi k_{rad}\rho}\frac{dT}{dr} - \frac{4\sigma T^3}{3\pi k_{cond}\rho}\frac{dT}{dr} = -\frac{4\sigma T^3}{3\pi k_{tot}\rho}\frac{dT}{dr} \tag{5.41}$$

where a total opacity k_{tot} is defined as

$$\frac{1}{k_{tot}} = \frac{1}{k_{rad}} + \frac{1}{k_{cond}} \tag{5.42}$$

The energy preferably flows through the mode of transport with the smaller opacity. An analogy between these two modes of energy transport and two electrical resistors in parallel may be made (the similarity of the equation above to that for two resistors in parallel is self-evident). In such an electrical circuit, the electrical current is larger in the branch containing the resistor with the smaller resistance. In most conditions found in stellar plasma, conductive opacity is very large as compared to radiative opacity and thus almost no energy is transported by conduction.

5.3.3   Convection

5.3.3.1   General Description of Convection

Stars are made up of plasma and are thus hydrodynamical objects. The plasma within a star is not static and macroscopic movements, flows or currents of gas can exist there. Since the theory of hydrodynamics is very complex and depends on physical values that are often not well established, a complete hydrodynamical theory for stellar plasma is not feasible. However, some hydrodynamic phenomena, such as convection can be modelled while making certain simplifying assumptions. This is the aim of this section.

Convection is the transport of energy by rising cells (or blobs) of matter within a star. For instance, if a cell of plasma is displaced within a star it can either fall back to its original position or may, under certain conditions, rise towards the surface due to buoyancy. If such conditions exist, the plasma is said to be unstable against convection and convective energy transport process can take place. For the discussion in the next section, it will be assumed that these cells do not exchange any energy with the surrounding plasma during their ascension (and it is therefore an adiabatic process) until they eventually dissolve in shallower (and therefore cooler) regions in the star. At this point, the excess heat of the cell is dispersed to the local medium and this process therefore transports energy from the inner to the outer regions of stars.

Convection may become important in stars when radiation cannot suffice to transport the energy from the central regions to the surface of stars. When the radiative flux, the opacity or both are very large, it leads to a large temperature gradient (see Eq. 5.18). In the next section, it will be shown that when the temperature gradient surpasses a certain limit, the medium becomes unstable and convection appears. A criterion can be established, called the Schwarzschild criterion, as to when this situation occurs. This criterion was first established by the German astrophysicist Karl Schwarzschild (1873–1916) and will be discussed in (the optional) Section 5.3.3.2.

Regions where partial ionisation of abundant elements (such as hydrogen or helium) occurs are prone to induce convective transport. As discussed in Chapter 1, when a given element is partially ionised, the excited atomic levels of the atoms of the less charged ion become relatively highly populated. This increases both the bound-bound and bound-free radiative opacity of the plasma. This increase in the opacity can then lead to a large temperature gradient that in turn, may render the medium convective. For example, convection is present in the ionisation zone of hydrogen of cool stars. In hotter stars, since the ionisation of hydrogen occurs in shallower layers where the temperature gradient is relatively small, no convection appears in that region.

Convective transport is present in the Sun and this phenomenon is apparent at its surface by the existence of granules (see Figure 3.14). These surface features are related to convective cells emerging at the solar surface, while cooler plasma cells sink towards the centre of the Sun. The solar granules have a diameter on the order of 1000 km. Convection is fundamentally a physical phenomenon that is hydrodynamical in nature. It is therefore very complex. However, a relatively simple description of convection called the mixing-length theory can be developed by making a number of assumptions and approximations. This theoretical framework of convection will be described in (the advanced optional) Section 5.3.3.3.

Similarly to the concept of radiative equilibrium, when all of the energy is transported by convection it is said to be in convective equilibrium (see (optional) Section 5.3.3.4). In reality, convective equilibrium is never completely attained in stars since there is always some radiative transport present. However, this approximation can sometimes simplify certain equations (see Section 5.4 for example).

5.3.3.2   The Schwarzschild Criterion for Convection[†]

To obtain a criterion under which convection can exist, several assumptions will be made. First, convection will be approximated by a process consisting of rising and falling cells of matter. If such a cell is displaced in the medium by a distance Δr, the cell rises towards the surface (due to Archimedes' principle[3]) if the density of the cell is smaller than the surrounding density. If the density of the cell is larger than the density of the surrounding medium, it falls back to its original position and no convective process takes place.

---

[3]The Archimedes principle states that a buoyancy force exists on any object that is immersed in a fluid (or gas). This force equals the weight of the fluid displaced by immersing the object. An object can therefore float in any fluid with a larger density than itself. In reality, the buoyancy is due to the difference of the pressure between the top and bottom of the object and is therefore due to the presence of a pressure gradient relative to depth in the fluid.

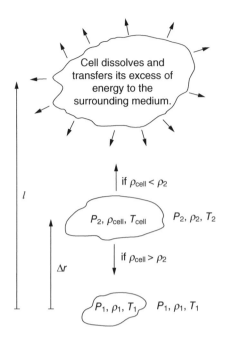

Figure 5.5 Illustration of the convection process. A convective cell found in the bottom part of this figure is displaced by a distance Δr. Initially the values of the pressure, density and temperature inside the cell are equal to the corresponding values found in the surrounding medium. During its displacement the pressure inside the cell is always equal to the pressure in the medium, however, its density and temperature change. After the displacement, if the density in the cell is larger than the density of the medium the cell sinks back toward its original position. If the density in the cell is smaller than the density in the surrounding medium, the cell is buoyant and travels a certain distance l (called the mixing length) before dissolving. It then transfers its excess of energy to the surrounding medium.

Another approximation made here is that no heat transfer occurs between the rising cell and the surrounding plasma during its rise in the star. Its excess energy is only dissipated after it travels a certain distance (see (optional) Section 5.3.3.3). The process is thus assumed to be adiabatic during the ascension of the convective cells. Also, it is assumed that the mean molecular weight of the particles in the plasma does not vary over the length on which the convective cells of matter travel.

Figure 5.5 illustrates a cell of plasma that is displaced in a star. It is assumed that the pressure inside the cell equals the pressure in the surrounding medium during its ascension. Therefore, the density inside the cell decreases during this displacement. If this density decreases faster than the decrease in the density of the surrounding medium (with respect to r), the cell becomes buoyant and convection ensues. The density within the cell changes by $\Delta\rho_{cell}$ when it is displaced by a distance Δr

$$\Delta\rho_{cell} = \left(\frac{d\rho}{dr}\right)_{adi}\Delta r < 0 \tag{5.43}$$

where $\left(\dfrac{d\rho}{dr}\right)_{adi}$ represents the gradient of the density during an adiabatic process. Since

$\Delta r > 0$ and $\left(\dfrac{d\rho}{dr}\right)_{adi} < 0$, this equation shows that the density of the cell decreases

($\Delta\rho_{cell} < 0$). Meanwhile, the density of the surrounding medium changes by $\Delta\rho_{med}$ on the distance Δr

$$\Delta\rho_{med} = \left(\frac{d\rho}{dr}\right)_{rad} \Delta r < 0 \tag{5.44}$$

where $\left(\dfrac{d\rho}{dr}\right)_{rad}$ represents the gradient of the density in the surrounding medium that is

assumed to be in radiative equilibrium (i.e. where all of the energy is transported by radiation). The density of the medium evidently also decreases along the trajectory of the rising cell.

According to the Archimedes principle, the criterion under which the convective cells rise is that the density of these cells must decrease more than the density of the medium: $|\Delta\rho_{cell}| > |\Delta\rho_{med}|$. By using Eqs. (5.43) and (5.44) and since the density gradients are negative, the criterion for convection to be present becomes

$$\left(\frac{d\rho}{dr}\right)_{adi} < \left(\frac{d\rho}{dr}\right)_{rad} \tag{5.45}$$

In other words, the above criterion stipulates that if the density gradient is smaller (i.e. more negative) inside a cell that is displaced upwards than the density gradient of the surrounding medium, the cell rises upwards due to buoyancy. However, this form of the convection criterion is not very useful. The rest of this section is dedicated to finding a different format for this criterion that is more physically insightful and that is expressed in terms of variables more commonly used in stellar astrophysics.

During an adiabatic expansion of an ideal gas, the results from the field of thermodynamics show that the gas pressure is proportional to a power law with respect to density

$$P \propto \rho^{\gamma} \tag{5.46}$$

where

$$\gamma = \frac{c_P}{c_V} \tag{5.47}$$

defines the ratio of the specific heats at constant pressure (c_P) and volume (c_V). The type of equation of state used above is called a polytrope and is often used to find analytical solutions to certain equations of stellar structure (see section 5.4 for more details). Since in this circumstance the pressure is proportional to a power of density, it is more useful to write the criterion given in Eq. (5.45) as

$$\left(\frac{d\ln\rho}{dr}\right)_{adi} < \left(\frac{d\ln\rho}{dr}\right)_{rad} \tag{5.48}$$

The gradient of $\ln \rho$ for an adiabatic expansion can be written as a function of $\ln P$ with Eq. (5.46) as

$$\left(\frac{d\ln\rho}{dr}\right)_{adi} = \frac{1}{\gamma}\left(\frac{d\ln P}{dr}\right)_{adi} \tag{5.49}$$

Meanwhile, in the radiative zone (i.e. outside the convective cells), the equation of state is assumed to be equal to that of an ideal gas or in other words $P \propto \rho T$. Therefore, $\ln P \propto \ln \rho + \ln T$ and the gradient of $\ln \rho$ in the radiative region is then

$$\left(\frac{d\ln\rho}{dr}\right)_{rad} = \left(\frac{d\ln P}{dr}\right)_{rad} - \left(\frac{d\ln T}{dr}\right)_{rad} \tag{5.50}$$

and the convection criterion can be written

$$\frac{1}{\gamma}\left(\frac{d\ln P}{dr}\right)_{adi} < \left(\frac{d\ln P}{dr}\right)_{rad} - \left(\frac{d\ln T}{dr}\right)_{rad} \tag{5.51}$$

During its ascent, it is assumed that the cell has the same pressure as its surrounding, and therefore, the gradients of $\ln P$ in the cell and in the radiative medium are equal

$$\left(\frac{d\ln P}{dr}\right)_{adi} = \left(\frac{d\ln P}{dr}\right)_{rad} \tag{5.52}$$

The convection criterion therefore becomes

$$\left(\frac{1}{\gamma}-1\right)\left(\frac{d\ln P}{dr}\right)_{rad} < -\left(\frac{d\ln T}{dr}\right)_{rad} \tag{5.53}$$

or taking into account that $\left(\dfrac{d\ln P}{dr}\right)_{rad}$ is negative

$$\left(\frac{\gamma-1}{\gamma}\right) < \frac{\left(\dfrac{d\ln T}{dr}\right)_{rad}}{\left(\dfrac{d\ln P}{dr}\right)_{rad}} = \left(\frac{d\ln T}{d\ln P}\right)_{rad} = \nabla_{rad} \tag{5.54}$$

where ∇_{rad} is the temperature gradient in the radiative zone. Assuming that the equation of state in the cell is also equal to that of an ideal gas that is expanding adiabatically, it must respect the following relations: $P \propto \rho T \propto P^{\frac{1}{\gamma}}T$ or $T \propto P^{\frac{\gamma-1}{\gamma}}$. The gradient of $\ln T$ with respect to $\ln P$ in the cell is therefore

$$\left(\frac{d\ln T}{d\ln P}\right)_{adi} = \left(\frac{\gamma-1}{\gamma}\right) = \nabla_{adi} \tag{5.55}$$

where ∇_{adi} defines the temperature gradient for an adiabatic process (in this case inside the convective cells). The so-called Schwarzschild criterion for convection can then be written as follows

$$\nabla_{rad} > \nabla_{adi} = \left(\frac{\gamma - 1}{\gamma} \right) \tag{5.56}$$

For an ideal monoatomic gas $\gamma = 5/3$ and $\nabla_{adi} = 0.4$. In this instance, the convection criterion is $\nabla_{rad} > 0.4$. Since stellar plasma is not monoatomic and radiation pressure is present, the value of γ is smaller than 5/3. Detailed calculations show that when radiation pressure completely dominates the Schwarzschild criterion is $\nabla_{rad} > 0.25$. Meanwhile, in partial ionisation zones the value of ∇_{rad} for which convection can take place can be lower than 0.25. Convective transport is prone to occur in regions where ionisation takes place.

More insight can be obtained by using the detailed equation for ∇_{rad} found in Section 5.2.3. The Schwarzschild criterion can be written

$$\nabla_{rad} = \frac{3k_R}{64\pi r^2 g} \frac{P}{\sigma T^4} L(r) > \nabla_{adi} = \left(\frac{\gamma - 1}{\gamma} \right) \tag{5.57}$$

This equation shows that regions where the opacity or the flux is large are prone to convection.

Finally, only ascending convective cells of matter carrying energy towards the stellar surface were discussed here. Of course, for each rising convective cell in the star, a cell that has an energy deficit descends toward its centre. This is necessary to conserve the amount of mass found at each depth of the star.

5.3.3.3   The Mixing-Length Theory[††]

In the previous section, a criterion was found under which convection can occur. However, it does not give any information on the efficiency of this transport process, or in other words, on the value of the flux of energy transported by it. In the presence of both radiative and convective flux, the total flux $H(r)$ at radius r is equal to

$$H(r) = \frac{1}{4\pi} \left(\frac{L(r)}{4\pi r^2} \right) = H_{rad}(r) + H_{conv}(r) \tag{5.58}$$

This section is dedicated to finding an approximate expression for the convective flux (H_{conv}) by using a simplified model for convection.

Contrarily to the assumption made in the previous section, a realistic convection process is not adiabatic. Therefore, the true temperature gradient in a convective cell (∇_{cell}) is larger than or equal to the gradient for an adiabatic process ($\nabla_{cell} \geq \nabla_{adi}$) because some energy may be lost during the ascension of the convective cells. Also, in the presence of convection, the medium is not in radiative equilibrium and the Schwarzschild criterion is slightly different from the one given above. In general, the temperature gradients $\frac{d \ln T}{d \ln P}$ for the different cases have the following relation to one another

$$\nabla_{rad} \geq \nabla_{med} \geq \nabla_{cell} \geq \nabla_{adi} \tag{5.59}$$

where ∇_{med} is the value of $\frac{d \ln T}{d \ln P}$ in the medium containing both radiative and convective transport. The Schwarzschild criterion then becomes

$$V_{\text{med}} > V_{\text{cell}} \qquad (5.60)$$

Since convection is a hydrodynamical process, a rigorous mathematical treatment of this physical phenomenon is not trivial. Complex numerical simulations of such hydrodynamical processes are sometimes employed in stellar astrophysics. However, a simpler description of convection that is often used in stellar structure or stellar atmosphere codes exists. It is the so-called mixing-length theory. The mixing-length theory is constructed by assuming that convection consists of rising and falling cells of matter. This physical process transports energy from the hotter to the cooler layers of the star. Since a series of (sometimes crude) approximations is necessary to simplify this convective theory, it is only an approximate depiction of convective flow. Even though it is customarily called a theory, the mixing-length theory is rather more a dimensional analysis than a rigorous theory. However, this 'theory' is very instructive to better grasp the concept of convective-energy transport.

First, it will be assumed that the convective cells of plasma rise a certain distance and then dissolve while transferring their excess of energy to the surrounding medium. A temperature difference

$$\Delta T = T_{\text{cell}} - T_{\text{med}} \qquad (5.61)$$

exists between the convective cell and the surrounding medium at the instant the cell dissolves. The energy per unit volume E transmitted from the convective cell to the medium is

$$E = \rho c_{\text{p}} \Delta T \qquad (5.62)$$

The specific heat at constant pressure is used here because it is assumed (see Section 5.3.3.2) that the pressure inside the cell is equal to the pressure in the surrounding medium.

The temperature difference ΔT for a convective cell travelling a distance Δr may be crudely approximated by the following expression

$$\Delta T \approx \left[\left(-\frac{\mathrm{d}T}{\mathrm{d}r} \right)_{\text{med}} - \left(-\frac{\mathrm{d}T}{\mathrm{d}r} \right)_{\text{cell}} \right] \Delta r \qquad (5.63)$$

Assuming that the average velocity of the convective cells is $\bar{V}$, the convective flux then becomes

$$F_{\text{conv}} = 4\pi H_{\text{conv}} = \rho c_{\text{p}} \bar{V} \Delta T = \rho c_{\text{p}} \bar{V} \Delta r \left[\left(-\frac{\mathrm{d}T}{\mathrm{d}r} \right)_{\text{med}} - \left(-\frac{\mathrm{d}T}{\mathrm{d}r} \right)_{\text{cell}} \right] \qquad (5.64)$$

It should be noted that the flux is equal to the energy density times the velocity that this energy travels through a given imaginary surface.

Now since

$$\frac{\mathrm{d}T}{\mathrm{d}r} = T \frac{\mathrm{d}\ln T}{\mathrm{d}r} = T \frac{\mathrm{d}\ln T}{\mathrm{d}\ln P} \frac{\mathrm{d}\ln P}{\mathrm{d}r} \qquad (5.65)$$

and assuming hydrostatic equilibrium

$$\frac{d \ln P}{dr} = -\frac{\rho g}{P} = -\frac{1}{H} \tag{5.66}$$

The variable H, which has units of length, is defined as the pressure scale height. In an isothermal gas, it is equal to the distance on which the pressure varies by a factor of e (see Example 2.1). Therefore, the derivatives of the temperature with respect to radius in the surrounding medium and inside a convective cell are, respectively,

$$\left(\frac{dT}{dr}\right)_{med} = -\frac{T}{H}\nabla_{med} \tag{5.67}$$

and

$$\left(\frac{dT}{dr}\right)_{cell} = -\frac{T}{H}\nabla_{cell} \tag{5.68}$$

A mixing length l may be defined as being equal to the average distance travelled by a convective cell before dissolving. When evaluating the convective energy flux at a given depth, the cells crossing a unitary imaginary surface have on average travelled a distance. $\Delta r = \frac{l}{2}$. These arguments and assumptions finally lead to the following expression for convective flux

$$H_{conv} = \frac{\rho c_P \overline{V} T}{8\pi}\left(\frac{l}{H}\right)(\nabla_{med} - \nabla_{cell}) \tag{5.69}$$

The parameter $\left(\dfrac{l}{H}\right)$ is a free parameter in this theoretical framework since it is not possible to calculate the average distance travelled by the cells. In typical astrophysical applications, its value is normally chosen to be a value on the order of unity. The convective flux is proportional to both the mixing length and to the average velocity of the convective cells. Also, and as expected from the Schwarzschild criterion, when $\nabla_{med} = \nabla_{cell}$ the convective flux is nil. To simplify the calculations, one may wish to assume $\nabla_{med} = \nabla_{rad}$ and $\nabla_{cell} = \nabla_{adi}$.

The average velocity $\overline{V}$ of the cells, which is up to now an unknown quantity in the equation above, must now be estimated. This will then lead to an equation for the convective flux of energy for which all of the variables may be calculated, with the exception of $\left(\dfrac{l}{H}\right)$. To evaluate $\overline{V}$, the work done on convective cells during their ascension will be calculated. The first step is to evaluate the total force F on a convective cell. This force is given by the following equation

$$F = F_{Arch} - m_{cell}g = \rho_{med}V_{cell}g - m_{cell}g = V_{cell}(\rho_{med} - \rho_{cell})g \tag{5.70}$$

where F_{Arch} is the Archimedes force on a cell of mass m_{cell} and volume V_{cell} caused by the displacement of the gas (of density ρ_{med}) of the medium. The total force f per unit volume acting on the cell is therefore

$$f = (\rho_{med} - \rho_{cell})g = -\Delta\rho g \tag{5.71}$$

where

$$\Delta\rho = (\rho_{\text{cell}} - \rho_{\text{med}})\qquad(5.72)$$

During the ascent of the convective cells, the pressure is assumed to be the same inside and outside the cell. By assuming an ideal gas ($P \propto \rho T$) the pressure difference ΔP between the inside and the outside of cells, which is nil, may be approximated by the following

$$\Delta P = 0 = \Delta\rho T + \rho\Delta T\qquad(5.73)$$

where $\Delta\rho$ and ΔT are, respectively, the density and temperature difference between the inside of cells and the surrounding medium. The equation above may be rearranged to give

$$\Delta\rho = -\frac{\rho}{T}\Delta T\qquad(5.74)$$

Using this result along with Eqs. (5.63), (5.67), (5.68) and (5.71), the force per unit volume on a convective cell is found to be

$$f = \left(\frac{\rho g}{H}\right)(\nabla_{\text{med}} - \nabla_{\text{cell}})\Delta r\qquad(5.75)$$

For the convective cells arriving at a given depth it may be assumed that they have on average been accelerated over a distance of $l/2$. The work per unit volume done on the cell can therefore be written as

$$W = \int_0^{l/2} f(y)\,dy = \frac{l^2}{8}\left(\frac{\rho g}{H}\right)(\nabla_{\text{med}} - \nabla_{\text{cell}})\qquad(5.76)$$

where the variable $y = \Delta r$ is used.

Since the convective cells must push aside other cells while ascending in the star, they lose part of their energy due to this kind of friction. To take this factor into account, it will be arbitrarily assumed here that only half of the energy obtained above is used to accelerate the cells. Therefore, the kinetic energy per unit volume of a cell arriving at the depth under consideration is

$$\frac{1}{2}\rho\bar{V}^2 = \frac{W}{2}\qquad(5.77)$$

and the average velocity of the convective cells is therefore

$$\bar{V} = \left(\frac{gH}{8}\right)^{1/2}\left(\frac{l}{H}\right)(\nabla_{\text{med}} - \nabla_{\text{cell}})^{1/2}\qquad(5.78)$$

The convective flux predicted by the mixing-length theory then becomes

$$H_{\text{conv}} = \frac{\rho c_P T}{8\pi}\left(\frac{gH}{8}\right)^{1/2}\left(\frac{l}{H}\right)^2(\nabla_{\text{med}} - \nabla_{\text{cell}})^{3/2}\qquad(5.79)$$

Here again, it is important to mention that the ratio of the mixing length to the pressure scale height $\left(\dfrac{l}{H}\right)$ is an unknown quantity in this theory. This free parameter can be adjusted by comparing the theoretical results to observations. However, because of the various approximations and assumptions made when deriving the mixing-length theory, a single value for this free parameter valid for all stars does not exist.

5.3.3.4   *Convective Equilibrium*[†]

In Section 5.2.3, the so-called energy-transport equation was found in the case where all the energy is transported by radiation or in other words when the plasma is in radiative equilibrium. A similar equation can be written when all the energy is transported by convection, a situation commonly called convective equilibrium.

Supposing that a star (or a portion of a star) is completely dominated by convection, it may be assumed that the stellar plasma is composed entirely of convective cells. In the schematic view of convection discussed previously, these cells are assumed to be rising and falling adiabatically. Therefore, the adiabatic equation of state should be valid everywhere in the region where convection completely dominates energy transport. The temperature gradient is then equal to (see Section 5.3.3.2)

$$\frac{d \ln T}{d \ln P} = \nabla_{adi} = \left(\frac{\gamma-1}{\gamma}\right) \tag{5.80}$$

and it is trivial to show that

$$\frac{dT}{dr} = -\left(\frac{\gamma-1}{\gamma}\right)\frac{\rho g T}{P} \tag{5.81}$$

This equation gives the temperature gradient (with respect to r) in a medium in convective equilibrium. As expected, since radiative transport is not present here the opacity is absent from this equation (as opposed to Eq. 5.18). It should be mentioned that in real situations, convective equilibrium is never strictly achieved since there is always some amount of energy transported by radiation in stars. However, convective equilibrium is a useful approximation to estimate the temperature stratification in convective regions of stars.

5.4   Polytropic Models

Solving the equations of stellar structure is an arduous task. To better appreciate this complexity, the link between the equations of stellar structure may be evaluated. The first two equations of stellar structure (Eqs. 5.1 and 5.4) describe the mechanical structure of stars. These two equations may be fused together by multiplying Eq. (5.1) by r^2/ρ and by deriving this result with respect to r

$$\frac{d}{dr}\left(\frac{r^2}{\rho(r)}\frac{dP(r)}{dr}\right) = -G\frac{dM(r)}{dr} \tag{5.82}$$

With the result of Eq. (5.4), the well-known Poisson's equation is obtained

$$\frac{1}{r^2}\frac{d}{dr}\left(\frac{r^2}{\rho(r)}\frac{dP(r)}{dr}\right) = -4\pi G\rho(r) \tag{5.83}$$

This equation is void of an explicit dependence on the thermal structure of the star and that is the reason it was mentioned that it portrays its mechanical structure.

Meanwhile, the other two equations of stellar structure (Eqs. 5.18 and 5.22) control the thermal properties or in other words the temperature dependence with respect to r. However, these four fundamental equations are coupled together. For instance, Eq. (5.1) depends on the temperature via the equation of state $P(\rho,T,X_i)$. Similarly, Eqs. (5.18) and (5.22) depend on the mechanical structure of the star via not only the equation of state (please note that ρ appears in Eqs. (5.18) and (5.22)), but also through the quantities $k_R(\rho,T,X_i)$ and $\varepsilon(\rho,T,X_i)$ that themselves also depend on the mechanical structure of the star.

In general, Eq. (5.83) cannot be solved independently from the energy-transport or the energy-conservation equations. However, there exists a type of equation of state that permits the decoupling of the mechanical structure from the thermal structure of a star and leads to an analytical solution for Eq. (5.83) (for certain cases). Such a decoupling between the equations just discussed above occurs for an equation of state where the pressure is independent of temperature. The equation of state in this case is customarily written

$$P = K\rho^{\frac{n+1}{n}} \tag{5.84}$$

where n is called the polytropic index and K is a constant. A star whose plasma obeys such an equation of state is commonly called a polytropic star. Using this type of equation of state allows the study of the mechanical structure of a star without having to deal with the intricacies of the equations reigning over its thermal structure. This approach is similar to the grey-atmosphere approximation seen in Chapter 3 that permitted the study of certain aspects of stellar atmospheres without having to calculate detailed monochromatic opacity spectra. It also leads to analytical solutions for certain cases as opposed to requiring complex numerical calculations.

There are several circumstances where the equation of state in a star can fittingly be approximated by a polytropic equation. The first example is a star in convective equilibrium. The ascent and descent of convective cells is typically assumed to be an adiabatic process. Thermodynamics shows that in such a situation the equation of state for an ideal monoatomic gas the equation of state is $P = K\rho^\gamma$, where $\gamma = 5/3$. The polytropic index for this case is $n = 1.5$. Other instances where the equation of state is polytropic are stellar regions where the pressure is dominated by a completely degenerate electron gas (see Section 5.6.3 for more details concerning degenerate gases). This situation can occur in the core of evolved stars and in white dwarf stars.

It is customary to write a dimensionless version of Poisson's equation shown above (Eq. 5.83) that leads to a well-known equation in stellar astrophysics called the Lane–Emden equation. To achieve this goal, a dimensionless function[4] $\theta(r)$ is defined such that the density for a polytrope of index n is

---

[4]In reality, this is a family of functions for each value of the polytropic index n. It is sometimes written with an n in subscript: $\theta_n(r)$. However, since a power n of this function appears in Eq. (5.85) and in order to avoid confusion, that subscript will not be used here.

$$\rho(r) = \rho_c \theta^n(r) \tag{5.85}$$

where ρ_c is the central density of the star and the function $\theta(r)$ is elevated to the power n. The value of this monotonically decreasing function is $0 \leq \theta(r) \leq 1$. By definition, it has the value 1 at $r = 0$ for all values of the polytropic index n. Also, since the density at the surface of the star is nil, $\theta(r=R_*) = 0$.

The pressure may then written as a function of $\theta(r)$

$$P(r) = K\rho_c^{\frac{n+1}{n}} \theta^{n+1}(r) = P_c \theta^{n+1}(r) \tag{5.86}$$

where $P_c = K\rho_c^{\frac{n+1}{n}}$ is the central pressure of the star. The Poisson equation found earlier (Eq. 5.83) then becomes

$$K\rho_c^{\frac{n+1}{n}} \frac{1}{r^2} \frac{d}{dr}\left(\frac{r^2}{\rho_c\theta^n(r)} \frac{d\theta^{n+1}(r)}{dr}\right) = -4\pi G\rho_c\theta^n(r) \tag{5.87}$$

and since

$$\frac{d\theta^{n+1}(r)}{dr} = (n+1)\theta^n(r)\frac{d\theta(r)}{dr} \tag{5.88}$$

this equation may be written

$$\frac{(n+1)P_c}{4\pi G\rho_c^2} \frac{1}{r^2} \frac{d}{dr}\left(r^2 \frac{d\theta(r)}{dr}\right) = -\theta^n(r) \tag{5.89}$$

Since $\theta(r)$ is dimensionless, the equation above requires that the quantity $\dfrac{(n+1)P_c}{4\pi G\rho_c^2}$ has the dimension of length squared. A new variable α that depends on the polytropic index n may be defined

$$\alpha^2 = \frac{(n+1)P_c}{4\pi G\rho_c^2} \tag{5.90}$$

Moreover, a new dimensionless variable ξ may also be defined for each polytropic index n

$$\xi = \frac{r}{\alpha} \tag{5.91}$$

The use of this variable can transform the Poisson equation into what is commonly called the Lane–Emden equation

$$\frac{1}{\xi^2} \frac{d}{d\xi}\left(\xi^2 \frac{d\theta(\xi)}{d\xi}\right) = -\theta^n(\xi) \tag{5.92}$$

which is now written as a function of dimensionless variables. A solution for $\theta(r)$ for each value of n may then be found by solving this differential equation. This equation is named after Jonathan H. Lane (1819–1880) and Robert Emden (1862–1940) who were, respectively, American and Swiss astrophysicists.

The Lane–Emden equation may also be written as

$$\frac{d^2\theta(\xi)}{d\xi^2} + \frac{2}{\xi}\frac{d\theta(\xi)}{d\xi} + \theta^n(\xi) = 0 \tag{5.93}$$

Since the central density at the centre of stars is finite, the solution of this equation (for a given value of n) must respect $\dfrac{d\theta(\xi)}{d\xi} = 0$ at $\xi = 0$ (that assures that the second term in Eq. (5.93) does not diverge) as well as the previously mentioned condition $\theta(\xi = 0) = 1$. Analytical solutions exist for three cases: $n = 0$, 1 and 5, while the solution for other values of n must be obtained numerically. The solutions for the three cases just mentioned are

$$n = 0: \quad \theta(\xi) = 1 - \frac{\xi^2}{6} \tag{5.94}$$

$$n = 1: \quad \theta(\xi) = \frac{\sin\xi}{\xi} \tag{5.95}$$

$$n = 5: \quad \theta(\xi) = \frac{1}{\left(1 + \dfrac{\xi^2}{3}\right)^{1/2}} \tag{5.96}$$

Example 5.3: Solve the Lane–Emden equation for $n = 1$ and show that the solution is

$$\theta(\xi) = \frac{\sin\xi}{\xi}.$$

Answer:

It can be shown that by applying a change of variable $\chi(\xi) = \xi\theta(\xi)$ the Lane–Emden equation becomes

$$\frac{d^2\chi(\xi)}{d\xi^2} = -\frac{\chi^n(\xi)}{\xi^{n-1}} \tag{5.97}$$

For $n = 1$, the solution to this equation is

$$\chi(\xi) = A\sin(\xi) + B\cos(\xi) \tag{5.98}$$

where A and B are integration constants. Therefore

$$\theta(\xi) = \frac{A\sin(\xi)}{\xi} + \frac{B\cos(\xi)}{\xi} \tag{5.99}$$

The values $A = 1$ and $B = 0$ for the integration constants respect the conditions $\theta(0) = 1$ and $\dfrac{d\theta(\xi)}{d\xi}\bigg|_{\xi=0} = 0$. The solution of the Lane–Emden equation for $n = 1$ is then

$$\theta(\xi) = \frac{\sin(\xi)}{\xi} \tag{5.100}$$

The solutions of the Lane–Emden equation for the three cases having analytical solutions are shown in Figure 5.6. The first zero (commonly written ξ_0) of the functions $\theta(\xi)$ corresponds to the stellar surface. For $n = 0$ (this polytropic solution corresponds to a hypothetical star with a constant density, see Eq. 5.85), $\xi_0 = \sqrt{6} = 2.449$ and for $n = 1$, $\xi_0 = \pi$. From Eq. (5.91), the radius of a polytropic star is

$$R_* = \alpha\xi_0 \qquad\qquad (5.101)$$

Meanwhile, all solutions for $n \geq 5$ do not possess a zero for finite values of ξ.

Figure 5.7 shows the density as a function of ξ (that is proportional to r, see Eq. (5.91)) for various polytropic models. The density is obtained by using the corresponding $\theta(\xi)$

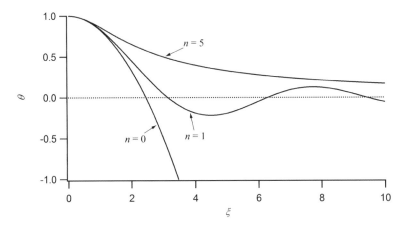

Figure 5.6   The dependence of $\theta(\xi)$ as a function of ξ for polytropic models with various indices.

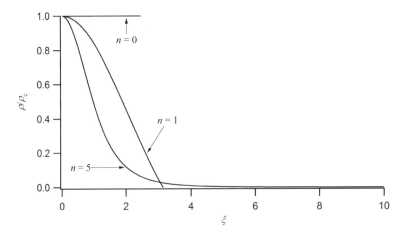

Figure 5.7   The dependence of ρ/ρ_c as a function of ξ for polytropic models of various indices.

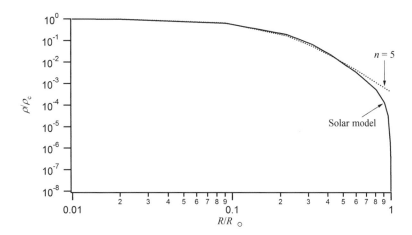

Figure 5.8   The dependence of ρ/ρ_c as a function of radius for a polytropic model with $n = 5$ as compared to a detailed numerical model of the Sun. The data for this theoretical model is found in Table 5.1 (Section 5.5).

functions (see Eq. 5.94, 5,95, and 5.96). In this figure, the model for $n = 0$ shown has a constant density up to the stellar surface where $\xi = \sqrt{6} = 2.449$, because as mentioned above, the $n = 0$ polytrope represents a hypothetical star with a constant density. To obtain such a configuration, the fluid must be incompressible. Since this is not the case for stars because they are made up of compressible gas, this solution is far from physical reality. For $n = 1$ the surface is found at $\xi = \pi$ (where the density is nil).

Figure 5.8 compares the density profile for a polytropic model with $n = 5$ to a detailed computation of the solar structure. The polytropic approximation in this case is reasonably good for approximately $r \leq R_\odot/2$ but is very far from reality in the outer regions. The divergence between the analytical and numerical solutions in the outer regions is not surprising since the analytical solution for $n = 5$ has an infinite radius (the radius in this case being defined as the distance from the centre to where the density becomes nil).

From Eq. (5.3), the mass of a star is

$$M_* = \int_0^{R_*} 4\pi r^2 \rho(r)\,dr \tag{5.102}$$

This equation may also be written as a function of the variables and function found in the Lane–Emden equation

$$M_* = 4\pi\alpha^3 \rho_c \int_0^{\xi_0} \xi^2 \theta^n(\xi)\,d\xi \tag{5.103}$$

Therefore, once the solution of the Lane–Emden equation is obtained for a given polytrope, the global mechanical properties of the star, its mass and radius (Eq. 5.101), can be calculated.

Example 5.4: Calculate the mass of a star with a polytropic equation of state with $n = 0$. Write this mass in terms of the quantities ρ_c and R_*.

Answer:

The mass of a star assuming a polytropic model is

$$M_* = 4\pi\alpha^3\rho_c \int_0^{\xi_0} \xi^2 \theta^n(\xi)\,d\xi \tag{5.104}$$

For $n = 0$, this equation becomes

$$M_* = 4\pi\alpha^3\rho_c \int_0^{\xi_0} \xi^2 \,d\xi = \frac{4\pi}{3}\alpha^3\rho_c\xi_0^3 = \frac{4\pi}{3}\rho_c R_*^3 \tag{5.105}$$

As expected, the mass in this case is simply that of a sphere of radius R_* with a constant density of value ρ_c.

5.5   Structure of the Sun

In the previous sections, the equations that need to be solved to calculate the structure of stars were discussed. Here, a case study, namely the structure of the Sun, is presented.

The Sun may be divided into six regions:

Core: The central regions where the thermonuclear reactions takes place. This zone extends from the centre to approximately 25 % of the solar radius.

Radiative zone: Region where energy transport is dominated by radiation. It extends to about 70 % of the solar radius.

Convection zone: Region above the radiative zone that extends to the surface and where energy transport is dominated by convection.

Photosphere: The surface region of the Sun from where the radiation escapes to outer space. This region is convective. Its thickness is only several hundred kilometres.

Chromosphere: Region extending approximately 2000 km above the photosphere and where the temperature rises up to $\sim 10^5$ K.

Corona: Region above the chromosphere that extends to several millions of kilometres farther and that has temperatures that reach 1 to 2 million degrees (see Section 3.6 for more details concerning the solar corona).

Figure 5.9 illustrates the solar interior. The solar interior is composed of the core, and the radiative and convective zones. The photosphere cannot be accurately calculated by a stellar interior model and is only correctly portrayed with a detailed stellar atmosphere model (see Chapter 4). Table 5.1 gives various physical quantities for the solar interior. In this table, the luminosity as a function of radius $L(r)$ is nil at the centre and it increases as a function of r due to the nuclear energy production rate within this radius (see Figure 5.3). It reaches its maximum value where nuclear reactions cease (i.e. at the surface of the

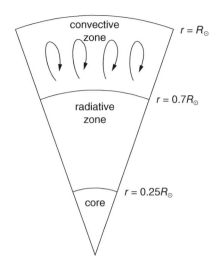

Figure 5.9   A cut-away diagram of the solar interior showing the radial extent of the core and the radiative and convective zones.

Table 5.1   Solar interior model[#].

r	$M(r)$	$L(r)$	T	ρ	$\log P$
$(R_\odot)$	$(M_\odot)$	$(L_\odot)$	$(10^6$ K$)$	(g cm^{-3})	(dyn cm^{-2})
0.007	0.00003	0.0002	15.7	150	17.369
0.02	0.001	0.010	15.6	146	17.355
0.09	0.057	0.361	13.6	95.73	17.177
0.22	0.399	0.966	8.77	28.72	16.525
0.32	0.656	1.000	6.42	9.77	15.724
0.42	0.817	1.000	4.89	3.22	15.324
0.52	0.908	1.000	3.77	1.05	14.722
0.60	0.945	1.000	3.15	0.500	14.322
0.71	0.977	1.000	2.23	0.177	13.721
0.81	0.992	1.000	1.29	0.0766	13.119
0.91	0.999	1.000	0.514	0.0194	12.119
0.96	0.9999	1.000	0.208	4.85×10^{-3}	11.118
0.99	1.0000	1.000	0.00441	2.56×10^{-4}	9.118
0.995	1.0000	1.000	0.00266	4.83×10^{-5}	8.118
0.999	1.0000	1.000	0.00135	1.29×10^{-6}	6.118
1.000	1.0000	1.000	0.00060	2.18×10^{-7}	4.918

[#]Reproduced courtesy of Cox, A.N., *Allen's Astrophysical Quantities*, Springer, New York (2004).

core, which is approximately found at $r = 0.25R_\odot$). The central temperature and density of the Sun are, respectively, found to have values of 15.7×10^6 K and 150 g cm^{-3}. The results given in Table 5.1 show that the pressure increases by more than twelve orders of magnitude from the surface to the solar centre.

5.6   Equation of State

5.6.1   Introduction

To calculate a stellar interior model, three main ingredients are needed. First, the nuclear production rate must be known. That physical quantity depends on the reaction rates of the various thermonuclear reactions that take place in stars. This topic will be discussed in Chapter 6. Secondly, the opacity of the stellar plasma must be evaluated. The opacity determines how energy is transported from the core to the outer layers of the star and affects its thermal structure. Opacities were discussed in Chapter 4. Finally, the appropriate equation of state must be known for the physical conditions found in the star under consideration. The aim of this section is to present some of the complexities that can arise in stellar interior model calculations relating to the equation of state.

Up to now, the ideal-gas approximation has been assumed to be valid in stars. An ideal gas is composed of noninteracting particles and leads to the well-known ideal-gas equation of state used previously throughout this book. An ideal gas also obeys the well-known Maxwell–Boltzmann statistics. However, this approximation breaks down when the interactions among the particles become important. Nonideal effects can then come into play.

Quantum effects must also be taken into account in certain circumstances. For example, in white dwarfs stars where the density is very large compared to normal stars, the pressure can be dominated by the so-called degenerate electrons. Since electrons are fermions, according to Pauli's exclusion principle two electrons cannot occupy the same quantum state. Pauli's principle is not of importance in low-density gases since only a small portion of quantum states is occupied. However, when the density of free electrons in stellar plasma becomes comparable to the density of states available for these particles, the electrons interact very differently and this leads to a very different equation of state from that for an ideal gas. It is therefore critical to properly evaluate the equation of state in such a case in order to appropriately model the structure of white dwarfs and the dense central regions of certain stars.

Another factor that can modify the equation of state is radiation pressure. The radiation field found in stellar plasma may in some cases lead to a large radiation pressure that can modify the structure of stars via the hydrostatic equilibrium equation (see (optional) Section 3.12.2). Another factor that is generally less important, namely magnetic pressure, can also have an effect when large magnetic fields are present.

Finally, relativistic effects can also intervene in certain circumstances and must be taken into account in the equation of state.

A complete treatment of the various equations of state that can be encountered in stars is outside the scope of this book. Only a brief presentation of this topic is presented here. In this section, some results for the pressure inside stars using the ideal-gas approximation will be given. A discussion about the pressure inside a degenerate electron gas will also be presented. The effect of radiation pressure will also be described.

5.6.2   The Ideal Gas

In most instances treated up to now in this book, the ideal-gas equation of state was used, namely

$$P = n_{tot}kT \tag{5.106}$$

where n_{tot} is the total number density of particles found in the gas. This equation of state is valid in most cases found in stars. It is also very easy to use and can help to understand the functioning of stars. In Chapter 1, the mean molecular weight μ (in units of m_H) of the particles in the gas was defined and it leads to a different form for the pressure

$$P = \frac{\rho kT}{\mu m_H} \tag{5.107}$$

The pressure may also be separated into two components: the partial pressure due to ions and that due to free electrons

$$P = P_{ions} + P_e = (n_{ions} + n_e)kT \tag{5.108}$$

Here, an equation for the pressure inside a star as a function of the mass fractions (X_i) of the various species that are present inside it will be found. This is useful because when computing stellar models, the mass fractions (or abundances) of the elements present are given as input. By definition, the mass fraction of species i is

$$X_i = \frac{\rho_i}{\rho} = \frac{n_i m_i}{\sum n_j m_j} \tag{5.109}$$

where ρ_i, n_i and m_i are, respectively, the mass density, number density and mass of the atoms of species i. The sum found at the denominator runs over all atomic species present. To achieve the aim of expressing the pressure as a function of mass fractions, it is possible to define a mean ionic weight μ_{ions} such that the ionic pressure is given by the following expression

$$P_{ions} = \frac{\rho kT}{\mu_{ions} m_H} \tag{5.110}$$

Since the contribution of free electrons to the mass density is negligible (i.e. $\rho \approx \rho_{ions}$), it is clear that μ_{ions} is the mean molecular weight of the ions in units of m_H.

A mean electron molecular weight μ_e such that the electronic pressure is

$$P_e = \frac{\rho kT}{\mu_e m_H} \tag{5.111}$$

may also be defined. However, the physical meaning of this quantity, which is given below, is not as straightforward as for μ_{ions}.

By Eqs. (5.107), (5.108), (5.110) and (5.111), the mean molecular weight is therefore equal to

$$\frac{1}{\mu} = \frac{1}{\mu_{\text{ions}}} + \frac{1}{\mu_e} \tag{5.112}$$

Expressions for both μ_{ions} and μ_e written as a function of the mass fractions present inside a given star will now be found.

The pressure due to ions can be written

$$P_{\text{ions}} = n_{\text{ions}}kT = \left(\sum n_j\right)kT \tag{5.113}$$

From Eq. (5.109), the number density of each species as a function of its mass fraction is

$$n_i = \frac{X_i}{m_i}\rho \tag{5.114}$$

The mass of each species may be approximated by

$$m_i = A_i m_{\text{H}} \tag{5.115}$$

where A_i is the number of nucleons in the nucleus of the species, and the pressure due to ions may then be written

$$P_{\text{ions}} = \left(\sum \frac{X_j}{A_j}\right)\frac{\rho kT}{m_{\text{H}}} \tag{5.116}$$

which leads to the following expression for the mean ionic weight

$$\frac{1}{\mu_{\text{ions}}} = \sum \frac{X_j}{A_j} \tag{5.117}$$

To obtain a similar expression for μ_e, complete ionisation of all of the species will be assumed. Although this approximation is not valid in all regions of a star, it is applicable for the central stellar regions. The use of such a simple expression can lead to an understanding of how pressure is modified when the relative abundances of hydrogen and helium change during the hydrogen burning phase in stars (see Chapter 6). In the complete-ionisation approximation, an atom for a given species j will furnish z_j electrons to the stellar plasma (where z_j is defined as the charge of the nucleus of the species under consideration) and the total electronic number density therefore becomes

$$n_e = \sum z_j n_j \tag{5.118}$$

Since the number density of each species is

$$n_i = \frac{X_i}{m_i}\rho = \frac{X_i}{A_i m_{\text{H}}}\rho \tag{5.119}$$

The two equations above may be combined to give the following expression for the electron number density

$$n_e = \frac{\rho}{m_H} \sum \frac{z_j X_j}{A_j}$$

(5.120)

and the electronic pressure is therefore

$$P_e = \frac{\rho k T}{m_H} \sum \frac{z_j X_j}{A_j}$$

(5.121)

From Eq. (5.111), it is easy to show that

$$\frac{1}{\mu_e} = \sum \frac{z_j X_j}{A_j}$$

(5.122)

and that the mean molecular weight is

$$\frac{1}{\mu} = \frac{1}{\mu_{ions}} + \frac{1}{\mu_e} = \sum \frac{X_j}{A_j} + \sum \frac{z_j X_j}{A_j}$$

(5.123)

The physical meaning of the mean electron molecular weight can be found by knowing that

$$P_e = \frac{\rho k T}{\mu_e m_H} = n_e k T$$

(5.124)

and therefore,

$$\frac{1}{\mu_e} = \frac{n_e}{\rho / m_H} = \frac{\text{number of free electrons}}{\text{number of nucleons}}$$

(5.125)

Consequently, $\dfrac{1}{\mu_e}$ measures the number of free electrons per nucleon in the plasma.

It can be shown (see Example 5.5) that for completely ionised plasma composed of an elemental mixture with a mass fraction of hydrogen equal to X and a mass fraction of helium equal to Y (and therefore, the mass fraction of the metals is equal to $Z = 1 - X - Y$) the mean molecular weight is approximately given by the following expression

$$\mu = \frac{2}{3X + \dfrac{Y}{2} + 1}$$

(5.126)

It can also be shown (see Exercise 5.11) that for plasma in which all atoms are neutral, the mean molecular weight is approximately

$$\mu = \frac{1}{X + \dfrac{Y}{4}}$$

(5.127)

Example 5.5: Find the mean molecular weight μ for completely ionised plasma containing metals with a mass fraction of hydrogen equal to X and a mass fraction of helium equal to Y.

Answer:

As seen above, the mean ionic weight and mean electron molecular weight are, respectively,

$$\frac{1}{\mu_{\text{ions}}} = \sum \frac{X_j}{A_j} = X + \frac{Y}{4} + \sum_{j>2} \frac{X_j}{A_j} \tag{5.128}$$

and

$$\frac{1}{\mu_{\text{e}}} = \sum \frac{z_j X_j}{A_j} = X + \frac{Y}{2} + \sum_{j>2} \frac{z_j X_j}{A_j} \tag{5.129}$$

Therefore,

$$\frac{1}{\mu} = \frac{1}{\mu_{\text{ions}}} + \frac{1}{\mu_{\text{e}}} = 2X + \frac{3Y}{4} + \sum_{j>2} \frac{(z_j+1)X_j}{A_j} \tag{5.130}$$

The sum $j > 2$ signifies that this sum is done for the metals only. Since for most elements, the number of protons in the nucleus is approximately equal to the number of neutrons there, $\frac{(z_j+1)}{A_j} \approx \frac{1}{2}$. Also, since the fraction of mass of the metals Z present in the plasma is by definition equal to

$$Z = \sum_{j>2} X_j \tag{5.131}$$

the mean molecular weight becomes

$$\frac{1}{\mu} = 2X + \frac{3Y}{4} + \frac{Z}{2} \tag{5.132}$$

By normalization, $Z = 1 - X - Y$, and the mean molecular weight for the plasma under consideration is given by the following expression

$$\mu = \frac{2}{3X + \dfrac{Y}{2} + 1} \tag{5.133}$$

During the main-sequence stage, hydrogen is transformed into helium in the stellar core. Therefore, Y increases with time. This leads to an increase of μ and a decrease of the pressure (see Figure 6.9). The decrease in pressure induces a slow contraction of the central regions of the star. Therefore, even while on the main sequence, the structure of a star changes (or evolves) with time. More details concerning the evolution of stars on the main sequence and beyond are given in Chapter 6.

5.6.3   Degeneracy

Degeneracy is a quantum-mechanical effect that modifies the equation of state of the gas. In classical physics, the states of free particles have a continuous energy spectrum and there is no limit on the number of particles that can be found in these continuous states. However, under certain conditions sometimes found in stars, quantum physics can come into play for the free-electron gas and this gas can then become degenerate. Degeneracy of such a gas is due to a combination of the Heisenberg's uncertainty principle and Pauli's exclusion principle. This section aims to explain degeneracy and describe its importance in stars.

Quantum physics shows that the smallest quantum cell that contains a single state in the six-dimensional phase space consisting of three space and three momentum components (x,y,z,p_x,p_y,p_z) is equal to

$$\Delta x \Delta y \Delta z \Delta p_x \Delta p_y \Delta p_z = h^3 \tag{5.134}$$

This result is consistent with Heisenberg's uncertainty principle. In stars, the degeneracy is due to the free electrons because of their relatively low momentum (see the equations below for a better understanding). Since electrons are fermions, they must obey Pauli's exclusion principle and there can only be two electrons (with opposite spins) in each of such quantum cells. Now, let us assume a unitary volume $\Delta x \Delta y \Delta z = 1 \text{ cm}^{-3}$, in spherical coordinates the equation above leads to

$$\Delta p_x \Delta p_y \Delta p_z = 4\pi p^2 \mathrm{d}p = 4\pi m_e^3 V^2 \mathrm{d}V = h^3 \tag{5.135}$$

where classical physics is assumed valid for the momentum of the free electrons ($p = m_e V$).

Since there can be two electrons in each state, the density of free electrons with momentum between p and $p + \mathrm{d}p$ defined as $n_e(p,p + \mathrm{d}p)$ must therefore respect the following relation

$$n_e(p, p+\mathrm{d}p) \leq \frac{8\pi p^2 \mathrm{d}p}{h^3} \tag{5.136}$$

where $\dfrac{8\pi p^2 \mathrm{d}p}{h^3}$ is the number of states per unit volume. When the number density of electrons is small (and clearly respects the inequality above), the statistics of the velocity distribution for the electrons are given by the Maxwell distribution and the equation of state is the one of an ideal gas. However, when the number density of electrons approaches this inequality (or in other words when the number density of free electrons becomes comparable to the density of quantum states), the velocity distribution differs strongly from the Maxwell distribution. The equation of state of a nonrelativistic degenerate gas is also quite different from that of an ideal gas. It is independent of temperature and only depends on density. Detailed theoretical considerations lead to a degenerate equation of state of the form

$$P \propto n_e^{5/3} \tag{5.137}$$

for a nonrelativistic degenerate gas of electrons. This equation may also be expressed in terms of the density via the definition of the mean electron molecular weight

$$P \propto \left(\frac{\rho}{\mu_e}\right)^{5/3} \tag{5.138}$$

Degeneracy is important in white dwarfs because of the large densities found there. It can also be important in the central regions of evolved stars. For example, soon after the Sun will leave the main sequence, its core will become degenerate. This will lead to a short phase where helium burns very fast. This is called the helium flash and will be discussed further in Section 6.8.3.

Figure 5.10 shows under what physical conditions degeneracy is important in stars. As expected, this figure shows that for a given temperature, stellar plasma becomes degenerate at high densities. At even higher densities, the degenerate gas becomes relativistic. The reason for this is due to the exclusion principle. Because of this principle, at such high densities more energetic levels become occupied and these high-energetic particles are relativistic. The equation of state for a relativistic degenerate electron gas is slightly different from Eq. (5.138) and is given by the expression

$$P \propto \left(\frac{\rho}{\mu_e}\right)^{4/3} \tag{5.139}$$

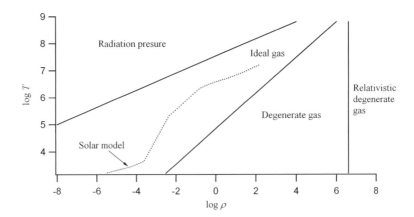

Figure 5.10 The approximate domains of the validity of the ideal-gas approximation, radiation pressure (see Section 5.6.4), degenerate and relativistic degenerate gases on a log T, log ρ diagram. The solid lines show the delimitations of the various regions. Also shown in this figure is the position of the solar model.

Also shown in Figure 5.10 is the log T – log ρ relation for the solar structure. This figure shows that inside the Sun, the ideal-gas assumption is valid.

5.6.4 Radiation Pressure

In stellar plasma, in addition to ions and free electrons, there are evidently an enormous number of photons. Since photons possess momentum, they can participate in the overall pressure. In (the optional) Section 3.12.2, the following expression was found for the radiation pressure

$$P_{\text{rad}} = \frac{4\sigma T^4}{3c} \tag{5.140}$$

where T is the local temperature of the plasma. As expected, Figure 5.10 shows that the radiation pressure can dominate gas pressure at higher temperatures. Also, for a given temperature, the radiation pressure becomes important at lower densities. This behaviour can be understood by comparing the energy density due to photons and that due to the gas. The energy density due to photons is proportional to T^4 while for gas it is proportional to ρT (see Exercise 5.12). This explains why even at relatively low temperatures, the energy due to photons can dominate the gas energy providing the density is sufficiently small.

When of importance, radiation pressure must be taken into account when solving the hydrostatic equilibrium equation. As discussed in (optional) Section 3.12, for massive stars, radiation pressure eventually leads to dynamical instability as the upper limit for stellar mass is reached. When this upper limit is reached, the effective temperature of the star becomes large enough to create sufficiently high radiation pressure to push out any mass exceeding the upper limit. The upper limit for the mass of stars is not a well-known quantity, but is approximately 120 $M_\odot$.

5.7 Variable Stars and Asteroseismology

5.7.1 Variable Stars

Variable stars are those that exhibit photometric or spectroscopic temporal changes. There exists a virtual zoo of variable stars, each of which often constitutes a research field by themselves. Variable stars can be divided into four categories: rotating, eruptive, explosive (sometimes called cataclysmic variables) and pulsating stars. Rotating variables are said to be extrinsically variables while the other three types are intrinsically variable stars.

Rotating variables can, for instance, be due to starspots (similar to sunspots) that come into and out of view during rotation. Other rotating variables can be due to their ellipsoidal shape, often caused by the tidal effects due to a companion star. A companion star can also cause an increase of the brightness on the side of the other star facing it by heating it with radiation. Variability can also be caused by eclipsing of a binary (or even multiple) star system commonly called eclipsing binaries. In this case, the variability is simply due to the fact that one star passes in front of the other (as seen by the observer) during their orbiting motion.

Eruptive variables are stars that can eject mass nonperiodically. Explosive variables are stars that exhibit explosive features either quasiperiodically (as for the case of recurrent novae, see Chapter 6 for more details) or nonperiodically (like supernovae). Pulsating stars possess photometric and/or spectroscopic variations due to generally periodic physical changes such as a change of their stellar radius and/or surface temperature. Even though throughout this chapter, equations stating the mechanical and energetic equilibrium were often discussed, stars may fluctuate around their true equilibrium state and pulsate. More information on these pulsations and how they may be measured by adequate observations will be discussed below.

Some pulsating stars are very useful in astrophysics because they can be used as distance indicators. Such pulsating variable stars possess a well-defined period–luminosity relation. Therefore, by measuring their period of variability, one can estimate their luminosity. Comparing the absolute and apparent magnitudes of such stars gives its distance from the observer. Such stars are sometimes referred to as standard candles since their brightness is well known. They can therefore be used as yardsticks to measure distances in our universe. For example, if such a pulsating star finds itself in a star cluster, the distance to this cluster can be determined. This distance is invaluable information for the study of the other stars found in this cluster.

Other kinds of variable stars such as novae or supernovae have well-established luminosities and can also be employed as standard candles. Once again, the observation of such phenomena can lead to an estimate of their distance from the observer.

Since the study of stellar pulsations may give information about the structure of stars, this category of variable stars is especially important. Pulsations may occur at different stages of evolution, or in other words in various parts of the H–R diagram. Figure 5.11

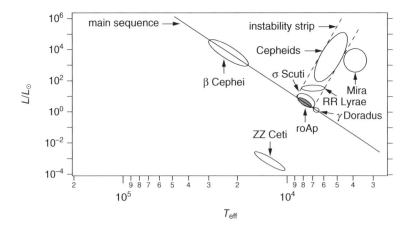

Figure 5.11 Approximate position of several types of pulsating stars in the H–R diagram. Also shown on the figure is the position of the main sequence (solid curve) and the instability strip (region between the dashed lines). Adapted with permission from Jörgen Christensen-Dalsgaard (private communication).

shows the position of several of types of pulsating stars in the H–R diagram. In this figure, the so-called instability strip is shown where a few types of pulsating stars are found. Stars that cross this region in the H–R diagram during evolution pulsate. Here is a nonexhaustive list of pulsating variable stars:

Type-I Cepheids (or classical Cepheids) are evolved stars with $T_{\text{eff}} = 6000$ to $8000\,\text{K}$ found in the instability strip. They possess a high metallicity (these are young population-I stars, see Chapter 6 for more details) and periods in the range from approximately 1 to 100 days. Classical Cepheids were named after the prototype of this type of pulsating stars: the star δ Cephei.

Type-II Cepheids (or W Virginis stars) are evolved stars found in the instability strip. Typically their masses are lower than Type-I Cepheids and are on the order of 0.5 $M_\odot$. They possess a low metallicity (these are old population-II stars, see Chapter 6 for more details) and periods from 1 to 50 days.

RR Lyrae stars are pulsating horizontal-branch stars with $T_{\text{eff}} = 6000$ to $7500\,\text{K}$. They are found in the instability strip of the H–R diagram. Their period of pulsation is typically in the range of 0.1 to 1 day.

Rapidly oscillating Ap stars (or roAp stars) are magnetic Ap stars on or near the main sequence that exhibit short periods of variability ranging from approximately 5 to 15 min. These stars are modelled by a star with a roughly dipolar magnetic field that is inclined with respect to its axis of rotation.

Beta (β) Cephei stars (or β Canis Majoris stars) are pulsating B-type stars on or slightly above the main sequence. Their periods of pulsations range from approximately 0.1 to 0.3 day. This class of stars is named after its prototype the star β Cephei. These stars should not to be confused with Cepheid stars.

Delta (δ) Scuti stars (or dwarf Cepheids) are pulsating A- or F-type stars on or near the main sequence with periods ranging from approximately 0.02 to 0.3 day. This class of stars is named after its prototype: the star δ Scuti.

Gamma (γ) Doradus stars are pulsating F0- to F2-type stars on or near the main sequence with periods ranging from approximately 0.4 to 3 days. This class of stars is named after its prototype: the star γ Doradus.

Mira variables are evolved (red giant) stars with periods ranging from approximately 100 to 1000 days. They are found just outside (on the cool side) of the instability strip. This class of stars is named after its prototype: the star Mira or Omicron Ceti. Figure 5.12 shows the observed light curve of the star Omicron Ceti over a decade. Mira variables are part of a larger class of variable stars called long-period variables (LPVs).

ZZ Ceti stars are pulsating white dwarfs stars of spectral type DA (or hydrogen-rich white dwarfs, see Section 6.10.1). These may also be written DAV, V standing for variable. These stars exhibit photometric variations of up to approximately 0.3 magnitude with pulsating periods ranging from approximately 100 to 1000 s. White dwarfs from other spectral types can also pulsate (i.e. DBV, DOV and DQV, see Section 6.10.1).

For a more thorough discussion of these types of variable stars (and others not mentioned here), the reader is referred to Percy, J.R., *Understanding Variable Stars*, Cambridge University Press, Cambridge (2007).

In addition to the list given above, there exist many other types of pulsating stars. For example, the Sun is a pulsating star. The Sun possesses a very large number of pulsation modes with characteristic periods on the order of 5 min. Solar pulsations have amplitudes

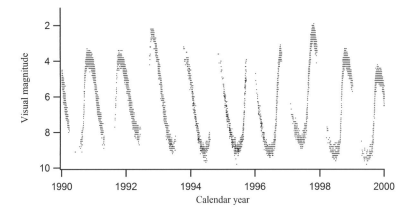

Figure 5.12 The observed visual magnitude for the prototype star for Mira variables, Omicron Ceti from 1990 to 2000. This star has a period of approximately 332 days, $M_* \approx 0.7\ M_\odot$ and $T_{\rm eff} \approx 3000\ {\rm K}$. Data courtesy of the American Association of Variable Star Observers (www. aavso.org).

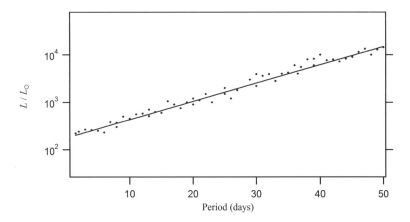

Figure 5.13 Illustration of the period–luminosity relation for classical (or Type-I) Cepheid stars. The dots represent individual classical Cepheid stars while the curve is the best linear fit of the data points giving the relation between the period and the luminosity.

that are much smaller than those shown above for Mira and below for a classical Cepheid (see Figure 5.14). Unlike these two types of pulsating stars, most stars do not have a dominant period of pulsation and a large number of pulsation modes can be present. Of course, the Sun also shows rotational (due to sunspots) and eruptive (solar flares) variability.

Since Type-I Cepheids are of great historical importance, this type of variable star merits further discussion. Both types of Cepheid stars have a well-known period–luminosity relation rendering these astronomical objects useful as distance indicators. This relation between the period T and the luminosity L is roughly given by a power law ($T \propto L^\alpha$). Figure 5.13 illustrates this relation for classical (or Type-I) Cepheids. Type-II Cepheids have a similar period–luminosity relation that lies underneath the curve shown in Figure

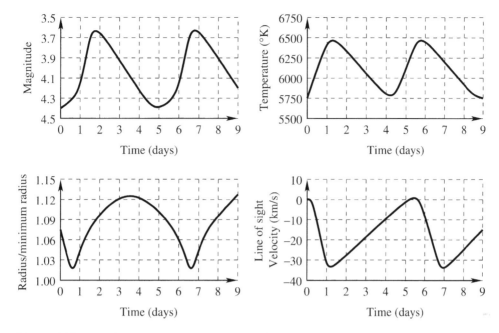

Figure 5.14 Magnitude, temperature, radius and line of sight velocity (or radial velocity) of its surface as a function of time for the classical Cepheid star δ Cephei (which is the prototype for this type of variable stars). The velocity given here is not corrected for the radial velocity of δ Cephei with respect to Earth, which has a value of approximately −16 km/s. Since the maximum of the velocity curve above is approximately found at 0 km/s, the velocities above −16 km/s represent contraction, while those below this value correspond to the phase of expansion. Reproduced with permission from Percy, J.R., Understanding Variable Stars, Cambridge University Press, Cambridge (2007).

5.13 because for the same period, Type-II Cepheids are less luminous than Type-I Cepheids. The period–luminosity relation for classical Cepheids was discovered by the American astronomer Henrietta Swan Leavitt (1868–1921) and has been a precious tool for estimating distances. This period–luminosity relation is sometimes called the Leavitt law. It should be noted that RR Lyrae stars are also often used as standard candles since their dominant pulsation period is on the order of 0.5 day and they are therefore easily identified. These stars are not only numerous but also have large amplitudes and well-defined absolute magnitudes.

Figure 5.14 shows the light curve of the prototype for classical Cepheids, namely the star δ Cephei. Its brightness varies by approximately 0.8 magnitude and its period of variability is approximately 5.4 days. Since the luminosity of a star depends on both its radius and effective temperature, the variation of these two quantities will determine the luminosity dependence as a function of time. Figure 5.14 shows that both δ Cephei's radius and surface temperature fluctuate with time. The maximum brightness of δ Cephei occurs near the maximum of the temperature curve and near the minimum of its radius. It is therefore the temperature changes that dominate the light curve for this star. The velocity of the outer layers of the star is also shown in Figure 5.14 and clearly demonstrates the dynamical pulsation of this star. This velocity can be obtained by studying the Doppler shifts of atomic lines as a function of time. The maximum brightness of this star occurs when its expansion velocity is near its maximum.

To sustain a pulsating motion, a driving mechanism must be present. If not, pulsations would be attenuated and disappear. This is similar to what happens to a classical oscillator (a mass hooked up to a spring for example) vibrating in a medium with friction. In this simple example, a driving force is necessary to maintain oscillations.

It was the British astrophysicist Sir Arthur S. Eddington (1882–1944) who suggested a possible explanation for the pulsation of classical Cepheids. He suggested that certain layers of the star, while it is in its compression phase of pulsation, might become quite opaque to radiation. The increase of opacity causes an accumulation of heat under these layers and eventually leads to an increase of pressure that pushes them outward. Once pushed out, the opacity of these layers decreases and permits the accumulated heat to flow out. Thereafter, these layers can fall back towards the centre since the pressure beneath them has decreased and a new cycle may then begin. This physical process acts as a valve that releases the accumulated heat and therefore decreases the pressure. The manifestation of this process is seen in Figure 5.14. When the star contracts, the surface temperature increases. Pressure then pushes the outer layers of the star outward. The maximum brightness occurs near where the aforementioned layers attain their maximum expansion velocity, which is also correlated to the maximum of the surface temperature.

The relatively large opacity needed for the mechanism described above in the case of Cepheids is due to the ionisation zone of HeII $\rightarrow$ HeIII. Helium, being the second most abundant element in most stars, can strongly contribute to the opacity. The increase of opacity is due to the fact that before total ionisation occurs, a growing number of atomic energy levels becomes populated and this increases the number of atomic lines present in their absorption spectrum. This hinders the radiative-transfer process and may then cause an accumulation of heat beneath the HeII $\rightarrow$ HeIII ionisation zone, which eventually leads to these layers being pushed out by the increased pressure. Once the ionisation zone is pushed out to layers where the temperature is lower, HeIII ions recombine with free electrons and the HeII ions become less excited and the opacity therefore decreases. This permits the accumulated heat to flow towards the exterior and leads to a contracting phase and a new cycle may begin. This is known as the *κ-mechanism* since the Greek letter κ often represents the radiative opacity (please note that in this book κ represents absorption only, while the total opacity is represented by the symbol k). The HeII ionisation zone is also believed to be the driving mechanism for RR Lyrae stars, while the hydrogen ionisation zone is most likely responsible for the pulsation of Mira variables, roAp and ZZ Ceti stars.

Meanwhile, the driving mechanism for the pulsation of β Cephei stars is thought to be due to an increase of the opacity of the stellar medium at depths where $T \approx 200\,000\,\mathrm{K}$ due mainly to the iron-peak elements. The increase in the opacity in this region is commonly called the Z-bump, Z standing for the metallicity (this feature was discussed in Chapter 3 and is illustrated in Figure 3.17). The opacity increase needed to sustain the pulsations could be due to an accumulation of iron at such depths. Since iron has a rich absorption spectrum, if its abundance is sufficiently amplified in that region, it can lead to a suitably large opacity to drive pulsations. In this scenario, the accumulation of iron is due to the atomic-diffusion process. Atomic diffusion is the migration of the various elements within a star due to the competition between gravity that causes settling of the heavy elements towards the stellar centre and radiative acceleration that can selectively push out elements towards the surface. Diffusion is only efficient in stable stellar regions where other mixing processes like convection or turbulence are not present. Atomic diffusion will be discussed in more detail in Chapter 7. Not all stars that are unstable due to the Z-bump need diffusion to pulsate. Some stars with uniform but relatively large abundances can possess instability due to the Z-bump.

Mira variables are also of particular historical importance since it is the prototype of this class of pulsating star that was the first variable star to be observed. It was discovered by the German theologian D. Fabricius (1564–1617) in 1596. Mira variables have a particularly large brightness variation. As seen in Figure 5.12, the prototype star for this type of pulsating stars' visual magnitude can vary by approximately 8 magnitudes (which is equivalent to a variation of the flux by more than a thousand!) which is much larger than for Cepheids for example. It is believed that when Mira variables are at maximum expansion (or minimum brightness), a large amount of TiO and other metallic oxide molecules form in their outer layers (see Reid, M.J. and Goldston, J.E., *The Astrophysical Journal*, 568, 931 (2002) for more details). These molecules obstruct a large amount of visible light leading to a relatively weak flux in the visible region of the spectrum and therefore to a large variation in the visual magnitude.

5.7.2   Asteroseismology[†]

Asteroseismology is the field of research that pertains to the pulsations (or oscillations) of stars in the aim of gaining information about their internal structure. The field of asteroseismology has had an exponential growth during the past several decades especially since it can probe the properties of the interior of stars as opposed to spectroscopy, for instance, that only samples the photosphere. Asteroseismology has therefore become a very powerful diagnostic tool. Meanwhile, the field of helioseismology relates to the study of the pulsations of the Sun.

Since stars can in general be considered as spheres composed of a compressible fluid, they may, under proper conditions, pulsate. For example, a star may have radial oscillations where its pulsations have a radial symmetry similar to those seen earlier for classical Cepheids. Figure 5.15 illustrates radial modes in a spherical object such as a star. These

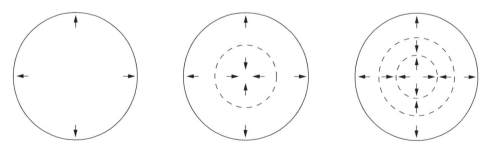

Figure 5.15   An illustration of radial pulsation modes in a spherical object. The dashed concentric shells are nodes of the standing waves. The fundamental mode is shown on the left in the expansion phase where the arrows represent the movement of the various portions of the star. The next two modes (sometimes called overtones) are also shown.

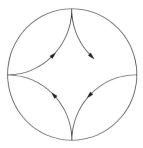

Figure 5.16 An illustration of the propagation of a nonradial mode of oscillation. The waves rebound between the surface and deeper layers due to refraction. The refraction is caused by the gradient of the speed of sound within the star that is due to the pressure gradient. The trajectory of such waves is represented by the arrowed path.

modes of oscillations are similar to those found for sound wave in a pipe of a musical instrument for instance.

Stars can also have some modes of oscillation that are nonradial. For instance, in the Sun acoustical waves travel are trapped within the star and rebound between the surface and the deeper layers (see Figure 5.16). These waves are reflected by the surface and the deeper layers because of refraction. In this instance, refraction is caused by the fact that the speed of sound varies with depth due to the pressure gradient. Refraction therefore modifies the direction of the waves travelling inward and eventually reflects them upwards. The recoil in this type of pulsation is caused by the pressure gradient and these waves are commonly called pressure waves (or p-modes). Since waves of different frequencies will be reflected at different depths, the study of the various modes present for a star can give information on the internal structure of the star. There are several million modes of pulsation that participate in the Sun's oscillation.

Another recoil process, namely buoyancy in the stellar medium can also cause stellar pulsations. Since buoyancy is caused by gravity, these are commonly called gravity waves and the related modes are called g-modes (g standing for gravity). The g-modes have smaller frequencies than p-modes.

Generally speaking, the luminosity variation of a pulsating star is caused by three factors: changes in the star's volume (i.e. general expansion or contraction), changes of the star's shape (stars when they pulsate may become nonspherical) and temperature variations over the stellar surface. Stellar pulsations can be quite complex since a large number of modes of oscillation along with their eigenfrequencies may be present simultaneously in a given star. Figure 5.17 shows a nonradial mode of pulsation for a ZZ Ceti star that is a white dwarf (see Section 6.10.1). The shades in this figure show the variation of the surface temperature over its surface. For ZZ Ceti stars it is the surface temperature variations that dominate their variability. A detailed and complex theoretical treatment such as the one leading to the results shown in Figure 5.17 may be used to extract information concerning the stellar structure by comparing the pulsation modes that are observed to those predicted by these numerical models. See Fontaine, G., Brassard, P., Charpinet, S., Quirion, P.-O. and Randall, S.K., ASP Conference Series (in press) for more details about these calculations.

Figure 5.17 Theoretical calculations showing the surface temperature of a nonradial mode of a ZZ Ceti star. The shades of grey get darker as the temperature increases. Figure courtesy of Gilles Fontaine.

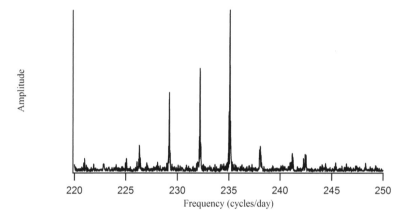

Figure 5.18 The pulsation frequencies observed for the roAp star HD 24712 (or HR 1217) by the MOST mission. This figure shows the intensity of the periodic signals as a function of frequency extracted from the photometric observations during a one-month period of the star under consideration. Data courtesy of Jaymie Matthews and Chris Cameron.

Pulsations lead to photometric and/or spectroscopic variations that may be measured with adequate observations. A common way of detecting pulsations is by photometric measurements. The method consists of taking a large number of measurements of the star of interest in a given photometric band. The integration time for each of these measurements must be smaller than the pulsation periods that are desired to be detected. A Fourier analysis, which is a mathematical method capable of obtaining the frequencies found in a given signal, is then applied to the observed data in order to extract the frequencies that are present in the astronomical data. An example of the type of scientific results obtained by the MOST satellite is shown in Figure 5.18. This figure shows the pulsation frequency spectra detected for the roAp star HD 24712 (or HR 1217).[5]

---

[5]The names of this star come, respectively, from the Henry Draper (HD) catalogue and the Harvard Revised (HR) photometry catalogue.

Special Topic – Asteroseismological Observations

The MOST mission:

The MOST (standing for Microvariability and Oscillations of STars) satellite is a Canadian observation mission that aims to study several types of astronomical objects including roAp and Sun-like stars. This satellite was launched in 2003 and is in a low-Earth orbit at 820 km altitude with an orbital period of approximately 100 min. This satellite is relatively small and contains a 15-cm reflecting mirror along with two CCD detectors (one for guiding and the other for data collection). This instrument has a broadband visual filter (3500 Å to 7000 Å) and a variable integration time from 1 to 60 s. It can observe a star for up to 58 days without interruption.

The relatively small size of this satellite is not indicative of its accuracy since it can detect a variability of only one part in a million. This is the equivalent of the variation of the amount of light received by an observer from a radiation source 1 km away as compared to when this observer moves closer to this source by only 0.5 mm!

Any naturally occurring periods such as the rotational period of the satellite must be accounted for when analysing the observational data. Space-based observatories such as MOST can detect small variabilities in stars that are not possible to detect from the surface of the Earth. Noise caused by the turbulence in the Earth's atmosphere would hinder such detections from Earth-based observatories. Figure 5.18 shows the type of results that can be extracted from MOST observations. These results are obtained by a Fourier analysis of photometric observations that finds the relative intensities of the periodic signals present, if any. In this figure, several pulsation frequencies are clearly found for this roAp star. For more information about these observations the reader is referred to Cameron, C., Ph.D. Thesis, U. of British Columbia (2009) and Cameron, C., Matthews, J., *et al.*, in preparation (2009).

The CoRoT mission:

The CoRoT (standing for Convection, Rotation and planetary Transits) mission consists of an orbiting telescope with a 27-cm diameter with four CCD detectors. This mission aims to study stellar seismology and search for exoplanets. Several types of pulsating stars will be studied with this instrument including RR Lyrae, δ Scuti, γ Doradus and β Cephei stars. This telescope was put into orbit at an altitude of 896 km in 2006. This mission is composed of a consortium of several countries spearheaded by France.

5.7.3   Basic Physics Behind Period–Luminosity Relations[†]

Figure 5.13 shows the period–luminosity relation for classical Cepheids. This relation demonstrates that Cepheids with larger luminosities have longer pulsation periods that can be explained by physical reasoning. Cepheids are evolved stars (see Figure 5.11) and the most luminous ones are supergiants with larger radii. It is therefore not surprising that these larger stars pulsate at smaller frequencies (or larger periods) than the smaller (and less luminous) Cepheids. A simple physical development undertaken below leads to the same conclusion.

It was seen that for the Sun, acoustical waves that are reflected from the surface to deeper layers are responsible for its pulsations. Therefore, for a given star, the period of pulsation may be approximated by the time its takes a sound wave to travel from its surface to its centre and back again. If $v_s(r)$ represents the speed of sound at distance r from the star's centre, the period of pulsation T may be approximated by

$$T \approx 2\int_0^{R_*} \frac{dr}{v_s(r)} \tag{5.141}$$

The adiabatic speed of sound is

$$v_s(r) = \sqrt{\frac{\gamma P}{\rho}} \tag{5.142}$$

where $\gamma = \dfrac{c_P}{c_V}$ is the ratio of the specific heats at constant pressure (c_P) and volume (c_V) and P and ρ are, respectively, the local pressure and density. The ratio P/ρ is not readily known without the calculation of the model of the stellar interior for the star under consideration. Since only general properties of the pulsation period are of interest here, simplifications may be made to estimate the integral found in the equation for the period of pulsation given above (Eq. 5.141). To make this integral evaluation analytically possible, the crude assumption of a constant density inside the star is made here. The pressure inside the star may then be found via the hydrostatic equilibrium equation

$$\frac{dP(r)}{dr} = -\frac{\rho G M(r)}{r^2} = -\frac{4\pi G \rho^2 r}{3} \tag{5.143}$$

since $M(r) = \dfrac{4\pi r^3 \rho}{3}$ for a sphere of constant density. The equation above can then be integrated from radius r to the surface and assuming the pressure at the surface is nil ($P(R_*) = 0$), it is easy to show (see Exercise 5.13) that this integration gives

$$P(r) = \frac{2\pi G \rho^2}{3}(R_*^2 - r^2) \tag{5.144}$$

This result may then be used to estimate the period of pulsation

$$T \approx 2\int_0^{R_*} \frac{dr}{v_s(r)} = 2\int_0^{R_*} \frac{dr}{\sqrt{\dfrac{2\pi G \gamma \rho}{3}(R_*^2 - r^2)}} = \sqrt{\frac{3\pi}{2G\gamma\rho}} \tag{5.145}$$

Therefore, in the approximation of a star with a constant density, the period of pulsation is inversely proportional to $\sqrt{\rho}$. This explains why more luminous (and therefore more voluminous and tenuous) Cepheids have larger pulsation periods than less-luminous (and denser) ones (see Figure 5.13). This equation also explains why the different types of pulsating stars seen in Section 5.7.1 have different pulsation periods. For example, the equation above makes it clear why compact stars such as ZZ Ceti star have periods of pulsation that are smaller than those of RR Lyrae stars that in turn pulsate more rapidly than Cepheids due to the relative position of these pulsating stars in the H–R diagram shown in Figure 5.11.

5.8   Summary

Equations of stellar structure:

$$\frac{dP(r)}{dr} = -\frac{\rho(r)GM(r)}{r^2} \tag{5.146}$$

$$\frac{dM(r)}{dr} = 4\pi r^2 \rho(r) \tag{5.147}$$

$$\frac{dT(r)}{dr} = -\frac{3k_R\rho}{64\pi r^2 \sigma T^3} L(r) \quad \text{(radiation only)} \tag{5.148}$$

$$\frac{dL(r)}{dr} = 4\pi r^2 \rho(r)\varepsilon(r) \tag{5.149}$$

Temperature gradient: $\nabla = \dfrac{d\ln T}{d\ln P}$ $\tag{5.150}$

Schwarzschild criterion: $\nabla_{\text{rad}} > \nabla_{\text{adi}} = \left(\dfrac{\gamma-1}{\gamma}\right)$ $\tag{5.151}$

Polytropic equation of state: $P = K\rho^{\frac{n+1}{n}}$ $\tag{5.152}$

Lane-Emden equation:

$$\frac{1}{\xi^2}\frac{d}{d\xi}\left(\xi^2\frac{d\theta(\xi)}{d\xi}\right) = -\theta^n(\xi) \quad \text{or} \quad \frac{d^2\theta(\xi)}{d\xi^2} + \frac{2}{\xi}\frac{d\theta(\xi)}{d\xi} + \theta^n(\xi) = 0 \tag{5.153}$$

where $\rho(r) = \rho_c\theta^n(r)$ and $\xi = \dfrac{r}{\alpha}$.

The six regions of the Sun are: core, radiative zone, convection zone, photosphere, chromosphere and corona.

Mean ionic weight: $\dfrac{1}{\mu_{\text{ions}}} = \sum \dfrac{X_j}{A_j}$ $\tag{5.154}$

Mean electron molecular weight: $\dfrac{1}{\mu_e} = \sum \dfrac{z_j X_j}{A_j}$ $\tag{5.155}$

Mean molecular weight: $\dfrac{1}{\mu} = \dfrac{1}{\mu_{\text{ions}}} + \dfrac{1}{\mu_e}$ $\tag{5.156}$

5.9   Exercises

5.1   Assume a star of mass M_* and radius R_* having a density profile equal to

$$\rho(r) = \rho_c \left(1 - \frac{r^2}{R_*^2} \right) \tag{5.157}$$

find the central density ρ_c in terms of M_* and R_*.

5.2   Show that when the energy transport is purely radiative

$$\nabla_{rad} = \frac{3k_R}{64\pi r^2 g} \frac{P}{\sigma T^4} L(r) \tag{5.158}$$

5.3   Assume a star of radius R_* having a density profile equal to

$$\rho(r) = \rho_c \left(1 - \frac{r}{R_*} \right) \tag{5.159}$$

and a nuclear production rate per unit mass equal to

$$\varepsilon(r) = \varepsilon_c \left(1 - \frac{r}{0.2R_*} \right) \quad \text{for} \quad r \le 0.2\,R_* \tag{5.160}$$

and

$$\varepsilon(r) = 0 \quad \text{for} \quad r > 0.2\,R_* \tag{5.161}$$

calculate the luminosity of the star at its surface in terms of R_*, ρ_c and ε_c.

5.4   Derive the four equations of stellar structure as a function of M given in the special topic found in Section 5.2.

5.5   Show that $\dfrac{dB_\nu}{dT} = \dfrac{2k^3 T^2}{h^2 c^2} \left[\dfrac{u^4 e^u}{(e^u - 1)^2} \right]$ $\tag{5.162}$

5.6   Find the value of u where the function $\Re(u)$ found in the expression for H_ν in stellar interiors (Eqs. 5.33 and 5.34) is at its maximum (numerical problem).

5.7   In the centre of cool white dwarfs composed of pure carbon, detailed calculations show that the conduction opacity is approximately

$$k_{cond} \approx 5 \times 10^{-7} \left(\frac{T}{\rho} \right)^2 \text{cm}^2 \text{ g}^{-1} \tag{5.163}$$

By assuming that radiative opacity is dominated by Thompson scattering, show that in the centre of such stars where $T \approx 10^7 \, \mathrm{K}$ and $\rho \approx 10^6 \, \mathrm{g/cm^3}$ conduction completely dominates energy transport.

5.8   Solve the Lane–Emden equation for $n = 0$ and show that the solution is $\theta(\xi) = 1 - \dfrac{\xi^2}{6}$.

5.9   Show that if $\chi(\xi) = \xi\theta(\xi)$ the Lane–Emden equation becomes

$$\frac{\mathrm{d}^2\chi(\xi)}{\mathrm{d}\xi^2} = -\frac{\chi^n(\xi)}{\xi^{n-1}} \tag{5.164}$$

5.10   Calculate the mass of a star with a polytropic equation of state with $n = 1$ in terms of ρ_c and R_*.

5.11   Show that in completely neutral plasma with a metallicity Z much smaller than X and Y, the mean atomic weight is approximately

$$\mu = \frac{1}{X + \dfrac{Y}{4}} \tag{5.165}$$

5.12   Find the temperature at which the radiation energy density is equal to the gas energy density when $\log \rho = -4$. Assume a radiation field with $I_\nu = B_\nu$ and plasma composed of pure hydrogen with an ideal-gas equation of state. Compare this result to the boundary found in Figure 5.10.

5.13   Show that the pressure inside a star of radius R_* with constant density ρ is

$$P(r) = \frac{2\pi G \rho^2}{3}(R_*^2 - r^2) \tag{5.166}$$

6

Nucleosynthesis and Stellar Evolution

6.1 Introduction

During their lifetime, stars dispense an enormous amount of energy. For instance, the Sun emits 3.9×10^{33} ergs (or 3.9×10^{26} J) every second, while stars at the supergiant stage irradiate up to millions of times this quantity of energy. Stars therefore need a source of energy to account for such a large power output.

Before the discovery of thermonuclear reactions, it was commonly thought that gravity was the energy source of stars. In Chapter 2, it was seen that when an astronomical body contracts, approximately half of the gravitational energy is emitted at its surface. If the gravitational energy of a given star of mass M_* and radius R_* is approximated by that of a body with a constant density (see Example 2.2), the energy emitted to interstellar space due to gravitational energy emitted since its formation may be approximated by the following expression (see Section 2.3)

$$\Delta U = -\frac{\Omega}{2} = \frac{3}{10} \frac{GM_*^2}{R_*} \tag{6.1}$$

By assuming that the star's luminosity L_* has been constant since its formation, a characteristic time called the Kelvin–Helmholtz time can be defined as

$$\tau_{KH} = \frac{\Delta U}{L_*} = \frac{3}{10} \frac{GM_*^2}{L_* R_*} \tag{6.2}$$

For the Sun $\tau_{KH} \approx 10^7$ yr, which gives a rough estimation of the age of the Sun assuming that gravity is the dominant source of energy for stars. However, its true age is approximately 4.5×10^9 yr, which can be estimated by geological dating of the oldest rocks on Earth or of meteorites. Since the age of the Sun is much larger than its Kelvin–Helmholtz time, gravity is not sufficient to generate the enormous amount of energy irradiated by the Sun since its formation.

Another possible energy source in stars is chemical energy. However, it is also easy to show (see Exercise 6.1) that energy coming from chemical reactions in stars is not sufficient to solve the energy problem for stars.

Another energy source must then be present. It wasn't until the beginning of the twentieth century that it was understood that the energy source responsible for the great luminosities of stars was nuclear in nature.

As mentioned previously in this book, a star begins its life by burning (or fusing) hydrogen in its core. As hydrogen is transformed into helium, the structure of the star readjusts. When all of the hydrogen in the stellar core is spent, the star evolves relatively rapidly. As will be seen later in this chapter, the star then evolves to other stages such as the red-giant phase, and depending on its mass, it will be able to burn (or fuse) heavier nuclei as time progresses. This process leads to the production of the various elements found in the Universe and is commonly called nucleosynthesis.[1] Nuclear burning is therefore an essential ingredient to understand how stars evolve.

In this chapter, the physics surrounding thermonuclear (or fusion) reactions and the energy they emit as well as their relation to the evolution of stars will be discussed. An optional section concerning nuclear models will be presented for the reader who wishes more information on the underlying physics behind nuclear fusion. For an even deeper understanding of nuclear fusion, an optional advanced section is also presented at the end of this chapter that discusses nuclear reaction cross sections and rates.

The various nuclear reactions taking place on the main sequence and at later evolutionary stages will be given. Examples of evolutionary tracks of stars in the H–R diagram will be shown. The relation of stellar evolution to stellar clusters, galaxies and stellar populations will then be described. The properties of the end states of stars, namely white dwarfs, neutron stars and black holes will be presented. The processes by which the elements heavier than iron are formed as well as the properties of novae and supernovae will also be discussed. However, before seeing these topics, general properties of thermonuclear reactions will be described.

6.2   Generalities Concerning Nuclear Fusion

Nuclear fusion is the union of light nuclei to produce heavier nuclei. Nuclear fusion reactions are exothermic and are the principal source of energy in stars. Energy is emitted by these reactions because the nuclei that are produced are more stable than those that fuse (see Figure 6.1). This figure shows that the average binding energy per nucleon generally increases up to the iron-peak elements (V, Cr, Mn, Fe, Co and Ni). The iron-peak elements have approximately the same binding energy per nucleon. Fusion can then occur up to the production of iron-peak elements. The excess energy, when light nuclei are fused into more tightly bound nuclei, is therefore released in the core of stars. This excess energy

---

[1]The discussion in this chapter relates to stellar nucleosynthesis. Shortly after the Big Bang, the light elements H, He and Li were formed. This is called primordial or Big-Bang nucleosynthesis.

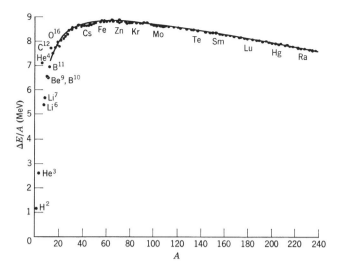

Figure 6.1   The average binding nuclear energy per nucleon $\Delta E/A$ as a function of the number of nucleons A in the various nuclei shown. The solid curve represents the results using the semiempirical mass formula (see (optional) Section 6.3.1). Reproduced with permission from Eisberg, R. and Resnick, R., *Quantum Physics of Atoms, Molecules, Solids, Nuclei and Particles*, John Wiley & Sons, Ltd, New York (1985).

(E) is related to the mass difference between the nuclei that are fused and those produced via Einstein's famous mass–energy equation

$$E = \Delta mc^2 = (m_{\text{initial}} - m_{\text{final}})c^2 \tag{6.3}$$

where Δm is the difference between the total mass of the nuclei being fused (m_{initial}) and the total mass of the particles produced by a given thermonuclear reaction (m_{final}). For thermonuclear reactions the energy given by the reaction above is positive. Energy may also be obtained when nuclei heavier than iron split into light atoms. This process is called nuclear fission and is the basis of nuclear power plants. It is of little or no importance for stars.

The average binding energy per nucleon shown in Figure 6.1 is simply the difference between the sum of the individual masses of the nucleons composing the nucleus and the mass of the constituted nucleus divided by the number of nucleons in it. For example, in the case of ^{4}He, which is composed of two protons and two neutrons, the average binding energy per nucleon is

$$\frac{\Delta E}{A} = \frac{1}{4}(2m_{\text{p}} + 2m_{\text{n}} - m_{^4\text{He}})c^2 \approx 7.07\,\text{MeV} \tag{6.4}$$

where $m_{^4\text{He}}$ is the mass of the ^{4}He nucleus and ΔE is the total binding energy of the nucleus and A is the total number of nucleons it contains.

A myriad of nuclear reactions takes place in stars during their various evolutionary stages. On the main sequence, stars fuse (or burn) hydrogen to produce helium. Section 6.5 is dedicated to the thermonuclear reactions on the main sequence. It will be seen later in this chapter that for less-massive stars, hydrogen is first fused to give deuterium (^{2}H) via the reaction (commonly called the proton–proton reaction)

$$^1\text{H} + {}^1\text{H} \rightarrow {}^2\text{H} + e^+ + \nu_e \tag{6.5}$$

In this reaction e^+ represents a positron (or antielectron) and ν_e is an electron neutrino. These two particles are respectively an antilepton and a lepton. Leptons are elementary particles that interact via the weak nuclear force or possibly the electromagnetic force (if they are electrically charged) but not the strong nuclear force. They have a spin of ½ (in units of $\hbar$) and are therefore fermions. Their name is derived from the Greek word *leptos* meaning thin because leptons are particles with relatively small masses. There are three so-called generations of leptons (see Table 6.1). Each generation contains two leptons, one of which is a neutrino. There are therefore three types of neutrinos known to exist in the Universe: electron, muon and tauon neutrinos. Neutrinos are electrically neutral. They are, respectively, associated to the following leptons: electron, muon and tauon that all possess the charge of the electron. Therefore, in the reaction above, an electron neutrino is emitted due to the presence of the positron (see below for the discussion on the conservation of the lepton number).

For each elementary particle, there exists their corresponding antiparticle. When a particle meets its associated antiparticle, they annihilate each other and their mass energy is transformed to energy in the form of radiation. This process must be taken into account when calculating the energy emitted by certain nuclear reactions.

Meanwhile, protons and neutrons are members of a class of elementary particles called baryons and are made up of three quarks. Baryons can interact not only via the weak nuclear force but more importantly also via the strong nuclear force that is responsible for keeping the protons and neutrons together in the atomic nucleus. The name baryon is derived from the Greek word *barys* meaning heavy. Similarly to leptons, there are three generations of quarks, each containing two different types of quarks. These six types of quarks (also called flavours) are named: up, down, charm, strange, top, and bottom (see Table 6.2). Quarks are charged particles with charges of $-1/3$ or $+2/3$ times the fundamental charge (see Table 6.3) and possess half-integral spins (quarks are therefore fermions).

Table 6.1   Leptons.

First generation	Second generation	Third generation
electron (e$^-$)	muon (μ$^-$)	tauon (τ$^-$)
electron neutrino (ν_e)	muon neutrino (ν_μ)	tauon neutrino (ν_τ)

Table 6.2   Quarks.

First generation	Second generation	Third generation
up (u)	charm (c)	top (t)
down (d)	strange (s)	bottom (b)

Table 6.3   Electric charge of quarks.

Quark	Charge (units of e)
up (u)	+2/3
down (d)	−1/3
charm (c)	+2/3
strange (s)	−1/3
top (t)	+2/3
bottom (b)	−1/3

A proton is composed of two up quarks and one down quark while a neutron is made up of two down quarks and one up quark. Quarks possess the peculiarity of interacting via all four fundamental forces.

Nuclear reactions must respect several conservation principles: conservation of electric charge, as well as the conservation of baryon and lepton numbers. The baryonic number for a proton or a neutron is +1. The lepton number for an electron or a neutrino is +1. These numbers are −1 for the corresponding antiparticles. For example, in the fusion reaction of two ^{1}H nuclei given above, the baryonic number is +2 and the lepton number is nil on each side of the reaction. The electric charge is also conserved during this reaction. The presence of the positron assures electric charge conservation, while the presence of the associated neutrino (i.e. the electron neutrino) assures the conservation of the lepton number.

On a more fundamental level, this thermonuclear reaction is the result of the following reaction: $p \rightarrow n + e^+ + \nu_e$. Since leptons are involved, it indicates that the weak nuclear force intervenes in this reaction. Protons are composed of three quarks: uud, while neutrons are made up of a udd quark trio. Therefore, on an even more fundamental level, the reaction $p \rightarrow n + e^+ + \nu_e$ is in reality a transformation of an up quark into a down quark via the reaction $u \rightarrow d + e^+ + \nu_e$. The reader who desires more information on the properties of quarks and other fundamental particles is referred to Martin, B.R. and Shaw, G., *Particle Physics*, John Wiley & Sons, Ltd, Chichester (2008).

Another similar reaction that is critical for the formation of the elements heavier than iron is the decay of neutrons into protons via $n \rightarrow p + e^- + \bar{\nu}_e$ where $\bar{\nu}_e$ represents an antielectron neutrino. On a more fundamental level, this reaction transforms a down quark into an up quark via the reaction $d \rightarrow u + e^- + \bar{\nu}_e$. The importance of this decay reaction relative to the formation of elements heavier than iron will be discussed in (optional) Section 6.12.

The energy emitted by the reaction ^{1}H + ^{1}H $\rightarrow$ ^{2}H + $e^+ + \nu_e$, while neglecting the energy due to the annihilation of the positron and that of the neutrino (more details are given below), is

$$E = \left(m_{\text{initial}} - m_{\text{final}}\right)c^2 = \left(2m_{^1\text{H}} - m_{^2\text{H}} - m_e\right)c^2 = 0.420\,\text{MeV} \tag{6.6}$$

where m_e is the mass of the positron (which is equal to the mass of the electron).

The masses found above are *nuclear* masses and not atomic ones. The atomic mass of several species is given in Appendix F. For calculations such as that done above, the total mass of the electrons found in the atom must be subtracted from the atomic mass to obtain the appropriate nucleus mass. To obtain exact nuclei masses, the binding energy of the electrons must be added. However, since such energies are on the order of 10 eV to 10 keV, except for the heavier nuclei they may be neglected when compared to typical nuclear masses and binding energies (which are on the order of 1 MeV or more).

Two other factors must be considered to precisely evaluate the total energy emitted by the reaction under consideration. First, the positron emitted in this reaction is quickly annihilated when it interacts with a free electron in the stellar plasma. This annihilation process emits $2m_e c^2 = 1.022$ MeV of energy. Secondly, since neutrinos interact very little with matter, most neutrinos emitted by stellar thermonuclear reactions can cross the upper layers of stars without interacting with the stellar plasma. These neutrinos take away energy and diminish the efficiency of such nuclear reactions. On average, the neutrino emitted by the nuclear reaction under consideration possesses an energy of 0.263 MeV (see Table 6.4 in Section 6.5.1). Therefore, the *total* energy emitted by the fusion of two ^{1}H nuclei is

$$E = 0.420\,\text{MeV} + 1.022\,\text{MeV} - 0.263\,\text{MeV} = 1.179\,\text{MeV} \tag{6.7}$$

To obtain the value of the nuclear energy produced per gram of matter per second $\varepsilon(r)$, the energy taken away by neutrinos is subtracted and counted as a loss of energy directly to interstellar space.

Example 6.1: Calculate the energy emitted by the reaction ^{3}He + ^{3}He → ^{4}He + $2\,^1$H (this reaction intervenes in the fusion of hydrogen via the proton–proton chains, see Section 6.5.1).

Answer:

The energy liberated by the reaction under consideration is found by using Einstein's mass–energy equation

$$E = \left(m_{\text{initial}} - m_{\text{final}}\right)c^2 = \left(2m_{^3\text{He}} - m_{^4\text{He}} - 2m_{^1\text{H}}\right)c^2 \tag{6.8}$$

The masses found in this equation are those of the nuclei. Appendix F gives the atomic masses for the neutral ionisation state. Even though these nuclei are completely ionised in stellar cores, the atomic masses can still be used in this case. Assuming that the reacting nuclei are neutral atoms, there is the same number of electrons (i.e. four) on either side of the nuclear reaction under consideration. The atomic masses may then be used because the electron masses cancel (as discussed above, the binding energy of the electrons may be neglected). Therefore, by using the data in Appendix F, the energy emitted is

$$E = \left(2 \times 3.0160293\text{u} - 4.0026032\text{u} - 2 \times 1.0078250\text{u}\right)c^2 = 0.0138054\text{u}c^2 = 12.860\,\text{MeV} \tag{6.9}$$

6.3   Models of the Nucleus[†]

The aim of this optional section is to give basic knowledge about the nucleus to the reader who is not well acquainted with nuclear physics and who desires more information on this topic. This is done by presenting two models for the nucleus: the liquid-drop model and the shell model. The results from these models can explain certain properties of the nucleus that are important for a deeper understanding of nuclear reactions.

6.3.1   The Liquid-Drop Model

The liquid-drop model of the nucleus can give insight on how and why the masses of nuclei vary with the number of protons (Z) and neutrons ($N = A - Z$) they contain. For this model, it is assumed that the nucleus has properties similar to those of a drop of liquid. For instance, Figure 6.1 shows that when A is large, the average binding energy per nucleon is approximately constant ($\Delta E/A \approx 8\,\text{MeV}$). For a drop of liquid, this property is equivalent to the fact that the heat of vaporization is proportional to its mass. Also, assuming that a nucleus is composed of nucleons coalescing in a spherical shape, the density of the nucleus may be assumed to be constant, just as it is inside a drop of liquid. Such a situation would arise if nucleons are approximated as mutually attracting hard spheres.

This model of the nucleus can lead to an equation estimating the mass of nuclei that contains several terms. This equation is called the semiempirical mass formula and was first established in 1935 by the German physicist Carl Friedrich Freiherr von Weizsäcker (1912–2007). It is written in units of atomic mass units u (u = $1.661 \times 10^{-24}\,\text{g} = 931.5$ MeV/c^2). Since the mass of the proton is 1.0072765u and that of the neutron is 1.0086649u, the mass of a nucleus with A nucleons (containing Z protons and N neutrons) is given by

$$m(Z, A) = 1.0072765Z + 1.0086649N - \frac{\Delta E}{c^2} \tag{6.10}$$

where ΔE is the binding energy of the nucleus in units of uc^2. This binding energy is due to the strong nuclear force that binds the nucleons together in the nucleus. The first two terms of this formula give the sum of the individual masses of the nucleons that constitute the nucleus in question. In the liquid-drop model, the binding energy comprises five terms

$$\Delta E = E_{\text{vol}} + E_{\text{surf}} + E_{\text{coul}} + E_{\text{asym}} + E_{\text{pair}} \tag{6.11}$$

namely the volume, surface, Coulomb, asymmetry and pairing terms. Each of these terms will now be described. The liquid-drop model is a mixture of a classical model (volume, surface and Coulomb terms) in which quantum effects are taken into account in a purely phenomenological manner (asymmetry and pairing terms).

In the hard-sphere approximation for the nucleons, the volume of the nucleus is directly proportional to A. As the number of nucleons increases, the total binding energy also increases. The volume term, which represents this portion of the binding energy, may be written as

$$E_{\text{vol}} = a_{\text{vol}} A \qquad (6.12)$$

where a_{vol} is a free parameter to be fitted to data that will be given below. This dependence is different from the gravitational potential energy, which is proportional to mass square (see Section 2.3). The reason for this is that the gravitational force has an infinite extent while the strong nuclear force only acts locally (at distances on the order of 10^{-15} m or less). A given nucleon in the nucleus mostly feels the attractive force of its neighbouring nucleons. This property explains why the total binding energy due to the volume term is proportional to the number of nucleons in the nucleus. The fact that nucleons near the surface have fewer near neighbours will be considered below in what is called the surface term. Since the strong nuclear force is attractive, the volume term in the semiempirical mass formula decreases the mass of the nucleus (i.e. $E_{\text{vol}} > 0$) by a value determined by Einstein's mass–energy relation.

In the depiction of a spherical nucleus, the nucleons at the surface have fewer near neighbours than those closer to the centre of the nucleus. These surface nucleons are therefore more loosely bound to the nucleus. Since the volume of the nucleus is proportional to A, its radius is proportional to $A^{1/3}$. The number of nucleons at the surface is proportional to the nucleus' surface ($\propto A^{2/3}$) and the average binding energy per nucleon is decreased by the so-called surface term

$$E_{\text{surf}} = -a_{\text{surf}} A^{\frac{2}{3}} \qquad (6.13)$$

This term renders the nucleus less stable and thus increases its mass.

The repulsive electric force among the protons also plays a role in the stability of the nucleus. The potential energy related to the Coulomb force is proportional to Z^2/R, where R is the radius of the nucleus ($R \propto A^{1/3}$). This decreases the stability of the nucleus and the following term in the semiempirical mass formula ensues

$$E_{\text{coul}} = -a_{\text{coul}} \frac{Z^2}{A^{\frac{1}{3}}} \qquad (6.14)$$

Experimentally, it is observed that stable isotopes of light elements lie near the line $Z = A/2$. There must then be a physical reason why isotopes lying far away from this line of stability are not stable and therefore cannot exist. This property is related to the shell model of the nucleus (the explanation will be given below in Section 6.3.2 describing the shell model of the nucleus). Therefore, a term in the semiempirical mass formula must be included to take this factor into account. A term with the following dependence is used in the semiempirical formula

$$E_{\text{asym}} = -a_{\text{asym}} \frac{(Z - A/2)^2}{A} \qquad (6.15)$$

Therefore, as Z deviates from the value $A/2$, this expression decreases the binding energy and eventually renders the isotope unstable. In reality, as A increases the number of neutrons become relatively large with respect to the number of protons (i.e. $Z < A/2$). The nuclear attraction force due to a larger number of neutrons serves to dilute the Coulomb repulsion and therefore somewhat destabilizes the nucleus. The factor $1/A$ in the term above, diminishes this correction factor for heavier elements and is present to assure that, as observed experimentally, a larger number of stable isotopes exists for such elements.

It is also observed that more than half of the stable nuclei have an even number of both protons and neutrons, while only four nuclei have an odd number of protons and neutrons (^{2}H, ^{6}Li, ^{10}B and ^{14}N). A pairing effect must therefore be present to take into account this preference of even-numbered proton and neutron nuclei (this property is also explained by the shell model). A pairing term of the type

$$E_{pair} = \begin{cases} \dfrac{a_{pair}}{A^{\frac{1}{2}}} & \text{if } Z \text{ and } N \text{ are even} \\ 0 & \text{if } Z \text{ is odd and } N \text{ is even, or if } Z \text{ is even and } N \text{ is odd} \\ -\dfrac{a_{pair}}{A^{\frac{1}{2}}} & \text{if } Z \text{ and } N \text{ are odd} \end{cases} \qquad (6.16)$$

fits the experimental data for the nuclear masses relatively well. This correction factor increases the stability of the nucleus when both N and Z are even and decreases it when both N and Z are odd. When only one of these numbers is odd, no correction factor is applied.

The following set of fitted values for the parameters used in the semiempirical mass formula (in units of uc^2) gives good results as compared to experimental nuclear masses: $a_{vol} = 0.01691$, $a_{surf} = 0.01911$, $a_{coul} = 0.000763$, $a_{asym} = 0.10175$ and $a_{pair} = 0.012$. These parameters can be used to estimate the mass of the nuclei and study the relative importance of the various terms found in the semiempirical mass formula.

Figure 6.2 shows the relative importance of the terms found in the semiempirical mass formula (with the exception of the pairing term). In the context of the liquid-drop model,

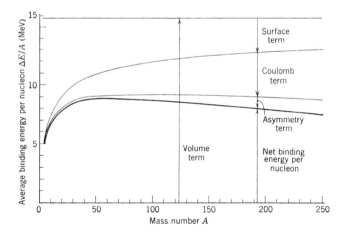

Figure 6.2 The relative importance of the volume, surface, Coulomb and asymmetry term of the semiempirical mass formula for the average binding energy per nucleon ($\Delta E/A$). Reproduced with permission from Eisberg, R. and Resnick, R., *Quantum Physics of Atoms, Molecules, Solids, Nuclei and Particles*, John Wiley & Sons, Ltd, New York (1985).

the combination of the surface, Coulomb and asymmetry terms conspires to give a maximum of stability for nuclei around iron. This fact is of critical importance for the evolution of stars (see Sections 6.7 and 6.8.4), since fusion is no longer efficient (i.e. exothermic) past this point of maximal stability.

6.3.2   The Shell Model

Since the nucleons inside a nucleus are found in a potential well due to the attractive strong nuclear force among protons and neutrons and since quantum mechanics applies to such systems, the nucleons are found in quantised energy levels similar to the electronic energy levels of atoms. For atoms, it is found that the ionisation energy varies as a function of Z and that it has a local maximum when electronic shells are full (see Figure 1.6). Similar shells are also present for the nuclear energy levels of the nucleus. This quantum-mechanical model of the nucleus is commonly called the shell model.

The numbers of nucleons necessary to fill energy shells are commonly called the magic numbers of the nucleus. Similarly to the inert gases found in the periodic table of the elements, nuclei with filled shells are particularly stable. Because of their high stability, these nuclei are generally less prone to fuse with other nuclei.

There are several manifestations of the nuclear energy shells. For example, the energy needed to extract a single neutron from a nucleus (which is equivalent to the extraction of an electron from an atom or an ionisation) varies greatly as a function of N. There are local peaks for this extraction energy (similarly to those seen in Figure 1.6 for the ionisation energy of atoms) for the values $N = 2, 8, 20, 28, 50, 82$ and 126. These quantities are the so-called magic numbers mentioned above. Also, elements that possess a number of protons equal to a magic number typically have a greater number of stable isotopes than the neighbouring elements.

It should be noted that the magic numbers are different from the number of electrons needed to fill electronic configurations of the atom. This should not be surprising because the strong nuclear force is very different from the electric force that is responsible for binding the electrons in the atom.

Additional evidence for the presence of magic numbers for the nucleus can be seen in Figure 6.1. The ^{4}He and ^{16}O nuclei that are doubly magic (i.e. both the number of protons and neutrons are magic numbers), have average binding energies per nucleon larger than their immediate neighbours.

Other supporting facts for the validity of the shell model of the nucleus can be observed by studying the relative abundances of the elements found in our Universe. Figure 6.3 shows the abundances of various elements found in the Sun. The abundances for fluorine ($Z = 9$) and scandium ($Z = 21$) have much lower abundances than their respective neighbours, namely oxygen and calcium. These last two elements possess a number of protons that equals a magic number (respectively, $Z = 8$ and 20). The reason for this difference in the abundances of these neighbouring elements is due to the high stability of nuclei with a magic number of protons. Once these nuclei are produced via nucleosynthesis, they are less reactive than their neighbouring elements and therefore less susceptible to destruction via nuclear reactions.

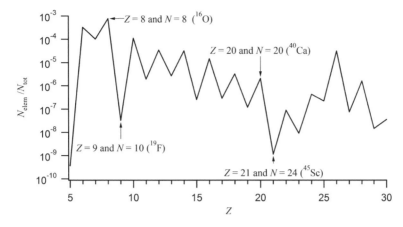

Figure 6.3   The abundance (in number of atoms of each species relative to the total number of atoms) of various elements in the Sun. The most abundant isotopes of O and Ca are doubly magic and are identified on the figure. The sole stable isotopes of F and Sc are also shown.

Also, the most abundant isotope for both O and Ca are doubly magic (i.e. ^{16}O and ^{40}Ca) or in other words both the number of protons and neutrons are magic. Meanwhile, fluorine and scandium only have a single stable isotope (^{19}F and ^{45}Sc) while other neighbouring elements have several. All of these properties for the abundance of the elements give more proof for the existence of a shell structure for the nucleus.

The asymmetry factor found in the semiempirical formula discussed in the previous section can also be explained by the shell model of the nucleus. Since protons and neutrons are fermions, they will independently fill up their respective nuclear energy levels, while respecting Pauli's exclusion principle. If there is a large asymmetry between the number of protons and neutrons, the energy of the nucleus becomes large and this gives an unstable configuration (see Figure 6.4). Such configurations where there are many more neutrons than protons can decay via the reaction $n \rightarrow p + e^- + \bar{\nu}_e$ (where $\bar{\nu}_e$ represents an antielectron neutrino). Such decays increase the number of protons in the nucleus (and therefore changes the nature of the element) and augment its stability. The importance of this decay process will be discussed in (optional) Section 6.12.

The pairing effect discussed earlier for the liquid-drop model is also clearly visible in Figure 6.3. Elements with an even number of protons are more abundant because these nuclei are more stable than their immediate neighbours in the periodic table with an odd number of protons. It is also found experimentally that the energy needed to extract a single neutron from a nucleus is larger when the number of neutrons in this nucleus is even as compared to when it is odd. Both of these experimental facts are manifestations of the pairing effect that can be explained by the shell model. In a classical-physics picture, the two paired identical nucleons have 'orbits' in opposite directions. Therefore, they are on average closer to one another than to other nucleons in the nucleus. This effect increases the binding energy of these nucleons.

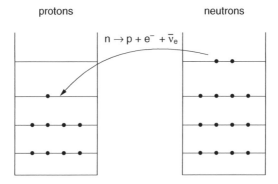

Figure 6.4   A schematic picture of the nuclear potentials felt by protons and neutron inside a nucleus. When the number of neutrons in the nucleus is much larger than the number of protons, one or several decays, commonly called β^- decays (n → p + e⁻ + $\bar{\nu}_e$), may occur that transform a neutron to a proton that finds itself on a lower energy level. This increases the stability of the nucleus.

6.4   Basic Physics of Nuclear Fusion

To properly understand the physics surrounding nuclear fusion, quantum-mechanical principles must be applied. Figure 6.5 shows schematically the approach of an incoming nucleus towards a second nucleus for which the nuclear potential is approximated by a square well. This well has the geometrical dimension on the order of the size of the nucleus (i.e. on the order of several fermis[2]). Besides the nuclear potential, there also exists a repulsive Coulomb force between the two interacting nuclei. In order for the incoming nucleus in Figure 6.5 to enter the nuclear potential of the other nucleus (and therefore fuse with it), it must penetrate the barrier potential that it encounters. Barrier-potential penetration is a well-known quantum effect (also called quantum tunnelling).

   Contrary to classical physics, quantum theory predicts that when a particle with energy less than the barrier potential it encounters, it has a nonzero probability of tunnelling through this barrier and finding itself on the other side. The probability that the incoming nucleus penetrates the barrier increases as the energy of the incoming nucleus increases, since the width of the encountered barrier decreases. Therefore, in stellar plasma, since the kinetic energy of particles increases with temperature, the nuclear reaction rate also increases with temperature. Also, because the Coulomb repulsion increases when the reacting nuclei are more highly charged, fusion for low-charged nuclei generally occurs at lower temperatures than for highly charged nuclei. This has important implications for stellar evolution, correctly predicting that the lowest charged nuclei are the first to burn. As the star evolves and its central temperature increases, heavier (and thus more highly charged) nuclei burn at later stages of evolution (more details are given later in this

---

[2]1 fermi = 10^{-15} m.

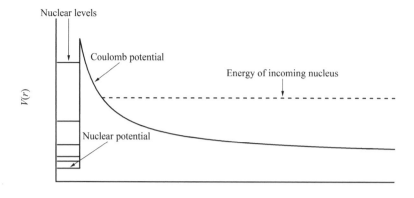

Figure 6.5   A schematic view of the potential of a nucleus (solid line). The potential due to the strong nuclear force is approximated by a square well. Outside this well the potential is due to the Coulomb repulsion force. The dashed line represents the energy of an incoming nucleus that is attempting to fuse with the nucleus, or to tunnel through the potential illustrated in the figure. Also shown in the figure are nuclear energy levels found inside the well.

chapter). There are exceptions to this general rule. For example, deuterium and lithium burn at a lower temperature than hydrogen. The reason for this is that for the proton–proton reaction ($^1H + {}^1H \rightarrow {}^2H + e^+ + \nu_e$), the weak nuclear force intervenes (which manifests itself by the presence of the positron and the neutrino), leading to a relatively small nuclear cross section. The burning of deuterium and lithium is discussed below in a special topic concerning brown dwarfs.

Generally, the energy of the incoming particle will not coincide with a bound nuclear level[3] of the nucleus inside the nuclear potential. These are called nonresonant reactions. This case is illustrated in Figure 6.5. The reactions occurring when the energy of the incoming particle is equal to a given nuclear energy level are called resonant reactions.

The energies at which fusion reactions occur in stars are relatively small as compared to, for example, the energies attained in laboratory (i.e. in particle accelerators). Even though during such a low-energy collision, the probability of tunnelling through the barrier potential is very small for the conditions found in stellar cores, the large number of nuclei found there leads to a large rate of nuclear energy production. The fact that the typical energies are much less than those that can be used to observe fusion in particle accelerators makes the gathering of experimental data of cross sections for stellar fusion reactions impossible. The rate of reaction is too low at the densities attained in particle accelerators for typical energies encountered in stellar cores. More details concerning nuclear reaction cross sections are given in (the advanced and optional) Section 6.13.

---

[3]For the reader who has chosen to skip Section 6.3, the nucleus has energy levels (and related shells) similar to those found for the electronic levels of atoms.

6.5   Main-Sequence Burning

To achieve stellar status, a collapsing interstellar cloud must have sufficient gravitational energy (or mass) to be able to heat up the interior of the protostar to attain the temperature needed for sustained hydrogen fusion. The lower limit of the mass for stars is approximately 8 % of $M_\odot$. Collapsed celestial bodies with masses just below this limit are called brown dwarfs (see special topic below). As mentioned previously, stars that burn hydrogen in their centre are found on the main sequence in the H–R diagram. This evolutionary phase of stars is the lengthiest and therefore merits special attention since, due to the length of this stage of evolution, the vast majority of stars observed are on the main sequence.

In this section, the detailed nuclear reactions that transform hydrogen to helium in the central regions of main-sequence stars will be discussed. Two sets of such reactions will be seen. The first, the proton–proton chains are responsible for energy generation for main-sequence stars with masses below approximately $1.5 M_\odot$. Meanwhile, the reactions of the CNO cycles dominate energy production for the stars with larger masses. A section pertaining to the duration of the lifetime of stars on the main sequence will also be presented. Finally, this section will conclude with an optional section on the famous solar neutrino problem.

Special Topic – Brown Dwarfs

Brown dwarfs are bodies with masses lying between those of large planets and stars. These substellar astronomical bodies have relatively small luminosities rendering their observation difficult. This is why they were first detected only in the mid-1990s. Since brown dwarfs have no sustaining energy source (i.e. sufficient thermonuclear reactions to account for their luminosity over time), eventually their luminosity diminishes with time after their formation. A search for these astronomical objects was undertaken in young stellar clusters so that they could more easily be detected because of their relatively large luminosity during their youth.

The upper limit for brown dwarf mass is $0.08 M_\odot$ or roughly 80 times the mass of Jupiter (M_{JUP}). In other words the mass at which hydrogen can burn. In reality, a small amount of hydrogen burns in the most massive brown dwarfs ($M > 60\ M_{JUP}$) during a short-lived period following the deuterium-burning phase (see below for details concerning deuterium burning). Detailed calculations show that hydrogen ceases to burn because the central temperature needed for this reaction to take place cannot be maintained. Since hydrogen burning is not sustained, these objects are not considered stars. The lower limit for brown dwarf mass is approximately 13 M_{JUP} (this is the mass where deuterium can burn, see below for more details). Below this mass, no fusion reactions take place.

Lithium can fuse with hydrogen (via the reaction $^7Li + {}^1H \rightarrow {}^4He + {}^4He$) at a lower temperature than that needed for hydrogen fusion. Brown dwarfs with masses between 60 M_{JUP} and 80 M_{JUP} burn their lithium in about 10^8 yr. Even though lithium only burns in the central regions of such brown dwarfs, since these objects are convective, the lithium found at their surface is eventually brought to the centre where it can be destroyed by nuclear fusion.

Massive brown dwarfs are not easily distinguishable from lower-mass stars. However, since brown dwarfs with masses less than 60 M_{JUP} never attain the temperature needed for Li burning,[4] which has an ignition temperature of approximately 3×10^6 K (while the critical temperature for burning hydrogen in low-mass stars is approximately 10^7 K, see Table 5.1), the detection of Li in the spectra of astronomical bodies with substellar masses gives an indication that these might be brown dwarfs (with masses below 60 M_{JUP}). An additional way of confirming the status of brown dwarfs is the presence of methane (CH_4) in their spectra. Methane molecules cannot exist at the surface of even the coolest stars, but such molecules can, however, exist at the surface of cool brown dwarfs.

The lower limit for the mass of brown dwarfs is defined by the mass below which deuterium cannot fuse with hydrogen via the reaction $^2H + {}^1H \rightarrow {}^3He + \gamma$. The critical temperature needed for this reaction to take place is approximately 1×10^6 K, which corresponds to the central temperature of brown dwarfs with masses of approximately 13 M_{JUP}. Astronomical bodies with masses smaller than this value are simply large exoplanets (assuming they are orbiting a star) and no thermonuclear activity exists there.

The effective temperatures of brown dwarfs are in the range from roughly 750 to 2200 K. This range of effective temperature is relatively uncertain because, for instance, since these astronomical bodies cool down as a function of time, cooler brown dwarfs than this lower limit should theoretically exist. Also, when they are young and are still burning their deuterium, their effective temperature can be larger than the upper limit given above. As mentioned earlier, brown dwarfs fade with time because they only have a limited amount of nuclear energy at their disposal due to the exothermic reactions of deuterium and lithium burning. After a certain time, the heat accumulated in brown dwarfs comes mainly through the contraction phase during their formation. Therefore, since the effective temperature of brown dwarfs is a function not only of their mass but also varies with time, it complicates their study. Four spectral classifications for brown dwarfs have been defined: M, L, T and Y. These spectral types are given as a function of decreasing temperature. The M-type brown dwarfs are hottest, while Y-type brown dwarfs are the coolest.

---

[4]The abundance of Li in the Universe can be predicted by the Big-Bang theory and may therefore be used as a test for this theory. However, since the fusion process can destroy Li in stars, when comparing the observed Li abundance in the Universe and the one predicted by the cosmological model, this destruction of Li in stars must be taken into account to properly interpret observations.

6.5.1   Proton–Proton Chains

The proton–proton chains consist of a series of thermonuclear reactions by which hydrogen is transformed into helium in stars. These chains are so-called because they begin by the fusion of two protons (or ^{1}H nuclei). There are three distinctive proton–proton chains given by the following reactions

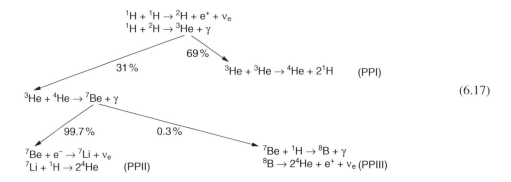

$$\text{(6.17)}$$

In the Sun's core, the ^{3}He nucleus has a probability of approximately 69 % of fusing with another ^{3}He nucleus (completing the PPI chain) and has a 31 % chance of reacting with a ^{4}He nucleus (leading to the PPII or PPIII chains). Meanwhile, the ^{7}Be nucleus reacts 99.7 % of the time with a free electron, thus leading to the PPII chain and it has a 0.3 % chance of reacting with a proton, thus completing the PPIII chain. Even though most of the thermonuclear energy produced in the Sun comes form the proton–proton chains, the CNO cycles (see below for more details) contribute almost a third of the total amount of hydrogen burning.

To gain more insight into these nuclear reactions, the PPI chain will be studied in more detail. In order to produce a ^{4}He nucleus, two ^{3}He nuclei must first be formed.[5] Therefore, two proton–proton fusion reactions must be followed by two proton–deuterium reactions

$$
\begin{aligned}
^1\text{H} + {}^1\text{H} &\to {}^2\text{H} + e^+ + \nu_e \\
^1\text{H} + {}^1\text{H} &\to {}^2\text{H} + e^+ + \nu_e \\
^1\text{H} + {}^2\text{H} &\to {}^3\text{He} + \gamma \\
^1\text{H} + {}^2\text{H} &\to {}^3\text{He} + \gamma \\
^3\text{He} + {}^3\text{He} &\to {}^4\text{He} + 2\,{}^1\text{H}
\end{aligned}
$$

$$\text{(6.18)}$$

The two ^{2}H and the two ^{3}He nuclei that are created by these reactions are later fused (and consequently destroyed). The net result is that these reactions fuse six ^{1}H nuclei (or protons) from the stellar plasma into a single ^{4}He nucleus and two ^{1}H nuclei, while also

---

[5]In reality, since ^{3}He nuclei were produced during the Big Bang, some exist within stars at their birth. However, to calculate the quantity of energy produced by a complete PP chain, it is assumed that all of the nuclei that are needed (except ^{1}H) are previously formed by thermonuclear reactions in the star.

Table 6.4 Neutrino energies for proton–proton chains.

Reaction	Average neutrino energy	Maximal neutrino energy
$^1H + {}^1H \rightarrow {}^2H + e^+ + \nu_e$	0.263 MeV	0.420 MeV
$^7Be + e^- \rightarrow {}^7Li + \nu_e$	0.80 MeV	0.862 MeV
$^8B \rightarrow 2{}^4He + e^+ + \nu_e$	7.2 MeV	14.02 MeV

creating two positrons and two electron neutrinos. Globally, the reactions of the PPI chain can be summarised as

$$6{}^1H \rightarrow {}^4He + 2{}^1H + 2e^+ + 2\nu_e + 2\gamma \tag{6.19}$$

or simply

$$4{}^1H \rightarrow {}^4He + 2e^+ + 2\nu_e + 2\gamma \tag{6.20}$$

It is easy to show that the PPII and PPIII chains also lead to similar global reactions that fuse four protons into a ^4He nucleus.

The energy emitted (neglecting positron annihilation and the energy carried away by the neutrinos, see below for more details) is

$$E = (m_{\text{initial}} - m_{\text{final}})c^2 = (4m_{^1H} - m_{^4He} - 2m_e)c^2 = 24.688 \text{ MeV} \tag{6.21}$$

The two positrons eventually interact with two free electrons in the stellar plasma leading to their annihilation. An amount of energy of $4m_e c^2 = 2.044$ MeV is released and must be added to the energy gain of this thermonuclear reaction giving a total energy of 26.732 MeV minus the energy carried away by the neutrinos. The energy of the neutrinos is subtracted here because they can easily leave the star without interacting with matter. Detailed calculations show that the average energy taken away by each of the neutrinos emitted in the PPI reaction is approximately 0.263 MeV. Meanwhile, the second neutrino emitted during the PPII and the PPIII chains, respectively, carry away approximately 0.80 and 7.2 MeV (see Table 6.4). In other words the proton–proton chains lose, on average, respectively 2.0, 4.0 and 27.9 % of their energy due to the neutrinos leaving the star.

6.5.2 CNO Cycles

The CNO cycles are made up of reactions in which protons are fused with C, N and O nuclei to produce helium. Since C, N and O nuclei are highly charged as compared to protons, the critical temperature needed for these reactions being able to take place is larger than for proton–proton chains. The reason for this is that the protons interacting with C, N and O nuclei must penetrate a stronger Coulomb potential and they therefore need larger velocities to fuse. This explains why the CNO cycles dominate energy generation in main-sequence stars only for masses larger than approximately $1.5 M_\odot$. It should be noted that both proton–proton and CNO reactions are present in such stars but their contribution to the overall energy production varies with their central temperature, which in turn depends on the stellar mass (see Figure 6.7 in the special topic presented below).

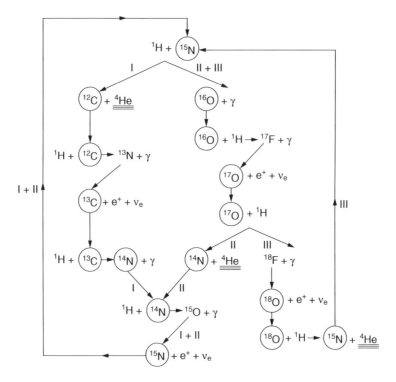

Figure 6.6 Illustration of the CNO cycles. The catalysts are circled. Figure reproduced with permission from Pearson, J.M., *Nuclear Physics: Energy and Matter*, Adam Hilger, Bristol (1986).

The reactions of the three CNO cycles are given below. The large arrows between the different cycles point to reactions that are in common among the various cycles. A more graphic way of illustrating the CNO cycles is shown in Figure 6.6.

CNOI   CNOII   CNOIII

$$^1H + {}^{12}C \rightarrow {}^{13}N + \gamma$$
$$^{13}N \rightarrow {}^{13}C + e^+ + \nu_e$$
$$^1H + {}^{13}C \rightarrow {}^{14}N + \gamma$$

$$^1H + {}^{14}N \rightarrow {}^{15}O + \gamma \;\rightarrow\; {}^1H + {}^{14}N \rightarrow {}^{15}O + \gamma$$
$$^{15}O \rightarrow {}^{15}N + e^+ + \nu_e \;\rightarrow\; {}^{15}O \rightarrow {}^{15}N + e^+ + \nu_e \tag{6.22}$$
$$^1H + {}^{15}N \rightarrow {}^{12}C + {}^4He \quad {}^1H + {}^{15}N \rightarrow {}^{16}O + \gamma \;\rightarrow\; {}^1H + {}^{15}N \rightarrow {}^{16}O + \gamma$$
$$^1H + {}^{16}O \rightarrow {}^{17}F + \gamma \;\rightarrow\; {}^1H + {}^{16}O \rightarrow {}^{17}F + \gamma$$
$$^{17}F \rightarrow {}^{17}O + e^+ + \nu_e \;\rightarrow\; {}^{17}F \rightarrow {}^{17}O + e^+ + \nu_e$$
$$^1H + {}^{17}O \rightarrow {}^{14}N + {}^4He \quad {}^1H + {}^{17}O \rightarrow {}^{18}F + \gamma$$
$$^{18}F \rightarrow {}^{18}O + e^+ + \nu_e$$
$$^1H + {}^{18}O \rightarrow {}^{15}N + {}^4He$$

Since the various C, N and O isotopes are produced and destroyed in equal amounts by fusion during a complete cycle of CNOI, II and III, globally each of these cycles may be simplified to the following reaction

$$4^1H + {}^{12}C \rightarrow {}^4He + {}^{12}C + 2e^+ + 2\nu_e + 3\gamma \tag{6.23}$$

or simply

$$4^1H \rightarrow {}^4He + 2e^+ + 2\nu_e + 3\gamma \tag{6.24}$$

In each of the CNO cycles presented above, four protons are transformed into a ^{4}He nucleus to produce energy. The energy emitted is the same as for the proton–proton chains except for the portion of the energy carried away by the neutrinos, which is different due to the differing nature of the reactions producing these leptons.

In these cycles, the CNO nuclei serve as catalysts. During a complete cycle, globally, no C, N, or O nuclei are either produced or destroyed. However, since the various reactions in the CNO cycles have different reaction rates, the amounts of C, N and O are modified until they achieve a steady state some time after the reactions of the CNO cycles are ignited. An example of such a steady state attained during nuclear reactions is given in (the advanced optional) Section 6.13.

It should be noted that in reality, there is an additional cycle not shown here. This fourth cycle fuses hydrogen with ^{19}F nuclei. This series of four cycles is commonly called the CNOF cycles. Since this fourth cycle is less prominent than the others, it is not discussed here.

Evidently, these cycles require the presence of CNO nuclei. However, since the heavy elements weren't formed during the Big-Bang nucleosynthesis, the first generation of stars formed in the Universe did not contain such elements. Therefore, the evolution of stars with no (or little) metals is different from the evolution of stars containing metals because only proton–proton chains may occur on the main sequence when no C, N and O atoms are present.

Special Topic – Energy Production Rates of the PPI Chain and CNO Cycles

Detailed calculations of the nuclear reaction rates for hydrogen burning show that the nuclear energy production rate per unit mass for the PPI chain and the CNO cycles are approximately given by the following expressions

$$\varepsilon_{PPI} = \frac{2.4 \times 10^4 \, \rho X^2 e^{\frac{-3.38}{T_9^{1/3}}}}{T_9^{2/3}} \tag{6.25}$$

$$\varepsilon_{CNO} = \frac{4.4 \times 10^{25} \, \rho X Z e^{\frac{-15.2}{T_9^{1/3}}}}{T_9^{2/3}} \tag{6.26}$$

where T_9 is the temperature in units of 10^9 K, X is the mass fraction of hydrogen and Z is the mass fraction of the metals. Since protons react with one another, the

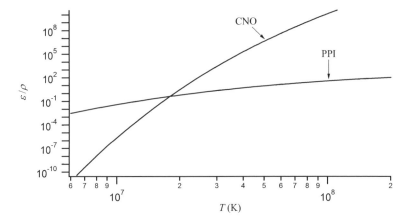

Figure 6.7 The temperature dependence of the nuclear energy production rate per unit mass (ε) for the PPI chain and the CNO cycles.

PPI rate depends on the square of X. Meanwhile, the CNO rate depends on XZ because protons react with CNO nuclei (here it is supposed that the CNO abundances scale with Z). Also, these rates increase linearly with density. The temperature dependence for the nuclear energy production rate is more complicated. Figure 6.7 compares these two rates and shows that the PPI chain dominates for lower temperatures, while the CNO cycle dominate energy production for higher temperatures. As discussed previously, higher temperatures (or in other words, higher velocities for the reacting nuclei) are needed for hydrogen fusion via the CNO cycles since a larger Coulomb barrier exists between the reacting nuclei. The crossover of the two curves shown in Figure 6.7 occurs for main-sequence stars with mass of approximately $1.5\,M_\odot$.

6.5.3   Lifetime of Stars on the Main Sequence

To use knowledge about stellar evolution in practical applications such as determining the age of stellar clusters (see Section 6.9.4), it is critical to be able to evaluate how long a star stays on the main sequence (or in other words how long hydrogen burning lasts in the centre of stars). The lifetime of stars on the main sequence can be estimated by comparing the available nuclear energy to its power output (or luminosity) on this branch of the H–R diagram. This is the aim of this section.

During the creation of a ^{4}He nucleus by either the proton–proton chains or the CNO cycles, a fraction of the mass F_m of the four fused protons is transformed to energy. This energy is eventually emitted at the star's surface. This fraction is approximately (while neglecting the energy taken away by the neutrinos)

$$F_m \approx \frac{26.7\,\text{MeV}}{4m_p c^2} \approx 0.007 \tag{6.27}$$

Since hydrogen only fuses at temperatures of approximately 10^7 K or more, only a portion of a star's mass is found in regions where the physical conditions needed for hydrogen fusion to occur exist. Therefore, only these stellar regions can participate in thermonuclear energy production. For a rough estimate of the duration of the hydrogen-burning phase, it is assumed that 10% of the mass of a star can be fused. Supposing that this mass is composed of pure hydrogen, the total energy E_{tot} available through hydrogen burning for a star of mass M_* can be estimated to be

$$E_{tot} \approx 0.0007 M_* c^2 \tag{6.28}$$

Assuming that the luminosity of the star L_* remains constant during the entire main-sequence phase, the time t_{ms} it will stay in this evolutionary phase is approximately

$$t_{ms} \approx \frac{E_{tot}}{L_*} \approx 0.0007 c^2 \frac{M_*}{L_*} \approx 10^{10} \left(M_* / M_\odot \right) (L_* / L_\odot)^{-1} \text{ yr} \tag{6.29}$$

This result predicts that the Sun will stay on the main sequence for 10^{10} years, which is close to more detailed evaluations for this quantity. Since the luminosity of stars and thus the rate at which they expend their energy increases much faster than stellar mass, this equation predicts that stars with large masses, even though they possess more nuclear fuel or hydrogen, have a shorter lifespan on the main sequence than less-massive stars (see Example 6.2). Massive stars therefore evolve much faster than less-massive ones. This is a very important result of stellar evolution and can be used as a tool to determine, for example, the age of stellar clusters (see Section 6.9.4).

Example 6.2: Calculate the lifespan on the main sequence for a B-type star with $M_* = 16 M_\odot$ and $L_* = 8000 L_\odot$ and an M-type star with $M_* = 0.1 M_\odot$ and $L_* = 8 \times 10^{-4} L_\odot$.

Answer:

Using the equation for t_{ms} given above

$$t_{ms} \approx \frac{16}{8000} \times 10^{10} \text{ yr} = 2 \times 10^7 \text{ yr for the B-type star} \tag{6.30}$$

and

$$t_{ms} \approx \frac{0.1}{8 \times 10^{-4}} \times 10^{10} \text{ yr} = 1.25 \times 10^{12} \text{ yr for the M-type star} \tag{6.31}$$

As expected, the hotter (or more massive) star has a shorter lifespan on the main sequence than the cooler one. It should also be noted that the lifespan of the M-type star studied here is larger than the age of the Universe (which is approximately equal to 13×10^9 yr). Therefore, the coolest stars that were formed shortly after the Big Bang are still on the main sequence today, while the high-mass stars formed at these early times have all had enough time to evolve away from the main sequence.

A final note concerning the results shown in this section is warranted. The portion of a star's mass that is transformed to energy during its evolution through the various evolutionary stages is negligible. The total mass of a star can then be considered as constant during its lifetime, with the exception that some mass loss at the surface often occurs. A star can, however, also eject considerable amounts of mass to interstellar space during explosive stages such as the supernova phase (see Section 6.11). On the other hand, stars in binary systems can gain mass through mass transfer from their companions. This modifies their evolution and renders the study of the evolution of binary systems quite complex.

6.5.4   The Solar Neutrino Problem[†]

As seen above, the thermonuclear reactions taking place inside the Sun emit a large amount of electron neutrinos. Due to their low probability of interaction, the vast majority of these neutrinos exit the Sun without being absorbed. Therefore, if astronomers were able to detect these neutrinos, such measurements could provide some direct information about the Sun's centre and the nuclear reactions taking place there. Gathering information directly from the inner portions of stars is impossible by normal astronomical observations that detect electromagnetic radiation since the radiation field emanates from the outer layers of stars.

Neutrinos emitted by the proton–proton reaction $^1H + {}^1H \rightarrow {}^2H + e^+ + \nu_e$ have a different energy spectrum from those emitted by the reaction $^7Be + e^- \rightarrow {}^7Li + \nu_e$ (the second neutrino emitted during the PPII chain) or those coming from the reaction $^8B \rightarrow 2\,{}^4He + e^+ + \nu_e$ (the second neutrino emitted during the PPIII chain). Therefore, the detection of the neutrinos and interpretation of these observations could serve to verify the relative reaction rates of the different nuclear reactions in the Sun. A method by which these neutrinos could be detected would be very useful to study the Sun's centre.

The first neutrino observatory was constructed in the US and is known as the Brookhaven Solar Neutrino Experiment. This observatory, which was built in a mine at a depth of approximately 1500 m, used ^{37}Cl as the target (or detection agent) and aimed to study neutrinos coming from the Sun. The nucleus of the ^{37}Cl atom can react with an electron neutrino via the reaction

$$\nu_e + {}^{37}Cl \rightarrow {}^{37}Ar + e^- \tag{6.32}$$

This reaction has a threshold energy of 0.814 MeV. It cannot detect the neutrinos emitted by the proton–proton reaction ($^1H + {}^1H \rightarrow {}^2H + e^+ + \nu_e$) since its neutrinos are not energetic enough (see Table 6.4). It is, however, sensitive to the neutrinos emanating from ^7Be and ^8B during, respectively, the PPII and PPIII chains. Since neutrinos do not interact strongly with matter, about 600 tons of C_2Cl_4 molecules were used as the target. Even with such a large amount of ^{37}Cl this neutrino observatory can only detect less than one neutrino interaction per day! The chief scientist for the Brookhaven Solar Neutrino Experiment who first suggested using ^{37}Cl to detect neutrinos, the American chemist Raymond Davis Jr. (1914–2006), shared the 2002 Nobel Prize in physics for the detection of the elusive electron neutrino and the first measurement of the solar neutrino flux that lead to the so-called solar neutrino problem (see below). The Japanese astrophysicist Masatoshi Koshiba

(b. 1926) also shared the 2002 Nobel Prize for confirming the solar neutrino problem with results from the Kamiokande neutrino observatory in Japan.

Neutrino observatories are built deep underground in order to avoid false detections from interactions with cosmic rays. Cosmic rays are blocked by the crust of Earth above the neutrino observatory. Meanwhile, neutrinos have no difficulty in traversing this crust.

Other atoms have also been used to detect neutrinos. For example, ^{71}Ga also reacts with electron neutrinos via the reaction

$$\nu_e + {}^{71}\text{Ga} \rightarrow {}^{71}\text{Ge} + e^- \tag{6.33}$$

This reaction has a threshold energy of 0.2332 MeV that is lower than for the ^{37}Cl reaction. Observatories using ^{71}Ga can therefore detect not only the neutrinos from ^{7}Be and ^{8}B but also those emitted by the proton–proton reaction. An example of such an observatory was the GALLEX experiment that was built in Italy and that collected data from 1991 to 1997. It consisted of about 30 tons of gallium as its target. Gallium was also used in the SAGE neutrino observatory in Russia. Water was used in the Kamiokande and Super-Kamiokande detectors in Japan and neutrinos were detected there through light produced by energetic electrons from elastic scattering of neutrinos from the Sun (more details given below).

The unit used to measure the quantity of neutrinos detected is the SNU standing for solar neutrino unit. The SNU unit measures the quantity of neutrino captures par target atom in units of $10^{-36}\,\text{s}^{-1}$. While using ^{37}Cl as a target, the predicted neutrino capture rate by using the theoretical production rates of neutrinos in the Sun according to the most precise solar models at hand is 7.9 ± 2.6 SNU. However, the results of the Brookhaven Solar Neutrino Experiment gave a value for neutrino captures of 2.1 ± 0.9 SNU. The observed rate is therefore much lower than the predicted rate and constitutes what is commonly called the *solar neutrino problem*. Too few neutrinos compared to predictions were also observed by Gallex, SAGE, Kamiokande and Super-Kamiokande experiments.

Several possible sources for this discrepancy have been suggested and are listed below:

Experimental errors: The lower observed rate could be due to experimental errors so that some detections are not counted, or due to the uncertainty of the reaction rate between the neutrinos and the target.

Uncertainties in nuclear reaction rates: If the nuclear reaction rates used for the theoretical calculations are wrong, or if other reactions take place, this would make the theoretical estimation of the predicted neutrino count uncertain.

Uncertainties in the solar model: If the model for the structure of the Sun is not accurate, this would also modify the theoretical estimation of the predicted neutrino count since nuclear reaction rates depend on the temperature, density and composition of the stellar plasma in the core of the Sun.

Neutrino absorption in the solar medium: If the solar medium captures a large number of neutrinos by some unknown reactions, the neutrino flux leaving the Sun would be lower than predicted.

Neutrino oscillation: Elementary particle theory or more specifically electroweak theory predicts that if neutrinos possess mass, they can change from one type to another. Consequently, the electron neutrinos emitted by the proton–proton chains in the Sun's central regions could change into tauon or muon neutrinos before reaching Earth. Since

the neutrino observatories discussed above solely or predominantly detect electron
neutrinos, this could explain the missing neutrino captures.

The first four of these five possibilities were gradually ruled out over the past decades
due to intensive experimental and theoretical work. Experimental errors sufficient to
explain the solar neutrino problem can be confidently eliminated since the results from
different neutrino experiments are consistent with one another. Also, more precise experimental estimations of the cross section of the target (^{37}Cl for example) relative to
neutrino absorption have been obtained over the past several decades rendering the
observations more reliable. The next three possibilities given above are unlikely to
explain the solar neutrino problem in the light of improvements in our understanding of
neutrino physics and solar physics (i.e. more precise experimental nuclear reaction cross
sections and better solar structure modelling). For example, the standard solar model
used to evaluate the solar neutrino flux was confirmed to be quite precise in predicting
the results of helioseismology. As discussed in (the optional) Sections 5.7.2 and 5.7.3,
helioseismology can probe the solar interior and give valuable information about its
physical structure. The standard solar model is also calculated with the most precise
nuclear and atomic data (including opacities) and physics available and is therefore
relatively accurate.

To verify if the last possibility, namely neutrino oscillation, can explain the solar neutrino problem, a neutrino observatory using heavy water as a target was constructed near
the city of Sudbury (Canada). This observatory is called the Sudbury Neutrino Observatory
(SNO). It is situated at a depth of about 2 km in an active nickel mine and contained
approximately 1000 tons of heavy water. It is illustrated in Figure 6.8. Heavy water is
composed of deuterium (^{2}H) instead of ^{1}H. The electron neutrinos from ^{8}B (or the PPIII
chain) can be detected via the reaction

$$\nu_e + {}^2\text{H} \rightarrow {}^1\text{H} + {}^1\text{H} + e^- \tag{6.34}$$

This reaction is commonly called the charged current reaction because charged W bosons
act as the intermediary for the weak nuclear force. It is also sometimes called a W exchange
process (or WEP). The reaction at the base of this interaction is $\nu_e + n \rightarrow p + e^-$, or more
fundamentally a down quark is transformed into an up quark via $\nu_e + d \rightarrow u + e^-$. However,
since this reaction is only sensitive to electron neutrinos, it is not by itself sufficient to
prove that a portion of solar electron neutrinos is transformed into the other two types.

The advantage of SNO is that it can also measure the total flux of the neutrinos (i.e.
due to all three types of neutrinos). This is done with deuterium because it can interact
with all three neutrino types via the reaction

$$\nu_x + {}^2\text{H} \rightarrow \nu_x + {}^1\text{H} + n \tag{6.35}$$

where $x = e$, τ or μ. This reaction is equally sensitive to all three types of neutrinos. It is
commonly called the neutral current reaction because in this instance neutral Z^0 bosons
act as the intermediary for the weak nuclear force. It is also sometimes called a Z exchange
process (or ZEP). Measurements of the appropriate interactions given in the two reactions
above can be used to determine the proportion of neutrinos coming from the Sun in the
form of electron neutrinos. If there is no neutrino oscillation, this fraction should be equal

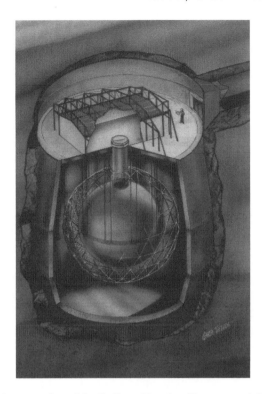

Figure 6.8   An artist's conception of the Sudbury Neutrino Observatory (photo courtesy of SNO). (See colour plate.)

to unity because no tauon or muon neutrinos are produced in the Sun. If this fraction is less than unity, it means that neutrino oscillation takes place.

Another interaction that is also sensitive to all three types of neutrinos is the elastic scattering of neutrinos on electrons via

$$\nu_x + e^- \rightarrow \nu_x + e^- \tag{6.36}$$

However, this reaction is much more probable for electron neutrinos than for the other two types. The electrons created by this process emit Cerenkov radiation that may be measured by detectors. Cerenkov radiation occurs when the scattered electrons have a velocity larger than the speed of light in the medium.

SNO operated from 1999 to 2006. In 2001, the scientific team of SNO found clear evidence of neutrino oscillation from a comparison of their results with those of the Super-Kamiokande neutrino detector in Japan. In 2002, with the help of the results from SNO related to the neutral current reaction seen above, it was shown that when the neutrino oscillation process is taken into account (assuming neutrinos possess mass), the total neutrino solar flux (including electron, tauon and muon types) is consistent with theoretical predictions. They found that only about one third of the electron neutrinos produced in the Sun's centre survive the trip to Earth, the other two thirds being transformed into the other two neutrino types. The mystery behind the missing neutrino counts of previous observational studies, such as those of the Brookhaven Solar Neutrino Experiment discussed above, had therefore finally been solved. The SNO observations answered one of the most important unresolved questions in the field of astrophysics of the late twentieth century. This discovery is a major triumph for modern science that has implications not only for our understanding of solar and stellar physics but also of particle physics.

6.6   Helium-Burning Phase

During the hydrogen-burning phase, four protons are transformed into ^{4}He nuclei and therefore the composition of the core gradually changes (i.e. the He abundance increases while the H abundance decreases). This process leads to an increase of the mean molecular weight in the stellar core. An increase of the mean molecular weight in the core leads to a decrease of gas pressure there. The weight of the layers above the core therefore leads to its contraction. The core progressively (and slowly) contracts during the core hydrogen-burning phase thereby increasing the density and the temperature (and therefore the pressure) there. Meanwhile, the global structure of the star readjusts and its radius increases. The star eventually becomes a red giant star. The reason it is called a red giant is that its effective temperature decreases slightly during this expansion and according to Wien's law (Eq. 1.7) it becomes redder. The position of the star in a Hertzsprung–Russell diagram changes during this process. This will be discussed in more detail in Section 6.8.

If the mass of the star is sufficiently large (approximately $M_* \geq 0.5 M_\odot$), the core will, following its contraction, attain the critical temperature ($\approx 10^8$ K) needed for the fusion of helium. Once the hydrogen contained in the core has burned, helium burning eventually dominates the nuclear energy production rate. It should be noted that stellar layers that were initially at temperatures too low to burn hydrogen may, after core contraction, possess the critical temperature needed for such reactions. Hydrogen may then burn in a certain layer (or a shell) outside the core. This process is commonly called shell burning.

Helium in the core of the evolved star can burn via the following chain of reactions[6]

$$^4\text{He} + {}^4\text{He} \rightarrow {}^8\text{Be} + \gamma$$
$$^8\text{Be} + {}^4\text{He} \rightarrow {}^{12}\text{C}* + \gamma \qquad\qquad (6.37)$$
$$^{12}\text{C}* \rightarrow {}^{12}\text{C} + \gamma$$

Here, the symbol ^{12}C* represents a carbon nucleus found in an excited nuclear energy state. The first nuclear reaction above is an endothermic reaction and requires an input of at least approximately 92 keV of energy (see Example 6.3). This is one of the reasons why the critical temperature for this reaction is relatively high compared to hydrogen fusion. At these temperatures, the average kinetic energy ($3kT/2$) of the particles in the gas is roughly the same order of magnitude as the energy input needed for the fusion of two ^{4}He

---

[6]The simplest way to burn helium would theoretically be to fuse it with protons. Unfortunately, no stable nuclei exist with five nucleons.

nuclei. There is therefore a considerable portion of particles with a kinetic energy larger than 92 keV. This input energy can then easily be found in the thermal bath of the plasma. The ^{8}Be nucleus is unstable to decay (into two ^{4}He nuclei) and has a short lifespan. However, this lifespan is large enough to allow for a second capture of a ^{4}He nucleus. Since the energy of the ^{8}Be + ^{4}He combination is almost equal to the total energy of the excited ^{12}C* nucleus, this fusion reaction is a so-called resonant interaction that has a relatively large cross section.

The global reaction of the chain given above is

$$3^4\text{He} \rightarrow {}^{12}\text{C} + 3\gamma \tag{6.38}$$

It is commonly called the triple-α reaction since three α particles (i.e. ^{4}He nuclei) fuse to create a carbon nucleus. Each triple-α reaction furnishes 7.275 MeV of energy to the star (see Exercise 6.4).

Example 6.3: Calculate the energy needed as input for the reaction ^{4}He + ^{4}He $\rightarrow$ ^{8}Be + γ.

Answer:

The energy can be calculated by using Einstein's mass–energy equation and the data given in Appendix F (note that since for this reaction, an equal number of electrons is present on each side of the equation for neutral ions, the atomic masses may be used instead of the nuclear masses)

$$E = \left(m_{\text{initial}} - m_{\text{final}}\right)c^2 = \left(2m_{4_{\text{He}}} - m_{8_{\text{Be}}}\right)c^2 = \left(2 \times 4.0026032\text{u} - 8.0053051\text{u}\right)c^2$$
$$= -9.87 \times 10^{-5}\ \text{uc}^2 = -91.9\ \text{keV} \tag{6.39}$$

Since the answer is negative it signifies that this reaction is endothermic and needs 91.9 keV of energy to take place.

The energy emitted by the triple-α reaction is smaller than the energy emitted by hydrogen burning. Also, since 12 nucleons are involved in this reaction, the mass fraction transformed into energy is much smaller than for hydrogen fusion. Moreover, since helium burning occurs at the red-giant phase, where the star has a much larger luminosity than on the main sequence, the helium-burning phase lasts much less than the main-sequence lifespan (see Exercise 6.7).

This tendency, in which the timescales of evolutionary phases related to the burning of nuclei with larger masses are shorter, can be partly explained by examining Figure 6.1. This figure shows that the difference of binding energy between the nuclei that are fused together from those formed becomes smaller as A increases. Also, since the advanced stages of nuclear burning occur when the stars are in a giant or supergiant state (i.e. stars with large radius and consequently large luminosities), the nuclear production rate is much larger than at earlier phases such as the main sequence. The higher rate of nuclear reactions, due to high central temperatures, explains why advanced nuclear-burning phases are of relatively short duration.

There are other ways in which helium may burn in stars. Similar to main-sequence burning, as helium is transformed into carbon, the core progressively contracts. As the carbon abundance increases, helium may also eventually burn via the reaction

$$^{12}\text{C} + {}^4\text{He} \rightarrow {}^{16}\text{O} + \gamma \tag{6.40}$$

Helium can also burn in several other ways including

$$^{16}\text{O} + {}^4\text{He} \rightarrow {}^{20}\text{Ne} + \gamma \tag{6.41}$$

$$^{20}\text{Ne} + {}^4\text{He} \rightarrow {}^{24}\text{Mg} + \gamma \tag{6.42}$$

At higher temperatures helium may also fuse with several other types of nuclei. However, due to the small reaction rates of these nuclear reactions, helium burning mostly creates carbon and oxygen.

6.7 Advanced Nuclear Burning

In this section, the major phases of nuclear burning and the related fusion reactions occurring during the advanced evolutionary phases of stars are presented. The reactions shown here are not only important for energy generation but also for the nucleosynthesis of some of the most abundant elements found in the Universe and on Earth. A multitude of secondary reactions, that are generally not important for energy production, but that do create other elements up to iron-peak elements, are unfortunately too numerous to be presented here. For a more comprehensive presentation of the thermonuclear reactions occurring in stars, the reader is referred to Iliadis, C., *Nuclear Physics of Stars*, Wiley-VCH, Weinheim (2007).

Big-Bang and stellar nucleosynthesis cannot explain the abundances of all the elements. For example, some light elements like Li, Be and B, whose abundances are much smaller than other light elements (see Appendix E), can be formed by a process called spallation. Astrophysical spallation is a process by which heavy nuclei found in the interstellar medium are hit by high-energy particles (cosmic rays composed mostly of protons) and are split into lighter nuclei. Since this topic is outside the field of stellar astrophysics, it will not be discussed any further.

Table 6.5 enumerates the elements produced (sometimes called the ashes) during each of the major phases of nuclear burning in massive stars. The central temperature at these phases and their duration for a $25\,M_\odot$ star are also given. As discussed previously, the duration of these nuclear-burning phases decreases as the evolution of the star advances, because less energy per unit mass is available during nuclear fusion and the

Table 6.5   Major phases of nuclear burning[#].

Burning phase	Elements produced	Central temperature	Timescale
H	He	$6.0 \times 10^7 \, \text{K}$	$7 \times 10^6 \, \text{yr}$
He	C, O	$2.0 \times 10^8 \, \text{K}$	$5 \times 10^5 \, \text{yr}$
C	O, Ne, Mg	$9.0 \times 10^8 \, \text{K}$	$600 \, \text{yr}$
Ne	O, Mg, Si	$1.7 \times 10^9 \, \text{K}$	$0.5 \, \text{yr}$
O	Si, S	$2.3 \times 10^9 \, \text{K}$	$6 \, \text{d}$
Si	Fe-peak	$4.0 \times 10^9 \, \text{K}$	$1 \, \text{d}$

[#]Theoretical results for a 25-$M_\odot$ star from Arnould, M. and Samyn, M., *La physique nucléaire en astrophysique*, *EDP Sciences*, Les Ulis (2002).

nuclear energy production rate and therefore the luminosity become larger. The timescale of the various phases vary enormously. For the case shown in Table 6.5, these timescales vary from $7 \times 10^6 \, \text{yr}$ for the main-sequence stage to one day for the silicon-burning phase. Each of these advanced nuclear-burning phases will now be discussed.

6.7.1   Carbon-Burning Phase

The next phase of nuclear burning after helium fusion is the burning of carbon, provided of course, that the mass of the star is sufficient to attain central temperatures that can ignite such reactions. In the previous section, it was mentioned that carbon may fuse with helium via the reaction

$$^{12}\text{C} + {}^4\text{He} \rightarrow {}^{16}\text{O} + \gamma \tag{6.43}$$

However, as the carbon abundance increases due to helium burning, carbon burning is eventually dominated by the reaction

$$^{12}\text{C} + {}^{12}\text{C} \rightarrow {}^{20}\text{Ne} + {}^4\text{He} \tag{6.44}$$

Carbon may also burn in several other ways including

$$^{12}\text{C} + {}^{12}\text{C} \rightarrow {}^{24}\text{Mg} + \gamma \tag{6.45}$$

Other reactions may also occur, although they are endothermic,

$$^{12}\text{C} + {}^{12}\text{C} \rightarrow {}^{16}\text{O} + 2\,{}^4\text{He} \tag{6.46}$$

$$^{12}\text{C} + {}^{12}\text{C} \rightarrow {}^{23}\text{Mg} + \text{n} \tag{6.47}$$

This last reaction also emits a free neutron. Such free neutrons may be absorbed by heavier nuclei to produce elements heavier than iron via the so-called s process (see discussion below in Section 6.7.4 and in (optional) Section 6.12).

6.7.2   Neon-Burning Phase

Carbon burning leads to the creation of oxygen, neon and magnesium. It would seem natural that the next phase of nuclear burning would be fusion reactions related to oxygen. However, the critical temperature for neon fusion with α particles, provided that α particles are present, is lower than for oxygen burning. At the high temperatures found at this stage of evolution, a sufficient number of energetic photons are present that can lead to photodisintegration of a portion of the neon nuclei

$$^{20}\text{Ne} + \gamma \rightarrow {}^{16}\text{O} + {}^{4}\text{He} \tag{6.48}$$

Photodisintegration is an endothermic reaction where a photon is absorbed by a nucleus that is broken into two or more parts. Such reactions are also sometimes called photonuclear reactions. This particular reaction takes away at least 4.73 MeV of energy from the star. However, these photodisintegration reactions create α particles that can then react with the remaining neon via the reaction

$$^{20}\text{Ne} + {}^{4}\text{He} \rightarrow {}^{24}\text{Mg} + \gamma \tag{6.49}$$

which emits 9.316 MeV of energy.

   Other secondary thermonuclear reactions can also occur during this stage. As magnesium accumulates, it can also fuse with α particles to produce silicon

$$^{24}\text{Mg} + {}^{4}\text{He} \rightarrow {}^{28}\text{Si} + \gamma \tag{6.50}$$

As the core contracts and the central temperatures increase, neon may also burn via the following reaction

$$^{20}\text{Ne} + {}^{20}\text{Ne} \rightarrow {}^{16}\text{O} + {}^{24}\text{Mg} \tag{6.51}$$

6.7.3   Oxygen-Burning Phase

The next major phase of nuclear burning is oxygen fusion. Oxygen accumulates in the core following the carbon- and neon-burning phases (see Eq. 6.43, 6.48 and 6.51). The most prevalent reactions by which oxygen may burn are

$$^{16}\text{O} + {}^{16}\text{O} \rightarrow {}^{30}\text{Si} + 2{}^{1}\text{H} \tag{6.52}$$

$$^{16}\text{O} + {}^{16}\text{O} \rightarrow {}^{28}\text{Si} + {}^{4}\text{He} \tag{6.53}$$

$$^{16}\text{O} + {}^{16}\text{O} \rightarrow {}^{31}\text{P} + {}^{1}\text{H} \tag{6.54}$$

Oxygen can also burn via a reaction that creates sulfur

$$^{16}\text{O} + {}^{16}\text{O} \rightarrow {}^{31}\text{S} + \text{n} \tag{6.55}$$

This last reaction also emits a free neutron that can then participate in the formation of the heavier elements via the s process (more details are given in the next section and in (optional) Section 6.12).

6.7.4 Silicon-Burning Phase

As seen above, photodisintegration (see Eq. 6.48 for instance) can create α particles in stellar cores. These α particles can then fuse with heavier nuclei. Silicon may then burn via the reaction

$$^{28}\text{Si} + {}^{4}\text{He} \rightarrow {}^{32}\text{S} + \gamma \tag{6.56}$$

This reaction can then be followed by successive captures of α particles

$$
\begin{aligned}
^{32}\text{S} + {}^{4}\text{He} &\rightarrow {}^{36}\text{Ar} + \gamma \\
^{36}\text{Ar} + {}^{4}\text{He} &\rightarrow {}^{40}\text{Ca} + \gamma \\
^{40}\text{Ca} + {}^{4}\text{He} &\rightarrow {}^{44}\text{Ti} + \gamma \\
^{44}\text{Ti} + {}^{4}\text{He} &\rightarrow {}^{48}\text{Cr} + \gamma \\
^{48}\text{Cr} + {}^{4}\text{He} &\rightarrow {}^{52}\text{Fe} + \gamma \\
^{52}\text{Fe} + {}^{4}\text{He} &\rightarrow {}^{56}\text{Ni} + \gamma
\end{aligned}
\tag{6.57}
$$

Since ^{56}Ni is unstable, it may disintegrate to give ^{56}Co, which in turn can disintegrate to produce ^{56}Fe via the reactions

$$
\begin{aligned}
^{56}\text{Ni} &\rightarrow {}^{56}\text{Co} + e^{+} + \nu_{e} \\
^{56}\text{Co} &\rightarrow {}^{56}\text{Fe} + e^{+} + \nu_{e}
\end{aligned}
\tag{6.58}
$$

The nuclei ^{56}Ni and ^{56}Co can also, respectively, produce ^{56}Co and ^{56}Fe by electron capture (i.e. $^{56}\text{Ni} + e^{-} \rightarrow {}^{56}\text{Co} + \nu_{e}$). As observed in Figure 6.1, ^{56}Fe is one of the most stable nuclei found in nature. Energy generation via nuclear fusion essentially ends at the creation of iron (and other iron-peak elements). The chain of reactions shown above explains why iron is among the most abundant elements in the Universe (see Table 1.6). The accumulation of iron in the stellar core leads to an energy crisis that can have dire results for stars. This energy crisis eventually induces the destruction of the star via an explosion (called a supernova). This calamity will be discussed in Sections 6.8.4 and 6.11.

There also exist many other reactions present at this stage that create the other elements (and their isotopes) not mentioned here up to the iron-peak elements. However, these reactions do not contribute significantly to the nuclear energy production rate and are therefore not of critical importance for studying the evolution of stars.

In addition to the nuclear reactions seen above, there is also another way in which silicon may burn and produce iron. This chain of reaction begins with the fusion of two silicon nuclei

$$^{28}\text{Si} + {}^{28}\text{Si} \rightarrow {}^{56}\text{Ni} + \gamma \tag{6.59}$$

which is followed by the same two successive disintegrations seen above, namely

$$
\begin{aligned}
^{56}\text{Ni} &\rightarrow {}^{56}\text{Co} + e^{+} + \nu_{e} \\
^{56}\text{Co} &\rightarrow {}^{56}\text{Fe} + e^{+} + \nu_{e}
\end{aligned}
\tag{6.60}
$$

or by electron capture. Because of the Coulomb barrier, the reaction $^{28}\text{Si} + {}^{28}\text{Si} \rightarrow {}^{56}\text{Ni} + \gamma$ is much less prevalent during the silicon-burning phase than the reactions in which α particles intervene. However, during the supernova phase (see Section 6.11), the conditions are such that this chain of reactions becomes important. The thermonuclear reactions taking place under such physical conditions are different from regular stellar nucleosynthesis and this process is commonly called explosive nucleosynthesis. Explosive nucleosynthesis does not only produce iron, but leads to the formation of other elements.

The various thermonuclear reactions taking place in stars can therefore create most elements up to the iron-peak elements. However, heavier elements are found on Earth and elsewhere in the Universe but cannot be accounted for by these nucleosynthesis processes. Another formation process must then take place to create these heavier nuclei. Some of these elements are formed when nuclei like iron absorb free neutrons. As the number of neutrons increases in the absorbing nuclei, they become unstable and neutrons can decay to produce protons via the reaction

$$n \rightarrow p + e^- + \bar{\nu}_e \tag{6.61}$$

This reaction therefore transforms the original element that absorbed these free neutrons into elements containing more protons.

Some elements are formed through a weak flux of neutrons. This is called the s (or slow) process. These neutrons are emitted during some of the nuclear reactions taking place during evolution, some of which were presented earlier in this section (i.e. $^{12}\text{C} + {}^{12}\text{C} \rightarrow {}^{23}\text{Mg} + \text{n}$). Other elements heavier than iron are produced when a large flux of neutrons exists (the so-called r (or rapid) process). Such large fluxes of neutrons can exist in later stages of stellar evolution namely when the central portion of a massive star is transformed into a neutron star (see Sections 6.8.4 and 6.10.2). The topic of r and s processes is covered in (optional) Section 6.12.

6.8   Evolutionary Tracks in the H–R Diagram

6.8.1   Generalities

When a star is born, it is situated on the so-called zero-age main sequence (or ZAMS). Therefore, the ZAMS curve represents the position of the stars in the H–R diagram at the onset of hydrogen fusion in their centre. While on the main sequence and as the hydrogen is progressively fused into helium in the star's central regions, the structure of stars readjusts and they slowly move away from the ZAMS. This structural change can be explained by the results found in Section 5.6.2. In the stellar core (where nuclear burning takes place), the mean molecular weight may be approximated by the molecular weight μ found in a completely ionised plasma (see Section 5.6.2) that is given by the following expression

$$\mu = \frac{2}{3X + \dfrac{Y}{2} + 1} \tag{6.62}$$

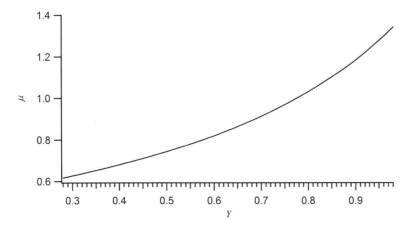

Figure 6.9 The variation of the mean molecular weight in completely ionised plasma as hydrogen is fused and the mass fraction of helium (Y) increases. A metallicity of $Z = 0.02$ was used here.

where X and Y are, respectively, the hydrogen and helium mass fractions. As hydrogen is progressively transformed into helium the mean molecular weight of the medium in the stellar core increases (see Figure 6.9). By using the ideal-gas approximation for the equation of state

$$P = \frac{\rho k T}{\mu m_{\mathrm{H}}} \qquad (6.63)$$

the pressure therefore decreases because of the increase of μ (assuming all else is equal). The core then slowly contracts due to the weight of the matter above the core. The local density increases because of this contraction, while the temperature increases due to the virial theorem. The structure of the star therefore readjusts. This partly explains why the observed main sequence has a certain width in the H–R diagram since stars with the exact same mass but with slightly different ages are found at slightly different positions in this diagram. Other factors such as different stellar composition can also account for scatter on the main sequence.

Figure 6.10 shows the partial (i.e. for a certain portion of a star's life) evolutionary tracks for stars of various masses in the H–R diagram obtained by the American astrophysicist Icko Iben Jr. (b. 1931). The numbers found in this figure represent different time steps. The point denoted as 1 defines the position of the zero-age main sequence. In this figure, the Sun at the present day is found between points 1 and 2 for the $1\,M_{\odot}$ curve.

At first, when a star arrives on the main sequence, its core slowly contracts as hydrogen is transformed to helium. As mentioned above, the density and temperature in the core increase with time. Since the nuclear energy production rate in a one-solar-mass star is approximately

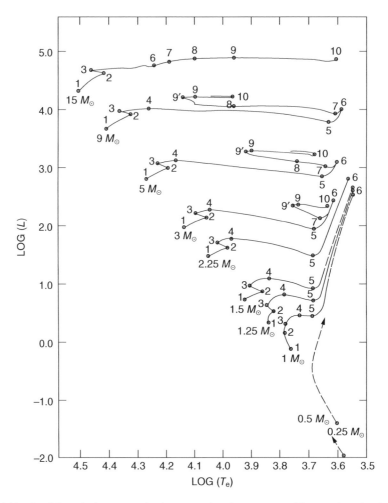

Figure 6.10 Partial evolutionary tracks for stars of various masses. The numbered points on the curves represent various time steps of the evolutionary process. Figure reproduced with permission from Iben, I., *Annual Review of Astronomy and Astrophysics*, 5, 571 (1967).

$$\varepsilon_{\mathrm{PPI}} = \frac{2.4 \times 10^4 \, \rho X^2 \mathrm{e}^{\frac{-3.38}{T_9^{1/3}}}}{T_9^{2/3}} \tag{6.64}$$

where T_9 is the temperature in units of $10^9\,\mathrm{K}$, the decrease of this quantity due to the diminishing hydrogen mass fraction X that occurs in the stellar core in the early stages of the main sequence will be more than compensated by the increase of the density and temperature. The nuclear production rate therefore increases as the core contracts and the luminosity of the star consequently increases. The surface temperature and the radius of the star gradually increase during this slow contracting phase of the core on the main sequence. Hydrogen can also start burning in layers above the initial core of the star as the temperature there becomes large enough to ignite these thermonuclear reactions.

Stars follow different evolutionary paths in the H–R diagram according to their mass. As stars with various masses evolve they have their individual particularities. For example, they can go through phases of sudden increases of nuclear burning (this is commonly called helium flashes when helium burns rapidly, or carbon flashes for carbon burning). They can also become pulsating stars if they, for example, cross the instability strip (see Section 5.7.1). They can also go through so-called dredge-up phases where the nuclear ashes (or synthesised material) can be brought to the surface of the stars by mixing. All of these details concerning the evolution of stars with various masses are outside the scope of this book. For a more complete discussion of the evolution of stars, the reader is referred to Kippenhahn, R. and Weigert, A., *Stellar Structure and Evolution*, Springer-Verlag, Berlin (1990).

To simplify the discussion of stellar evolution, stars may be divided into three evolutionary groups. Stars with approximately $M_* \leq 0.5 M_\odot$ can only burn hydrogen before evolving into white dwarfs. These stars do not possess sufficient gravitational energy to heat the core to the temperature necessary for helium fusion. Stars within the range $0.5 M_\odot \leq M_* \leq 10 M_\odot$ first burn hydrogen and then go on to burn helium (and possibly other elements depending on their mass). Stars with approximately $M_* \geq 10 M_\odot$ are able to burn all elements up to the production of iron-peak elements, eventually leading to supernovae. Following the evolution of these massive stars, they leave either a neutron star or a black hole as remnants. The exact lower limit for the mass of stars where iron production occurs is somewhat uncertain. Here, it is assumed to be equal to $10 M_\odot$.

Since nuclear fusion requires high temperatures, stellar nucleosynthesis takes place in the central regions of stars. As time passes, the metallicity in the core is enhanced due to fusion reactions occurring there. Generally speaking, this increase of metallicity is not observable since it occurs in the core of stars. However, after leaving the main sequence, certain stars have phases where the medium enriched by nucleosynthesis can be dredged-up to the stellar surface and can therefore be detected by spectroscopy. An example of such stars is classical carbon stars that exhibit a large abundance of this element at their surface. The carbon found at their surface has been dredged-up from the central regions during a phase of evolution called the asymptotic branch phase (see Figure 6.13 and discussion below for the position of the asymptotic-giant branch in the H–R diagram). Such details concerning the evolution of stars will not be discussed here.

Also, it should be noted that in this section only isolated (or singular) stars are considered. The evolution of binary systems is more complicated since mass transfer can occur during their evolution. An example of such a case leading to a nova or supernova will be discussed in (optional) Section 6.11.

Since the evolution of stars for each mass value has its specific features, a limited number of cases will be discussed here. Since the Sun is of particular interest to humans, the details surrounding its evolution will be described below. The evolution of a typical high-mass star that fuses the elements up to iron and eventually produces a supernova will also be presented. But first, a short description of the evolution of very low-mass stars is given.

6.8.2   Evolution of Low-Mass Stars ($M_* \leq 0.5\,M_\odot$)

In this section, the evolution of very low-mass stars will be discussed briefly. The lowest-mass stars ($M_* \approx 0.08M_\odot$) have a lifetime on the main sequence of approximately 10^{13} yr, while those with $M_* \approx 0.25M_\odot$ will stay approximately 10^{12} yr on the main sequence. Since the length of the lifetime of stars with $M_* \leq 0.5\,M_\odot$ is longer than the age of the Universe, none of these stars have had time to leave the main sequence since their formation. It is then impossible to obtain observational data to study the evolution of such stars, and this field of study is purely theoretical. However, since a relatively large portion of stars found in the Universe are low-mass stars (see Figure 2.4), these stars merit their own study.

Following the hydrogen-burning phase, very low-mass stars do not have sufficient gravitational energy to heat up their core to temperatures capable of helium fusion. Stars with masses lower than approximately $0.5\,M_\odot$ can therefore never achieve helium burning. Detailed calculations show that very low-mass stars are completely convective for a large portion of their lifetime. Since hydrogen from the surface layers can be brought to the core, a large portion of the hydrogen in these low-mass stars can be burned during their journey on the main sequence. This lengthens their stay on the main sequence. For instance, when a $0.08\,M_\odot$ star leaves the main sequence, its hydrogen mass fraction will only be about 0.11.

Another peculiarity of very low-mass stars is that some of these astronomical objects never achieve red-giant status (see Section 6.8.3). Figure 6.11 shows the radius of very low-mass stars as a function of time. These evolutionary models show that stars with

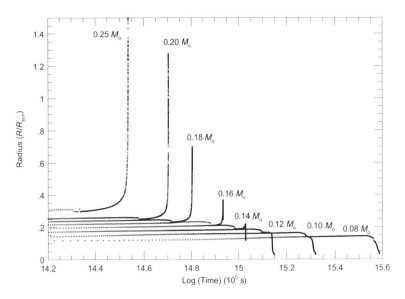

Figure 6.11   The radius of stars with masses between 0.08 and $0.25\,M_\odot$ as a function of time during their evolution. In this figure, a maximum radius is reached for all stars except for the $0.25\,M_\odot$ star. Reproduced by permission of the AAS from Laughlin, G., Bodenheimer, P. and Adams, F.C., *The Astrophysical Journal*, **482**, 420 (1997).

masses lower than $M_* \approx 0.16 M_\odot$ do not appreciably expand and go from the main sequence directly to the white-dwarf branch without becoming a red giant. The stars shown in Figure 6.11 will eventually produce white dwarfs composed mostly of helium. White dwarfs are dense stellar remnants where nuclear energy is no longer produced. These objects will therefore cool off with time. As mentioned in Section 5.6.3, white dwarfs are supported by the pressure due to a degenerate electron gas. More details concerning white dwarfs will be given in Section 6.10.1.

The vast majority of stars has relatively low mass (see Figure 2.4) and end their life as white dwarfs. This has dire effects on galaxies. On a very long timescale (i.e. much longer than the age of the Universe), once all stars of a given galaxy have evolved and not enough interstellar matter exists for further star formation, that galaxy will eventually have a very low luminosity. It is estimated that the luminosity of such a galaxy will be on the order of a single typical star like the Sun. This is much lower than the characteristic luminosity of average galaxies. For example, the luminosity of our galaxy, The Milky Way, has a value of approximately $3.6 \times 10^{10} L_\odot$.

6.8.3   Evolution of a $1 M_\odot$ Star: Our Sun

It is natural for humans to seek to understand how the central star in our planetary system will evolve with time. This is the aim of this section. However, it is clear that since the evolutionary times of stars (e.g. see Section 6.5.3) are much longer than the times of interest for humanity, such studies will have no direct impact on our society.

Similarly to all other stars, a $1 M_\odot$ star begins its life on the main sequence while burning hydrogen in its core. Hydrogen is depleted faster in the centre of the star than in the outer core because of the dependence of nuclear reaction rates on temperature. Figure 6.12

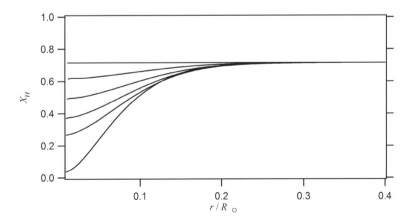

Figure 6.12   Hydrogen mass fraction (X_H) as a function of distance from the centre of the star (written in units of $r/R_\odot$) for times of 0.0, 1.39, 3.02, 4.53, 5.75 and 8.07 Gyr after the onset of hydrogen burning for a $1 M_\odot$ star. Data courtesy of Mathieu Vick, Jacques Richer and Georges Michaud obtained with the Montréal stellar evolution code.

shows the depletion of hydrogen as it is progressively transformed into helium during the evolution of a $1 M_\odot$ star. This figure shows that approximately 8 Gyr after the onset of hydrogen burning, all of the hydrogen in the star's centre is depleted. The curve for 4.53 Gyr after the onset of hydrogen burning, shown in this figure, approximately represents the Sun at its present age.

The slow evolution taking place on the main sequence has already been discussed in Section 6.8.1. It was shown that as hydrogen is fused into helium, the core of the star slowly contracts because of the increase of the mean molecular weight there. This contraction leads to an increase of the central temperatures so that hydrogen then burns at a faster rate and may also start burning in regions outside the core where before this physical readjustment of the star, the temperature was too low to sustain fusion. This process leads to an increase in the luminosity of the star.

As the amount of hydrogen is progressively depleted in the core of the star, hydrogen starts burning in a growing shell outside the core. A larger portion of the star's mass therefore participates in the nuclear processes. When hydrogen is completely exhausted in the central region of the star, the temperature there is not yet sufficient to go to the next phase of nuclear burning, namely helium fusion. The critical temperature for helium fusion is approximately 10^8 K. However, hydrogen is still burning at an ever-increasing pace in a growing shell outside the core. The energy produced in this shell-burning phase causes the outer portion of the star to expand, leading eventually to a red-giant star. During this expansion, the surface temperature of the star decreases so that the star becomes redder. This decrease of the effective temperature is due to a readjustment of the temperature gradient in the star that maintains proper energy transport. The star will first move to the subgiant branch and then become a red-giant star (see Figure 6.13). The red-giant phase lasts for approximately 10^9 yr for a 1-$M_\odot$ star.

Meanwhile, the core continues contracting, which causes an increase of its temperature. For the star under consideration here (i.e. $M_* = 1 M_\odot$), when the critical temperature for

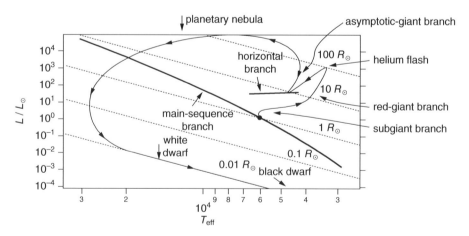

Figure 6.13   Illustration of the approximate evolutionary track of a $1 M_\odot$ star in the H–R diagram. The dotted lines show the position for various values for the radius.

helium fusion is reached, detailed modelling shows that the equation of state in the core describes degenerate matter. The degenerate free-electron gas now dominates the pressure there. This state of degeneracy leads to a strange phenomenon. In normal plasma, where the equation of state is that of an ideal gas, the contraction of the core leads to higher temperatures and a higher nuclear energy production rate that is self-regulated in the following way. The increase of the energy production in the core leads to a higher temperature that increases the pressure and eventually stops the contraction of the core. This process regulates the energy production rate, since the core collapse ceases when sufficient energy is produced.

However, the situation is quite different when the gas is in a degenerate state. As seen in Section 5.6.3, the gas pressure in a degenerate gas is independent of temperature. Therefore, when the core contracts, the temperature increases, which leads to more nuclear energy production from helium fusion. However, the temperature increase due to this production of energy does not affect pressure, and the core continues to collapse, increasing even more the central temperature and the energy production rate. This has a runaway effect that leads to a very large rate of helium fusion that is commonly called the helium flash. During this helium flash the luminosity of the star is greatly increased (see Figure 6.13). The helium-flash phase is short lived. Detailed study shows that the temperature increase in the central regions eventually lifts the degeneracy of the stellar plasma. The core and the star as a whole then stabilize on what is commonly called the horizontal branch (its name coming from the fact that this branch of the H–R diagram is approximately horizontal). On this branch, the star burns helium in its core while hydrogen is still burning in a shell outside the helium-burning core. The horizontal branch is the equivalent of the main-sequence branch for helium burning in low-mass stars.

After a stay of approximately 10^8 yr on the horizontal branch (or only approximately one hundredth the time it stayed on the main sequence), the star runs out of helium in its core that is now composed mostly of carbon nuclei. Figure 6.14 illustrates the internal structure of a 1-$M_\odot$ star near the end of its nuclear energy production phase. The core is carbon rich and is surrounded by a helium-burning shell. Above this shell, there is a

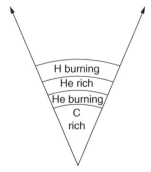

Figure 6.14 Illustration of the central region of a 1-$M_\odot$ star near the end of its nuclear-burning life. It is composed of a carbon core and nuclear-burning shells (He- and H-burning shells) separated by an inert helium-rich shell. This drawing is not to scale.

helium-rich and nuclear inert shell where the temperature is not sufficient for helium burning. Above this inert region there is a hydrogen-burning shell. An energy crisis is at hand. The core contracts (both helium and hydrogen continues to burn in distinctive shells outside the core) while the radius of the star increases. The star then climbs the asymptotic-giant branch in the H–R diagram (see Figure 6.13) and becomes a supergiant. It will stay approximately 2×10^7 yr on this branch. However, a 1-$M_\odot$ star is not massive enough to contract the core and increase the central temperature sufficiently to fuse carbon. The end of the nuclear-burning phase of the star is nearing. The core once again becomes degenerate and the pressure from the free electrons eventually stops the contraction of the core. A white dwarf composed mainly of carbon is forming in the centre of the star.

The star then enters a phase in its life called a planetary nebula. The outer layers of the star continue to expand, therefore permitting an observer to see deeper (i.e. T_{eff} increases, see Figure 6.13) in the star. The central white dwarf eventually becomes visible. The mass of the white dwarf left after the evolution of a 1-$M_\odot$ star is approximately $0.6 M_\odot$. The exact value of this mass depends on the mass-loss rate during the various phases of evolution and the quantity of mass expelled during the planetary nebula phase. A planetary nebula is a white dwarf that is surrounded by a tenuous shell of gas expelled during the evolutionary process. An example of a planetary nebula, namely M57 is shown in Figure 6.15. The expanding shell shown in this figure has a temperature of approximately 10 000 K. The typical value for the velocity of the expanding matter from a planetary nebula is on the order of 10 km/s. It should be noted that planetary nebulae are not an explosive phenomena like supernovae. The planetary phase lasts approximately 5×10^4 yr.

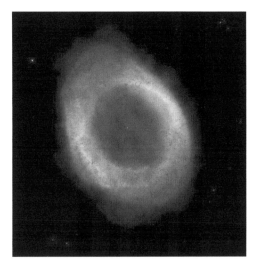

Figure 6.15   M57 (also called the Ring Nebula) is a well-known planetary nebula. The central white dwarf is the pinpoint body visible in the middle of the expanding shell of matter. This object has a diameter of approximately 2.4 ly and is situated 2300 ly from Earth. (NASA/courtesy of nasaimages.org). (See colour plate.)

Since the radius of a white dwarf is relatively small, the luminosity of the star plummets as it moves to the white-dwarf branch. White dwarfs are the stellar remnants from the evolution of low- to intermediate-mass stars and will be discussed in more detail in Section 6.10.1. Also, since white dwarfs do not produce any nuclear energy, they will cool off on a very long timescale and become cold inert objects commonly known as black dwarfs (see Figure 6.13).

White dwarfs possess an upper limit for their mass whose value is approximately $1.43 M_\odot$. Stellar remnants above this mass collapse to form a neutron star. Meanwhile, neutron stars with a mass above approximately $3 M_\odot$ become black holes. It is important to realise that since stars lose mass during evolution (during the planetary nebula phase for example), stars with masses above $1.43 M_\odot$ can become white dwarfs as long as they lose sufficient mass to bring the remnant below this upper limit.

6.8.4 Evolution of Massive Stars ($M_* \geq 10 M_\odot$)

Above a certain mass, stars are able to fuse elements up to iron as described by the nuclear reactions seen in Section 6.7. The exact value of this mass is not known with precision, but for our purposes here will be estimated as $10 M_\odot$. Such massive stars go through successive core-burning phases. As with all other stars, they begin their lives on the main sequence while burning hydrogen. When the hydrogen found in the core is spent, the star evolves and then burns helium to produce carbon and oxygen (see Sections 6.6 and 6.7 for details surrounding the various nuclear reactions at play). Once this burning phase is finished, the core contracts and then successively burns carbon, neon, oxygen and silicon. During each of these burning phases, the fusion reaction from the previous phase occurs in a shell outside the core. Near the end of the nuclear-burning period of massive stars, the star has an onion-like structure made out of various burning shells (see Figure 6.16)

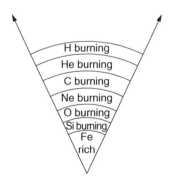

Figure 6.16 Onion-like structure of a massive star ($M_* \geq 10 M_\odot$) near the end of its life. The buffer regions (or nuclear inert shells) where no or little nuclear burning takes place that are found between the nuclear-burning shells are not shown here (like in Figure 6.14). Calculations show that at the late stages of the evolution of a $25 M_\odot$ star, over 50 % of its mass is found within the boundaries of the nuclear-burning process. This drawing is not to scale.

and nuclear inert shells. At this stage, the star is a supergiant, and its radius can attain more than $1000 R_\odot$ and its luminosity up to approximately $10^6 L_\odot$ (the exact values depend on the mass of the star).

As shown in Table 6.5, each successive burning phase is of shorter extent. For example, a 25-$M_\odot$ star stays 7×10^6 yr on the main sequence, while the silicon-burning phase only lasts 1 day! The evolution of stars therefore accelerates as they advance in the nuclear burning sequences. The reason for the shortening of the successive phases is that when stars evolve their central temperature increases, thereby increasing the nuclear reaction rates. Since the nuclear fuel burns at a very high rate, it shortens the lifetime of each successive burning phase. From an observational point of view, since the luminosity during advanced stages of evolution is very large and the quantity of energy per nucleon that can be liberated is small when fusing heavier elements (see Figure 6.1), the nuclear reaction rate must then be very large at these advanced stages. Therefore, the lifetime of these advanced burning stages is relatively small.

A critical situation occurs when the core of the star is composed of iron following the silicon-burning phase. Since the iron nucleus is one of the most stable nuclei of the periodic table of the elements, it cannot be further fused to produce energy. At this point of the evolution of massive stars, an energy crisis occurs in the core. When silicon burning ceases, the core contracts and the central temperatures increase. However, contrary to previous phases of evolution, its content (i.e. iron) can no longer produce thermonuclear energy. Contraction therefore continues until the temperatures in the central regions are so large that there exist photons of energy sufficient to destroy iron nuclei by photodisintegration reaction in which the energy of the photons is used to break up iron. An example is the following reaction

$$^{56}\text{Fe} + \gamma \rightarrow 13\,^4\text{He} + 4\text{n} \tag{6.65}$$

This process takes energy away from the radiation field found in the star and leads to an acceleration of the core collapse that in turn increases the local temperature even more. When the temperature is sufficiently large, even helium can be photodisintegrated via the reaction

$$^4\text{He} + \gamma \rightarrow 2\text{p} + 2\text{n} \tag{6.66}$$

At this point, the core of the star is composed mainly of protons, neutrons and free electrons. As the core continues its collapse, the densities become so large that even protons and electrons may fuse to give neutrons via the reaction

$$\text{p} + \text{e}^- \rightarrow \text{n} + \nu_e \tag{6.67}$$

This is commonly called the neutronization of the medium. The disappearance of the free electrons takes away their contribution to the pressure in the medium and once again, leads to an acceleration of the core collapse. A new type of astronomical body called a neutron star (composed mainly of neutrons) is therefore created by the collapsing core of massive stars. If the mass of the remnant is more than approximately $3\,M_\odot$, the neutron star becomes a black hole. This upper limit for the value for neutron stars, which is not precisely known, is called the Tolman–Oppenheimer–Volkoff limit. It is similar to the upper limit for the mass for white dwarfs (called Chandrasekhar's limit, see Section 6.10.1). Black holes are stellar remnants that are so dense that not even light can escape their gravitational pull, therefore rendering them black (see Section 6.10.3). The term black hole for describing these dense objects was first used by the American physicist John Archibald Wheeler (1911–2008) in 1967.

During the core collapse, the iron nuclei that were created in the most central parts of the star are destroyed by photodisintegration. However, not all the iron is destroyed; the iron found outside the region where photodisintegration takes place survives.

The death of massive stars is much more violent than the demise of lower-mass stars. When the core collapses toward becoming a neutron star, the core eventually rebounds when its density surpasses the typical density of nuclei. Since neutrons are fermions they obey Pauli's exclusion principle and this explains why the core cannot attain an infinite density. This core bounce produces a shock wave that propagates to the outer layers of the star. This shock wave violently ejects the outer layers of the star. This process is called a supernova. The energy generated by a supernova can also trigger what is commonly called explosive nucleosynthesis in the layers outside the core. It is believed that some of the neutrinos emitted during the neutronization process can be absorbed and transfer energy to the stellar medium to maintain the shock wave propagating in the star. Also, the large number of free neutrons present in the centre of exploding stars can create elements heavier than those of the iron peak. The procedure of producing heavy elements is called the r process (r standing for rapid) and is discussed in (optional) Section 6.12. Finally, numerical simulations of the final collapse and explosion of the stellar core causing a supernova show that this is a very rapid process that lasts on the order of a second. This period of time is extremely small as compared to the evolution timescales of stars at the various stages of nuclear burning (see Table 6.5).

During the supernova, the star's luminosity becomes as large as that of an entire galaxy (on the order of $10^{10} L_\odot$)! Since supernovae have similar maximum brightness they can be employed to estimate the distances to distant galaxies. It should also be noted that there are several types of supernovae that will be discussed in (optional) Section 6.11.

To confirm the presence of a supernova, neutrino observatories are employed since a large number of neutrinos are emitted during the neutronization of the core. More information concerning neutrino observatories was given in (optional) Section 6.5.4.

Supernovae are rare phenomena because massive stars are not common. For example, less than 0.1 % of stars have a mass larger than $15 M_\odot$. Supernovae are, however, very important for the chemical evolution of galaxies since they eject matter enriched in heavy elements. This enriches the metallic content of the interstellar medium. Therefore, future generations of stars (i.e. formed at later times) have a larger metallicity. This enrichment of the metallicity as a function of time will be discussed in Section 6.9.1.

An example of a well-known supernova, namely M1 or the Crab nebula, is shown in Figure 6.17. This supernova was recorded by Chinese and Arab astronomers in the year 1054. In its centre, a pulsar is observed. A pulsar is composed of a rotating neutron star with a large magnetic field (see Section 6.10.2 for more information concerning these astronomical objects).

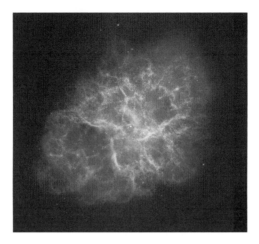

Figure 6.17 The Crab nebula (M1), which is a supernova remnant. This object is composed of a central pulsar surrounded by the matter ejected during the supernova. It has a diameter of approximately 10 ly and is found at a distance of 6300 ly from Earth. It is expanding at a velocity of 1800 km/s. (NASA/courtesy of nasaimages.org). (See colour plate.)

6.9   Stellar Clusters

Even though only a small portion of stars in our galaxy are found in stellar clusters, they are important astronomical objects that can be very useful for gaining a better understanding of stars. For example, since stellar clusters contain stars of various masses they are excellent 'laboratories' where stellar evolution theory may be verified. There are two types of stellar clusters: open (sometimes also called galactic clusters) and globular clusters. Open clusters are relatively young, while globular clusters are older astronomical objects. More information about the properties of these clusters is given below. However, before discussing stellar clusters, it is important to examine in more detail stellar populations that were very briefly introduced in Chapter 1. Also, since galaxies contain stars, it is instructive to study them to gain insight about certain aspects of stellar astronomy. Therefore, the various types of galaxies found in the Universe will be described.

6.9.1   Stellar Populations, Galaxies and the Milky Way

Stars can be divided into three types of populations: I, II and III. These populations are related to the quantity of metals they contain. Population-I stars have large metallicities while population-II stars have relatively low metallicities. Population-III stars theoretically have a nil metallicity (i.e. $Z = 0$). The Sun is a population-I star.

   The metallicities of the various stellar populations are related to their age. During the Big Bang, only hydrogen, helium and a small amount of Li was created. Therefore, the first generation of stars (or population-III stars) that formed in our Universe did not contain metals (expect for the small amount of Li formed during the Big Bang mentioned above). These stars did not contain elements such as C, N, O, … Fe. As time elapsed, the more-massive stars of this first generation evolved to create the various elements and ended up as supernovae. During their explosion, elements formed during nucleosynthesis were expelled to the interstellar medium. Therefore, as the Universe ages, more metals are found in the interstellar medium and the following generations of stars formed contain a growing amount of metals. Consequently, stars formed recently (i.e. population-I stars) contain more metals than older stars (i.e. population-II stars). Since only the outer layers of stars are probed during spectroscopic observations, the metallicity detected in stars, even those that have left the main sequence (with a few exceptions[7]), is determined by the metallicity of the primordial cloud in which these stars were formed.

No stars have yet been detected with zero metallicity. One reason for this anomaly could be that population-III stars might have been contaminated with enriched interstellar matter that has been accreted by these stars. For instance, a neighbouring star that has gone through the supernova phase and expelled enriched material could be the source of such contamination.

There are three types of normal (as opposed to active) galaxies in the Universe: elliptical, spiral and irregular galaxies (see Figure 6.18). The classification of galaxies is primarily based on their morphological properties. The colour of galaxies depends strongly on the quantity of gas they contain. Generally, the proportion of gas in galaxies relative to their total mass varies among these three types of galaxies. Elliptical galaxies generally contain very little gas, while irregular galaxies are those that contain the larger proportion of gas among the three types of galaxies. Since interstellar gas is necessary for star formation, more star-formation events take place in irregular galaxies than in spiral galaxies. Very little or no star formation occurs in elliptical galaxies. It is believed that for these galaxies, a very rapid star-formation rate early in their lifetime gobbled up the whole amount of gas they contained.

Irregular galaxies are generally bluer than spirals, which are in turn bluer than elliptical galaxies. The reason is that early-type stars are still present in irregular and spiral galaxies because they contain interstellar gas so that star formation continues taking place. In elliptical galaxies, the more-massive stars have already evolved and since these galaxies contain no or little gas, star formation has ceased. Therefore, elliptical galaxies contain only lower-mass stars and are less blue than other types of galaxies.

The different types of galaxies also differ by the type of stellar populations they contain. Elliptical galaxies contain population-II stars (i.e. relatively old stars), while irregular and spiral galaxies contain both population-I and population-II stars (i.e. both young and old stars).

Our galaxy, the Milky Way, is a spiral galaxy. Spiral galaxies can be divided into three parts: the central bulge, the disk (containing the spiral arms) and a halo (see Figure 6.19). The disk contains both gas and population-I and -II stars. The halo is composed of population-II stars (most of which are found in globular clusters, see Section 6.9.3 below).

Table 6.6 shows the general approximate properties of population-I and -II stars in the Milky Way. Population-I stars have a relatively low scale height (horizontally from the

---

[7]For instance, there is a class of stars called carbon stars that contain a large abundance of carbon at their surface. This carbon comes from helium fusion and is brought to the surface by convection during its evolution.

Figure 6.18 Examples of the various types of normal galaxies: (a) a spiral galaxy seen face on: the Whirlpool galaxy (M51) along with its companion galaxy, (b) a spiral galaxy seen edge on: the Sombrero galaxy (M104), (c) an elliptical galaxy: #NGC1132 and where a host of other galaxies can also be seen on this picture, (d) an irregular galaxy: M82 also known as the cigar galaxy. (NASA/ courtesy of nasaimages.org). (See colour plate.)
#NGC stands for the New General Catalogue that contains deep sky objects.

disk) and are therefore found in or near the galactic disk. Population-II stars are on average farther away from the galactic plane and form the halo of the Milky Way. Also, their average velocities perpendicular to this plane (σ_w) are relatively large. This may be explained by the fact that halo (or population-II) stars are older and have had time to be subjected to a large number of gravitational interactions with other stars for instance. These old stars (and old clusters), that were originally in orbit in the plane of our galaxy have been deviated from their initial trajectory and now orbit the galaxy in a plane inclined from the galactic disk. Since globular clusters (composed of population-II stars) are older than open clusters (see Section 6.9.2 below), they are distributed spherically with respect to the centre of our galaxy. Meanwhile, open clusters (composed of population-I stars) are

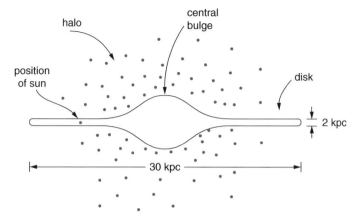

Figure 6.19   The three components of the Milky Way galaxy: the central bulge, the disk (containing the spiral arms) and the halo. The halo is composed of globular clusters and individual stars that have a spherical distribution. The position of the Sun is also shown. The diameter of the disk is approximately 30 kpc, while its thickness is 2 kpc. The unit pc standing for parallax-second (see Appendix C) will be defined in Section 6.9.5. This drawing is not to scale.

Table 6.6   Typical properties of population-I and -II stars in the Milky Way.

	Population I		Population II	
	very young	young	old	very old
Scale height (kpc)	60	100	500	2000
σ_w (km s^{-1})	8	10	25	75
Z	>0.02	0.01	0.005	<0.002
Age (relative to the Universe)	<0.05	0.25	0.75	1
Distribution	generally found in aggregates		spherical distribution	

found in the galactic disk and are generally found in aggregates mostly in the spiral arms of the Milky Way.

6.9.2   Open Clusters

Open clusters are typically comprised of roughly 10^2 to 10^3 loosely bound stars. Their dimension is up to approximately 10 pc. They are relatively young objects composed of population-I stars. These clusters may therefore contain early-type stars because these stars haven't had time to evolve yet. In our galaxy, they are mainly found in the spiral arms of

Figure 6.20   The open cluster M45 commonly called the Pleiades. It is found at a distance of approximately 440 ly from Earth. (NASA/courtesy of nasaimages.org). (See colour plate.)

Figure 6.21   The globular cluster M4. It is found in the Scorpio constellation at a distance of approximately 7200 ly from Earth. (NASA/courtesy of nasaimages.org). (See colour plate.)

the galactic disk. Figure 6.20 shows an example of a very well-known open cluster, the Pleiades or M45. There are over 1000 known open clusters in our galaxy.

6.9.3   Globular Clusters

Globular clusters typically contain on the order of 10^5 stars. Their dimension is up to approximately 100 pc. They are much more tightly bound than open clusters leading to their spherical shape (see Figure 6.21) with a much larger concentration of stars in their central region. They are very old objects that no longer contain early-type main-sequence

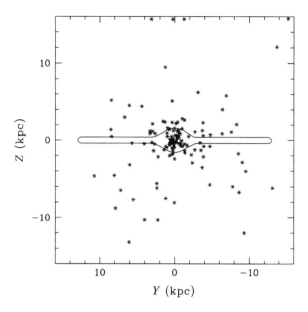

Figure 6.22 The distribution of globular clusters in the Milky Way. This drawing of the Milky Way is not to scale. Figure reproduced with permission from Cox, A. N., *Allen's Astrophysical Quantities*, Springer, New York (2004).

stars. They are therefore composed of population-II stars. Globular clusters have a spherical distribution with respect to the centre of our galaxy (see Figure 6.22) and are consequently mostly found in the halo. Detailed accounting shows that there are probably less than 200 globular clusters in our galaxy.

6.9.4 Age of Stellar Clusters

The age of stellar clusters may be estimated with the theoretical concepts of stellar evolution seen previously in this chapter. As a basic premise, it is assumed that all of the stars composing a cluster were created at roughly the same time. Generally, stars with various masses make up stellar clusters. The time a star may stay on the main sequence depends on its mass. The early-type (or more massive) stars have a shorter stay on the main sequence than late-type stars. Therefore, as time elapses, stars with progressively smaller masses leave the main sequence (see Figure 6.23). The age of the stellar cluster may be approximated as the time the most massive stars of this cluster still observed to be on the main sequence stay at this evolutionary phase. According to Eq. (6.29), a cluster for which the most massive stars that remain on the main sequence are stars with $M_* = 5M_\odot$, is approximately 10^8 yr old.

Figure 6.24 shows a colour-magnitude diagram for the M3 globular cluster. In this case, the turn-off point where stars are about to leave the main sequence is found at $M_* \approx 1M_\odot$ giving an estimated age of 10 Gyr according to Eq. (6.29).

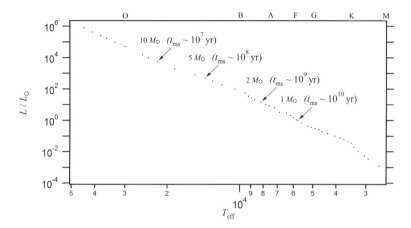

Figure 6.23   The time stars spend on the main sequence according to Eq. (6.29) for various stellar masses. The dots are main-sequence stars of various masses.

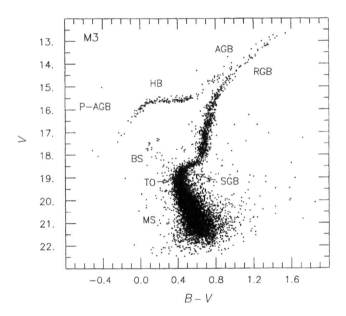

Figure 6.24   Colour-magnitude diagram for the M3 globular cluster. Shown in the figure are the main-sequence (MS), the turn-off point (TO), the subgiant (SGB), red-giant (RGB), horizontal (HB), asymptotic-giant (AGB) and postasymptotic (P-AGB) branches. Also seen in the figure are blue stragglers (BS). Figure reproduced with permission from Renzini, A. and Fusi Pecci, F., *Annual Review of Astronomy and Astrophysics*, vol. 26, 199 (1988).

In this figure, one may notice that there is a small number of stars still on the main sequence above the turn-off point. Such stars are called blue stragglers. The presence of these stars that should have theoretically left the main sequence according to the estimated age of the cluster pose a problem for the standard evolutionary theory of stars. However, there are several possible circumstances that can explain the presence of these stars without casting doubt on our understanding of the evolution of stars. The most probable explanation is that these stars are in binary systems. In such a system, a low-mass star could have gained mass by accretion of matter from the other member of the binary system. Therefore, a blue straggler could have started its life as a lower-mass star and stayed on the main sequence longer than predicted for an isolated star with a mass equal to the new mass of the blue straggler. Another possibility that could explain blue stragglers is that these stars might have formed more recently than the other stars found in the cluster.

6.9.5   Distance to Stars and Stellar Clusters

Since the Earth revolves around the Sun once a year, the stars near us move back and forth in the sky during the year (relative to very distant stars that remain fixed). The distance to these stars can be obtained by measuring the angle (related to the angle of parallax, see below) by which they move over a six-month period. This angle is obtained by measuring the angle by which a telescope's direction must change to keep the star in its sight over that six-month period. This assumes that the angular displacement of the star under consideration due to its tangential velocity of its proper motion is negligible. For any star, a six-month period can be found where an isosceles triangle can be defined with the star under consideration at the peak of the triangle and the position of the Earth at the beginning and end of the six-month period forming its base (see Figure 6.25). In this figure, the angle of parallax α that is used below is that of half of the angle at the summit of the isosceles triangle.

Since the distance between the Sun and the Earth is a known quantity (i.e. 1 AU), the distance d to the star is found by simple trigonometry. Assuming that d is given in AU,

$$\sin \alpha = \frac{1}{d} \tag{6.68}$$

where α is defined as the angle of parallax. Since the distance to stars is much larger than the astronomical unit, the angle of parallax is very small. It is easy to show (see Exercise 6.6) that for small angles (i.e. $\sin \alpha \approx \alpha$, where α is in radians), when α is expressed in arcsec the distance d in units of parsecs is

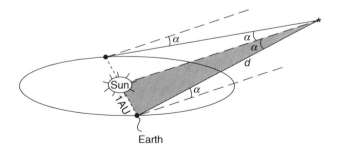

Figure 6.25   Illustration showing the angle of parallax α of a star. The angle by which a telescope's direction on Earth must move over a six-month period in order to keep the star in its sight is equal to 2α.

$$d = \frac{1}{\alpha} \tag{6.69}$$

A parsec (name taken from parallax-second or the distance for a parallax angle of one arcsec) is equal to 3.26 ly or 206264 AU. This method of measuring distances to stars is called the parallax method. Stars near the Earth have a larger angle of parallax than those farther away.

However, this method of measuring distances to stars has its limitations. Since telescopes have a limited resolution, they cannot measure arbitrarily small angles. The distance to stars far away from the observer that possess angles of parallax too small to be detected due to the limit of resolution of the telescope cannot be measured with this method. In other words, such stars do not move observably in the sky during a six-month period.

As discussed in Chapter 1, with the knowledge of the absolute magnitude M of a given star and by measuring its apparent magnitude m, its distance d (in units of pc) can be obtained by using the equation for the distance modulus

$$m - M = 5 \log \left(\frac{d}{10} \right) \tag{6.70}$$

This method of measuring distances to stars is especially useful for stellar clusters. With proper photometric measurements, stars of a cluster can be positioned in a H–R diagram. Meanwhile, the absolute magnitude of zero-age main sequence stars can be calculated theoretically. The difference between the absolute magnitude of the ZAMS and the apparent magnitude of main-sequence stars of the star cluster gives its distance by using the distance modulus equation given above (see Figure 6.26). This method of measuring the distance to stars is commonly called the spectroscopic parallax method. The distance to stars with well-known luminosities such as certain pulsating stars that possess a well-

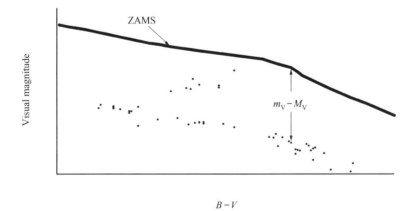

Figure 6.26 Illustration of the difference between the absolute magnitude of the zero-age main sequence (ZAMS) and the apparent magnitude of the main sequence of a star cluster.

determined period–luminosity relation (see Figure 5.13), can also be obtained with this method. In this specific case, the measurement of the dominant pulsating period gives the luminosity of the star in question and consequently its absolute magnitude.

6.10   Stellar Remnants

6.10.1   White Dwarfs

White dwarfs are stellar remnants of low to intermediate-mass stars. The name of these stars is related to the fact that they are relatively hot (or white in colour) and compact. Well over 90 % of stars will end up as white dwarfs. However, since less-massive stars live longer than their more massive counterparts, most of them have not yet become white dwarfs, and therefore white dwarfs now only account for a few per cent of the stars found in our galaxy. White dwarfs are compact objects with a typical radius on the order of 1 % of the solar radius (or roughly equal to the Earth's radius) and they have an average density of roughly 10^6 g/cm^3. This density is six orders of magnitude larger than the average density of the Sun. White dwarfs have relatively small luminosities because of their size and are found on the lower portion of the H–R diagram (see Figures 1.11 and 1.14). The surface gravity of white dwarfs is several orders of magnitude larger than for typical main-sequence stars. This leads to large values of pressure at their surface and to wide atomic lines in their spectra. These large atomic line widths simplify the observational identification of white dwarfs.

Table 6.7 gives the fundamental parameters of the well-known white dwarf Sirius B. In reality, the star Sirius is a binary-star system. Sirius A is the star visible with the naked eye while Sirius B is its companion, which happens to be a white dwarf. Sirius B is too faint to be seen with the naked eye.

The internal pressure in white dwarfs is due to a degenerate electron gas. The equation of state of such a gas leads to a pressure that depends on the number density of the free electrons in the stellar plasma but that is independent of local temperature. Detailed calculations show that a white dwarf's radius decreases as a function of its mass. It is found that a white dwarf with a mass of approximately $1.43 M_\odot$ has a radius that theoretically tends towards zero. This mass, which is called Chandrasekhar's limit, is the upper limit for the mass of white dwarfs. In the late stages of evolution of massive stars, if the degenerate core's mass reaches this value, it collapses and leads to the neutronization of matter where protons and electrons fuse to generate neutrons via the reaction

Table 6.7   Fundamental parameters of the white dwarf Sirius B.

M_*	$1.05 M_\odot$
T_{eff}	27 000 K
L_*	$0.03 L_\odot$
R_*	$0.008 R_\odot$
$\log g$	8.65

$$p + e^- \rightarrow n + v_e \tag{6.71}$$

This process consequently produces a neutron star (see Section 6.10.2).

The core of white dwarfs is mostly formed of He or C and O. As seen in Section 6.8.2, stars with very low masses can only burn hydrogen since they cannot attain the central temperature needed for He burning. However, since the lifetime of stars with very low masses on the main sequence is larger than the age of the Universe, these stars are still on the main sequence at the present time and cannot account for white-dwarf cores made of helium. Another way of obtaining a white dwarf with a helium core is via the evolution of a binary-star system. A star that has burned the hydrogen in its core and is in the process of becoming a red giant may have its outer shells stripped by the gravitational pull of its companion star. If this mass loss occurs before helium has started to fuse, it can stop further evolution of the star and leave a white dwarf made up mostly of helium. As mentioned previously, the evolution of binary systems can strongly modify the way stars in such a system evolve. Meanwhile, the evolution of an intermediate-mass star such as the Sun, leaves behind a white dwarf with a core made mostly of C and O. Its outer layers may contain hydrogen and/or helium depending on the quantity of mass lost during its evolution.

White dwarfs are divided into spectral classes that are related to the composition of their outer layers (see Table 6.8). The system of classification of white dwarfs is symbolised by a D (standing for degenerate) followed by second letter that defines the spectral features. Contrarily to main-sequence stars classification that is directly related to T_{eff}, the classification of white dwarfs are associated to their composition. Therefore, the T_{eff} range for certain spectral classes can overlap. Some white dwarfs pulsate and are therefore variable and in this case a V is added to their spectral class such as DAV stars (these stars are also called ZZ Ceti stars after the name of the first of this type of star discovered). Some white dwarfs also show hybrid spectral characteristics. For example, some have hydrogen-rich atmospheres but also show neutral helium lines in their spectra and are classified as DAB white dwarfs.

Magnetic fields are detected for a small portion (a few per cent) of white dwarfs by the Zeeman splitting of their spectral lines. Some of these magnetic white dwarfs have field strengths of up to 10^9 G. These magnetic fields are enormous compared to the average field at the surface of the Sun (or the Earth) which is equal to approximately one G, while it is on the order of a couple of kG in sunspots. It is believed that the magnetic fields in white dwarfs are remnants from the evolution of magnetic ApBp stars (see Section 7.2.2).

Table 6.8 Spectral classification of white dwarfs.

Spectral Class	Spectral characteristics
DA	Hydrogen-rich atmosphere (strong HI lines)
DB	Helium-rich atmosphere (with strong HeI lines)
DC	No strong lines present
DO	Helium-rich atmosphere (with strong HeII lines)
DQ	Carbon-rich atmosphere
DZ	Metal-rich atmosphere (e.g. CaI, MgI, FeI lines)

Magnetic fields at the surface of ApBp stars can reach values of more than 10 kG. When the stellar core contracts to form such white dwarfs, the magnetic field lines follow the condensing plasma that intensifies the value of the magnetic field found in the original main-sequence stars.

Since white dwarfs have no thermonuclear energy source, they cool down with time. They will eventually become very faint objects commonly called black dwarfs. Theory predicts that these black dwarfs will ultimately crystallize. Since most black dwarfs are made predominantly out of carbon, these objects are sometimes called the Universe's diamonds.

6.10.2   Neutron Stars, Pulsars and Magnetars

As discussed in Section 6.8.4, neutron stars are created by the collapse of the cores of stars with relatively high masses. The lower limit for the mass of neutron stars is equal to the Chandrasekhar limit since below this mass a white dwarf is obtained. The upper limit for the mass of neutron stars is more uncertain, mostly because of the limited knowledge of the equation of state inside neutron stars, but this upper limit is approximately $3M_\odot$. Above this upper limit, the core remnant collapses to form a black hole (see Section 6.10.3 for properties of such bodies).

The typical radius of neutron stars is on the order of 10 km and these objects are therefore much denser than white dwarfs. The average density of neutron stars is on the order of 10^{14} g/cm^3 (this value is estimated below from simple physical considerations in a special topic) and their central density may reach 10 times this value. To put the average density of neutron stars in perspective, one cubic centimetre of neutron-star matter contains as much mass as all the humans on our planet!

The outer crust of neutron stars is made of solid matter composed of nuclei (most likely iron formed during the advanced stages of evolution) and free electrons. Some theoretical models predict that the core of neutron stars might be composed of quarks. Since observational and experimental data for the inner structure of neutron stars is not available, it is difficult to confirm the validity of theoretical models for these astronomical objects.

A pulsar is composed of a fast rotating neutron star with an intense and largely dipolar magnetic field that is inclined with respect to the axis of rotation of the neutron star (see Figure 6.27). A typical magnetic field in pulsars is on the order of 10^{14} G.

Pulsars emit a large amount of radiation via a physical process called synchrotron radiation. Free electrons with high velocities (near the speed of light) surrounding the neutron star have curved trajectories in the magnetic field due to the Lorentz force. The magnetic-field lines therefore accelerate these relativistic particles that in turn causes synchrotron radiation (see Figure 6.28). The emission spectrum of the synchrotron process emits more radiation energy in the short- (X-ray) and especially the long- (radio) wavelength portions of the electromagnetic spectrum as compared to blackbody (or thermal) radiation. Pulsars therefore emit a relatively large amount of radio radiation. Two cones of radiation, similarly to lighthouse beacons, sweep across space as the neutron star rotates. Pulsars may be observed if one of the paths of these cones of radiation crosses the position of the Earth (see Figure 6.27). Pulses of radiation can then be detected, which is why these astronomical objects are called pulsars.

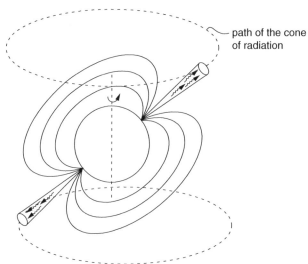

Figure 6.27   Schematic model of a pulsar. It is composed of a rapidly rotating neutron star that has a strong dipolar magnetic field that is inclined with respect to the axis of rotation of the neutron star. A large amount of radiation is emitted in a cone emerging from both magnetic poles.

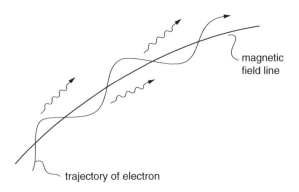

Figure 6.28   Illustration of the deviation of a free relativistic electron's trajectory by the presence of a magnetic field. This process emits radiation that is commonly called synchrotron radiation.

The rotation periods of the neutron stars that make up pulsars, measured as the periods of the observed pulses, vary from about 1 ms to 10 s. As pulsars emit energy, their rotational velocity decreases. The detection of the rate at which their rotational period increases can be compared to the predicted rate given by theoretical models in order to extract information on pulsars.

The type of pulsars discussed above is often called radio pulsars. There also exist X-ray pulsars. The large amount of X-radiation emitted by these objects is thought to be due to the accretion of matter from a binary companion.

The first pulsar was discovered in 1967 by the British astronomers Anthony Hewish (b. 1924) and Jocelyn Bell Burnell (b. 1943) with observations in the radio region of the electromagnetic spectrum. Up to now, over 1500 pulsars have been identified. Most of these pulsars have been detected in the radio region of the electromagnetic spectrum, while some were detected with X-rays. The best-known pulsar is the one at the centre of the Crab nebula (M1) found in the constellation Taurus (see Figure 6.17). M1 is the remnant of a supernova observed in 1054 and recorded by Chinese and Arab astronomers.

Special Topic – Average Density of Neutron Stars

Simple physical considerations can lead to an estimation of the minimum average density of a neutron star. A spherical neutron star of mass M and radius R with a rotational frequency $\omega = 2\pi/T$ (where T is the period of rotation), will be torn apart unless the gravitational acceleration is larger than the centrifugal acceleration

$$G\frac{M}{R^2} > \omega^2 R \qquad (6.72)$$

The minimum average density of the neutron star needed to keep matter at its surface from being flung out into space by the centrifugal force so that the neutron star is stable against rotation is therefore

$$\bar{\rho} > \frac{3\pi}{GT^2} \qquad (6.73)$$

where the average density is $\bar{\rho} = \dfrac{M}{\dfrac{4}{3}\pi R^3}$. Since the fastest pulsing pulsars have a

period on the order of 1 ms, this means that the central neutron star rotates up to approximately 1000 times a second. Using this value as an upper limit, the above equation gives $\bar{\rho} \approx 10^{14}\,\text{g/cm}^3$. This is in agreement with the typical average densities of neutron stars given by theoretical models.

In 1979, the first of several sporadic gamma-ray signals were observed. These were named soft gamma-ray repeaters (SGR) and some of them were eventually found to be associated with supernova remnants. However, it was only in 1992 that the American astrophysicist Robert Duncan (b. 1955) and his Canadian counterpart Christopher Thompson (b. 1961) theorised the existence of a new class of astronomical objects, namely magnetars, which could be responsible for SGRs. Magnetars are pulsars that possess magnetic fields up to 1000 times stronger than normal. It is hypothesised that magnetars are born with very short rotation periods with a dynamo process inside the neutron star that produces these large magnetic fields.

The large magnetic fields in these dense objects causes stress on their surface that may sporadically lead to cracks. This process causes seismic tremors commonly called star-quakes. The magnetic field is deformed and charged particles can be accelerated and emit the gamma-rays observed. It is believed that roughly 10% of supernovae caused by the evolution of massive stars might lead to a magnetar remnant. The rotation rate of magnetars decreases relatively quickly and after a period of approximately 10^4 yr, these objects are no longer as active. Due to the short timescale for their activity, it is not surprising that only a few SGRs have been detected.

For a portion of the magnetars that have been associated with supernova remnants some are not exactly where the centre of the corresponding supernovae remnant is expected to be. It is hypothesised that during a stellar explosion, the strong magnetic field present can redirect the energy preferentially in the direction of one of the magnetic poles, thus giving the star a velocity 'kick' so that the magnetar is not always found in the centre of the supernova remnant. Observations show that the velocity of the central remnant may reach 1000 km/s.

Since magnetar astronomy is a relatively new field of study and only a small number of these objects have been detected, much uncertainty still remains concerning the physical processes involved.

6.10.3   Black Holes

Black holes are the densest and least voluminous of the three types of stellar remnants. Since black holes possess large gravitational fields, the general relativity theory of gravitation proposed by the renowned German-born physicist Albert Einstein (1879–1955) must be employed to study these objects since classical gravity fails there due to the large gravitational fields involved. This theory treats gravity not as an attractive force between particles as in Newton's classical formulation, but as a distortion of spacetime. This theory correctly predicts (see special topic on spacetime curvature below) that even photons are affected by gravity due to the curvature of spacetime.

In this section, classical mechanics will be used to obtain certain properties of black holes. Even though classical physics is not valid under the physical conditions encountered near black holes, it gives the same exact results as general relativity for certain aspects because some relativistic terms cancel in certain theoretical developments.

Consider a spherical object of mass M and radius r from which a mass m at its surface is trying to escape at velocity V. The total energy of mass m is

$$E = \frac{1}{2}mV^2 - G\frac{Mm}{r} \tag{6.74}$$

The mass m is able to escape to infinity when $E \geq 0$, or in other words when the kinetic energy is larger than or equal to the potential energy due to gravitation. The escape velocity V_{esc} is obtained by assuming $E = 0$, which gives

$$V_{esc} = \sqrt{\frac{2GM}{r}} \tag{6.75}$$

If the mass M is progressively compressed into a shrinking volume, the escape velocity can eventually equal the speed of light. At that point, the radius is defined as the so-called Schwarzschild radius

$$R_{Sch} = \frac{2GM}{c^2} \qquad (6.76)$$

Since special relativity imposes a speed limit of c on all particles, a particle at the surface of a body that is compressed within the Schwarzschild radius can never escape. Such a dense object is called a black hole since nothing is able to escape once it enters the horizon defined by the surface of the sphere of radius R_{Sch}. The horizon is the point of no return for an object approaching a black hole.

In general relativity, the horizon defines what is called a singularity. This singularity exists because terms in certain equations of general relativity cannot be defined there due to the fact that a nil value appears at the denominator. For example, the Schwarzschild metric that measures the distance ds surrounding a massive spherical body of mass M is given (in spherical coordinates) by the equation

$$ds^2 = \left(1 - \frac{2GM}{rc^2}\right)c^2 dt^2 - \left(1 - \frac{2GM}{rc^2}\right)^{-1} dr^2 - r^2\left(d\theta^2 + \sin^2\theta d\varphi\right) \qquad (6.77)$$

This metric has a singularity for $r = \dfrac{2GM}{c^2}$ (i.e. for the Schwarzschild radius) since the second term on the right-hand side gives an infinite value.

Current physical theories cannot give any information about the inside of black holes (i.e. inside the horizon). Therefore, it is not known, for example, to what point the mass inside a black hole collapses and its internal structure is therefore completely unknown. Only global properties can be defined for black holes. Black holes have three independent fundamental properties: mass, angular momentum and electric charge.

Special Topic – Spacetime Curvature and Gravitational Redshift

The theory of general relativity predicts that gravity distorts the geometry of spacetime. One of the ways to experimentally verify the presence of spacetime curvature is to observe stars near the limb of the Sun during a total solar eclipse. During such an eclipse, stars can be seen in the sky due to the darkness that ensues. It has been verified that the known position of stars changes slightly when these stars are found near the limb of the Sun. Figure 6.29 shows the path of photons coming from a star that is found behind the Sun (and therefore should not be observable). The trajectory of photons is bent due to the curvature of spacetime near the solar surface caused by the gravity of the Sun. This beam of light can then be observed on Earth during a total solar eclipse (which is needed to darken the sky) and its apparent position in the sky is slightly shifted. The measurement of such shifts confirms the presence of spacetime curvature predicted by the theory of general relativity.

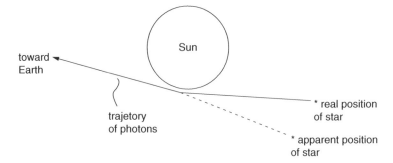

Figure 6.29 Illustration of the apparent displacement of the position of a star observed near the limb of the Sun. Such observations are made during a solar eclipse in order for the stars near the solar limb to be visible. This illustration is not to scale and the angle of deviation is greatly exaggerated.

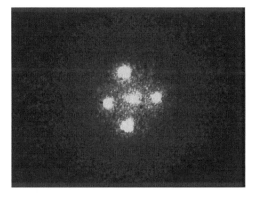

Figure 6.30 An image of Einstein's cross. This system is made up of a distant quasar that is found behind another galaxy that is seen in the centre of this image. This central galaxy serves as a gravitational lens that bends the light coming from the quasar and giving four distinct images of the distant quasar. (NASA/courtesy of nasaimages.org). (See colour plate.)

There exists other observational proof of spacetime curvature. For example, a distant quasar[8] that is found behind another less-distant galaxy is seen as four distinct images in the astronomical object called Einstein's cross (see Figure 6.30). The central galaxy acts as what is called a gravitational lens by bending the light coming from the distant quasar. The geometry of the system therefore gives four images of the distant quasar. If for instance, both the lensing and the distant galaxy were spherical and perfectly aligned, the image of the quasar would be a circular disk seen around the central galaxy. Some galaxies that are lensed by galaxies for which their position is slightly offset from the line of sight are observed as being

---

[8]Quasars are active galaxies that are young and extremely luminous. The energy source responsible for their power output is believed to be related to a supermassive black hole (with a mass up to approximately $10^{10} M_\odot$) at their centre.

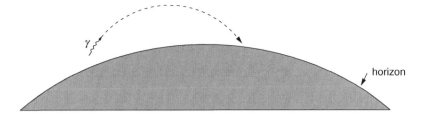

Figure 6.31   Illustration of the curved path of a photon (γ) emitted (by an atom for instance) near the horizon of a black hole. The effect of gravity on photons illustrated here is relativistic in nature and does not occur in classical physics.

arced or stretched and their image resembles a crescent. The curvature of spacetime caused by astronomical objects such as galaxies and black holes can therefore distort astronomical observations.

Since photons are affected by the gravitational distortion of spacetime, those emitted near the surface of a black hole do not follow a linear path. The photons follow the curvature of spacetime and can fall back towards the surface (see Figure 6.31). Even photons having a trajectory perpendicular to the black hole surface are not able to escape. General relativity predicts that a photon of wavelength λ_0 at a distance r from a spherical mass M is redshifted when travelling in a radial direction away from the mass according to the following equation

$$\lambda_{\text{obs}} = \lambda_0 \left(1 - \frac{2GM}{rc^2}\right)^{-1/2} = \lambda_0 \left(1 - \frac{R_{\text{Sch}}}{r}\right)^{-1/2} \tag{6.78}$$

where λ_{obs} is the wavelength observed when the photon is far away from the mass M and R_{Sch} is the Schwarzschild radius of this mass. Photons therefore lose energy when exiting a gravitational field and are redshifted. This redshift is negligible for photons exiting a normal star, but is important when photons try to exit the strong gravitational fields around black holes. This equation shows that the redshift is infinite when the photon is emitted at $r = R_{\text{Sch}}$. Therefore, a photon that attempts to exit a black hole perpendicularly to the surface loses all of its energy and does not exit the black hole, even if the direction of its trajectory is not affected by gravity.

Table 6.9 gives values of the Schwarzschild radius and the average density for various masses having reached black-hole status. The density for less-massive black holes is larger than for more-massive ones (see Exercise 6.12). For example, to produce a black hole from a proton, it must be compressed within a radius on the order of 10^{-52} cm and to a density of approximately 10^{130} g/cm^3! There exists no natural process, except possibly under the conditions found in the very early Universe, which can create such mini black holes. Furthermore, if such mini black holes were created in the early Universe they would very quickly evaporate (see below). A natural process that can create black holes with

Table 6.9 Schwarzschild radius and average density for several masses of black holes.

Mass	R_{Sch}	Average density
m_p	2.5×10^{-52} cm	3×10^{130} g/cm^3
1 kg	1.5×10^{-25} cm	7×10^{76} g/cm^3
$M_\oplus$	0.89 cm	2×10^{27} g/cm^3
$3 M_\odot$	8.9 km	2×10^{15} g/cm^3
M_{Galaxy}	0.4 ly	1×10^{-8} g/cm^3

masses larger than approximately $3M_\odot$ does exist, namely the evolution of massive stars (see Section 6.8.4). These stars possess enough gravitational energy to compress their central regions sufficiently to form black holes. It should also be noted that it is very probable that a supermassive black hole (with a mass up to approximately $10^{10} M_\odot$) exists at the centre of all galaxies. The presence of such supermassive black holes can, for instance, explain the large amount of energy emitted by a class of galaxies called active galaxies. Recent observations seem to confirm that a supermassive black hole exists in the centre of the Milky Way. Contrary to black holes formed by the evolution of massive stars, supermassive black holes have very low average densities (see Table 6.9).

In addition to stellar black holes and supermassive black holes it is theorised that a third type exists, namely, primordial black holes. Primordial black holes are those that could have been formed in the extreme conditions following the Big Bang. Since these black holes are not formed by the collapse of stars, they can have masses much lower than $3 M_\odot$.

Several strange phenomena occur near black holes. For example, if a body ventures near the horizon of a black hole it will be stretched by tidal forces. Tidal forces responsible for the tides on Earth are due to the gradient of the gravitational acceleration due to the Moon from the side of our planet facing it to the opposite side of the Earth. The stretching of physical bodies approaching a black hole may be illustrated by a simple example using classical gravity. At the surface of the Earth, the gravitational force acting on a person standing is approximately the same at his or her feet as at the top of his or her head. If the height of this person is represented by h, the relative change of gravitational force felt from his or her feet to his or her head is

$$\frac{\dfrac{1}{R_\oplus^2} - \dfrac{1}{(R_\oplus + h)^2}}{\dfrac{1}{R_\oplus^2}} \approx \frac{2h}{R_\oplus} \quad (\text{if } h \ll R_\oplus) \tag{6.79}$$

where $R_\oplus$ is the radius of the Earth (6378 km). Since $h \approx 1.5$ m, this proportion is extremely small ($\approx 10^{-7}$) and therefore the gravitational acceleration can be considered constant over this length. When nearing a black hole, the distance to which an astronaut (for instance) could approach before crossing the horizon is on the order of 100 km. The relative change given by an equation similar to Eq. (6.79) in this case is on the order of 10^{-5}, which is still quite small. However, since the gravitational acceleration in this case is very large (for example for a black hole with a mass of $10 M_\odot$, the acceleration is, according to classical physics, 10^{13} cm/s^2) and therefore a small proportion of this acceleration is enough to stretch and destroy any body that approaches a black hole. Therefore, no astronaut could survive such a journey near the horizon of a black hole.

Even though it was shown earlier that nothing can escape a black hole, in the discussion leading to this conclusion quantum effects were neglected. In 1974, the renowned British physicist Stephen Hawking (b. 1942) predicted that black holes can emit energy and eventually evaporate (see Hawking, S.W., *Nature*, 248, 30 (1974) for more details). This may be explained by quantum tunnelling. Therefore, there is a nonzero probability for a particle trying to escape a black hole to emerge outside by quantum tunnelling. This phenomenon may be explained in another way. In the vacuum surrounding a black hole, quantum theory predicts that particle–antiparticle pairs can appear and quickly annihilate themselves. As long as the time Δt they exist is shorter that what is prescribed by the Heisenberg uncertainty principle (here ΔE is the energy of the particles created)

$$\Delta t = \frac{h}{4\pi\Delta E} \tag{6.80}$$

such pair creation remains unobservable. In this case these particles are called virtual particles. In other words, there exist fluctuations in a quantum vacuum. For example, for an electron–positron pair, $\Delta E = 2m_e c^2$ and $\Delta t < 3 \times 10^{-22}$ s. When such a phenomenon occurs near the horizon of a black hole, one of the two particles may fall into the black hole and the second one can sometimes escape. This is equivalent to a quantum tunnelling of a particle coming from inside the black hole. A black hole can consequently evaporate. Stephen Hawking associated a temperature with such an evaporating black hole and developed a theoretical framework commonly called the thermodynamics of black holes. The value of this temperature depends on the mass M of the black hole according to the following equation

$$T = \frac{hc^3}{16\pi^2 kGM} \tag{6.81}$$

Black holes emit energy known as Hawking radiation with a blackbody spectrum related to this temperature. Since the temperature of black holes is inversely proportional to their mass, the timescale on which black holes produced by stellar evolution evaporate is much larger than the age of the Universe (see Exercise 6.13). The temperatures of such black holes are very low (see Example 6.4).

Example 6.4: Calculate the temperatures of black holes with a mass equal to m_p, 1000 kg, $M_\oplus$ and $5\,M_\odot$.

Answer:

Since the temperature of a black hole is

$$T = \frac{hc^3}{16\pi^2 kGM} \tag{6.82}$$

simple substitution finds respective temperatures of approximately 7×10^{49}, 1×10^{20}, 0.02 and 1×10^{-8} K for the four values of the masses given above. This leads us to conclude that mini black holes evaporate much faster than those produced by stellar evolution. Also, since the temperature of black holes from stellar remnants (i.e. those with masses on the order of several $M_\odot$) is so small, the energy they emit is negligible and thus not observable.

Since the temperature of black holes produced by stellar evolution is very low, the radiation they emit is too weak to detect. It is thus impossible to directly observe such black holes. However, when black holes are accreting mass (for instance when they are found in a binary system), the matter falling in emits X-rays and this radiation may be detected. The first black-hole candidate, Cygnus X-1, was detected in the mid-1960s. This is a binary system composed of a B-type supergiant star and a second component that emits no radiation in the visible part of the spectrum but does emit a large amount of X-rays. By estimating the mass of the supergiant star and by measuring the rotation period of the system, the mass of the invisible component was estimated to be between 5 and $10 M_\odot$. The invisible component is therefore too massive to be a neutron star. All of these observational facts may lead us to conclude that the companion of the supergiant star is a black hole towards which matter coming from the supergiant is being accreted.

6.11   Novae and Supernovae[†]

The word *nova* is derived from a Latin word meaning new. Novae and supernovae are stars that suddenly become visible due to a sudden increase of their luminosity. Such astronomical phenomena have been detected and registered by human beings for the past several millennia. In the modern era, when supernovae are discovered, they are named SN followed by the calendar year and then by one or two letters in order of their discovery. For example, the well-known supernova SN 1987A was the first supernova detected in 1987. It took place in the Large Magellanic Cloud that is a neighbour galaxy to our own Milky Way. This supernova is a Type-II supernova (see below for the definition of the various types of supernova) and is believed to be the product of the evolution of a $20 M_\odot$ star. It was first discovered by visual observations. The neutrinos emitted by the supernova were also detected as an initial burst of neutrinos lasting less than three seconds. This gives an indication of the timescale of the final phase of a massive star's life, leading to its explosion.

In Section 6.8.4, it was seen that supernovae are created by the evolution of massive stars that explode at the end of their life. However, a supernova can also occur when a white dwarf explodes. In this section, several types of supernovae will be discussed. Various types of supernova light curves (i.e. their luminosity as a function of time) will also be shown. But first, the physical processes leading to nova will be revealed.

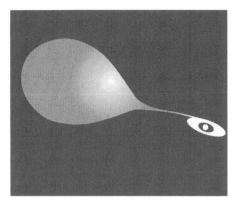

Figure 6.32   Illustration of mass transfer from a giant star to a white dwarf. An accretion disk is shown around the white dwarf. Such mass loss can lead to nova or supernova.

Up to now, only the evolution of isolated (or single) stars has been considered. The evolution of binary stars is more complicated since mass transfer can occur from one companion to the other. Two particular cases of evolving binary stars consisting of a white dwarf and a companion will be discussed here. The first case described below gives what is called a nova, while the second case seen later in this section leads to a Type-Ia supernova.

A nova is an astronomical event that occurs when an accretion disk is formed around a white dwarf in a binary system (see Figure 6.32). When relatively small amounts of matter (not enough for the white dwarf to approach the Chandrasekhar limit as for the case of SN Ia, see below) falls from the companion star onto the white dwarf, it causes a flare up of thermonuclear reactions that emits a large amount of radiation thereby giving a nova. A nova does not lead to the destruction of the system and can occur on a periodic basis within the same system. Typically, the luminosity of a nova is on the order of $10^4 L_\odot$.

Supernovae are astronomical events that emit an extremely large amount of energy (on the order of 10^{51} erg or more) on a very short timescale. This energy is emitted in the form of neutrinos, radiation and kinetic energy of the matter ejected by the explosion. A large portion of the neutrinos are emitted during the first seconds of this phenomenon, while radiation is emitted over several months or years following the explosion. A considerable portion of the exploding star is ejected at speeds on the order of 10^4 km/s. These speeds are detected by the Doppler shifts of the atomic spectral lines of supernovae. For example, some recent observations of SN 2008D show expansion speeds of more than 10^4 km/s. This supernova was discovered by NASA's Swift satellite in the X-ray portion of the electromagnetic spectrum. This satellite was observing another supernova in the galaxy NGC 2770 (situated 88 Mly away from Earth) that had exploded weeks earlier when it detected SN 2008D. The probability of having two supernova events in a host galaxy separated by such a short period is extremely small (estimated to be less than 1 in 10 000). This fortuitous discovery of a supernova explosion at the early stage of the process should shed light on the physics of these fascinating astronomical events. Normally, supernovae are detected in the visible part of the spectrum. The maximum visible brightness typically

Table 6.10   Types of supernovae.

Type	Precursor	Spectral characteristics
SN Ia	White dwarf in binary system	No H lines and strong Si lines
SN Ib	Evolution of a massive star	No H lines and no Si lines
SN Ic	Evolution of a massive star	No H lines, no Si lines and no He lines
SN II	Evolution of a massive star	Strong H lines

occurs a few weeks after the explosion and therefore, such detection does not give direct information about the early phase of the supernova phenomenon.

The ejected matter causes shock waves in the interstellar matter surrounding the supernova. This process can heat the interstellar matter up to approximately 10^8 K, which leads to the emission of a large amount of X-rays. Since the interstellar medium is much more tenuous than the ejected mass, the ejected matter will at first progress with little impediment. After about 10^2 yr, the shock wave will be slowed down considerably and the local temperature will also decrease.

Supernovae can be divided into two spectral types related to the strength of the hydrogen lines observed in their spectra. Type-I (or SN I) supernovae show no hydrogen lines in their spectra while Type-II (or SN II) supernovae have strong hydrogen lines (see Table 6.10). SN II supernovae are due to the explosion of massive stars at the end of their lives (see Section 6.8.4). At the time of their explosion, the supernovae of this type have a large amount of hydrogen in their outer layers, which explains the strong hydrogen lines.

Type-I supernovae come in three types named SN Ia, Ib and Ic. SN Ib and Ic supernovae are also thought to be due to the explosions of massive stars. It is believed that the outer layers of these stars are lost either by strong stellar winds or to stripping of these layers by a binary companion. Type-Ib supernovae have lost most of their hydrogen in their outer layers but still have a considerable amount of helium in their envelopes. Meanwhile, SN Ic are believed to have lost most of their hydrogen and helium before exploding.

Type-II, -Ib and -Ic supernovae are due to the explosion of relatively young (or population-I) stars. A clue to the population of these stars is also obtained by knowing that these types of supernovae are never observed in elliptical galaxies. As seen in Section 6.9.1, elliptical galaxies are composed of old (or population-II) stars and therefore, all massive stars have already evolved and died giving neutron stars or black holes as remnants following their explosion. No massive stars are available for supernovae production in such galaxies.

Type-Ia supernovae are considerably different from the other types discussed above. They are due to the explosion of a white dwarf found in a binary-star system. A contraction of the white dwarf occurs when mass transfer from the companion star to the white dwarf is sufficient to lead to a mass approaching the Chandrasekhar limit. This contraction leads to runaway thermonuclear reactions and the ensuing supernova. Since little or no hydrogen is found in white dwarfs, no hydrogen atomic lines are observed in their spectra. Also, because of the nucleosynthesis taking place during the explosive stage of Type-Ia supernovae, a considerable amount of Si is produced leading to the presence of strong Si lines in their spectra (while Si lines are absent for

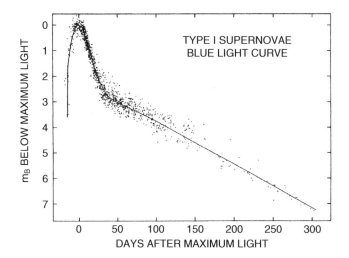

Figure 6.33 The light curve of SN I showing the magnitude in the blue part of the spectrum below maximum brightness as a function of time. This figure is composed of a large number of Type-I supernovae. Reproduced by permission of the AAS from Doggett, J. B. and Branch, D., *The Astronomical Journal*, 90, 2303 (1985).

Type-Ib and -Ic supernovae, see Table 6.10). Contrary to the other types of supernovae, during the explosion, the white dwarf is completely destroyed leaving no central remnant. The heavy elements produced during the thermonuclear reactions in SN Ia are ejected to interstellar space and this contributes to the enrichment of the metallicity of interstellar matter.

In addition to the spectral types of supernovae, it is also very instructive to study their light curves. Figure 6.33 shows the light curve for SN I. After the initial increase of luminosity, two distinctive downward slopes are observed and can be explained by the decay of the elements produced during the explosive nucleosynthesis that takes place in the shock wave that disrupts the star. Several radioactive elements are formed during the supernova and their decay determines the shape of the light curves of supernovae. The most important one is ^{56}Ni. This element can decay via two modes. It can decay via the emission of a positron or a so-called β^+ decay (with a half-life of 6.1 d)

$$^{56}\text{Ni} \rightarrow {}^{56}\text{Co} + e^+ + \nu_e + \gamma \tag{6.83}$$

^{56}Ni can also be destroyed by electron capture

$$^{56}\text{Ni} + e^- \rightarrow {}^{56}\text{Co} + \nu_e \tag{6.84}$$

Detailed calculations show that during supernovae the decay of ^{56}Ni is completely dominated by electron capture. The emission of energy by this process explains the first downward slope of the light curve of SN I.

As ^{56}Ni is progressively transformed into ^{56}Co, this nucleus, which is also unstable, can in turn decay via the following reaction

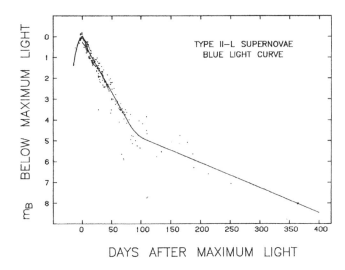

Figure 6.34 The light curve of SN II-L showing the magnitude in the blue part of the spectrum below the maximum brightness as a function of time. This figure is composed of a large number of Type-II-L supernovae. Reproduced by permission of the AAS from Doggett, J. B. and Branch, D., *The Astronomical Journal*, 90, 2303 (1985).

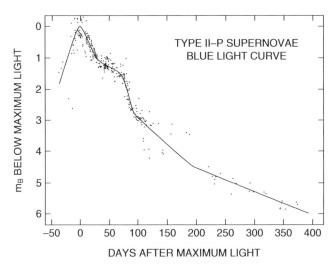

Figure 6.35 The light curve of SN II-P showing the magnitude in the blue part of the spectrum below maximum brightness as a function of time. This figure is composed of a large number of Type-II-P supernovae. Reproduced by permission of the AAS from Doggett, J. B. and Branch, D., *The Astronomical Journal*, 90, 2303 (1985).

$$^{56}\text{Co} \rightarrow {}^{56}\text{Fe} + e^+ + \nu_e + \gamma \tag{6.85}$$

^{56}Co has a half-life of 77.7 d. Again, calculations show that electron capture dominates the decay of ^{56}Co in supernovae (^{56}Co + e^- → ^{56}Fe + ν_e). The decay of ^{56}Co explains the presence of the second (and less steep) of the two slopes observed on the light curves of SN I (see Figure 6.33).

The light curves of SN II come in two distinctive types. Type II-L (L standing for linear) shows a similar time dependence to those of SN I (see Figure 6.34). Meanwhile, a second type of light curve for SN II is found (see Figure 6.35) where a sort of plateau is observed on the ^{56}Ni slope. This type of supernova is called SN II-P (P standing for plateau). The plateau is due to the nontransparency of the envelope to outgoing radiation. The mass loss in Type-Ib and -Ic supernovae makes it easier for radiation to exit the system and no plateau exists for their light curves. A certain amount of mass loss for SN II-L is assumed to have taken place and to have sufficiently diminished the opacity of its outer layers. For SN II-P, the change of slope (from ^{56}Ni to ^{56}Co decay) occurs about 200 d after the explosion, while it occurs in less than 100 d for SN II-L. Finally, Type-Ia supernova light curves are similar to those of Type-II-L because the medium of the outer portions of the exploding white dwarf is relatively transparent to radiation. No plateau therefore exists in light curves of SN Ia.

Supernovae are rare occurrences. It is estimated that in our galaxy, about three SN I and two SN II per century take place. Only a small number of supernovae have been visible to the naked eye in recorded history. One of the best-known is the Crab nebula Type-II supernova shown in Figure 6.17. Another example of a supernova discovered in early times is the Type-I supernova detected by the famous Danish astronomer Tycho Brahe (1546–1601) in 1572.

As for other astronomical objects classifications, the classification of supernovae is not an exact science. Some supernovae with peculiar properties cannot be explained by the classification system described above. More theoretical and observational investigation is therefore needed to completely grasp the behaviour of supernova and the related physics.

6.12   Heavy Element Nucleosynthesis: s, r and p Processes[†]

6.12.1   The Slow and Rapid Processes

As discussed earlier, the elements up to the iron peak can be produced in stars by nuclear fusion. Another physical process is therefore needed to explain the existence of the elements heavier than those of the iron peak. Most of the isotopes of these heavier elements can be explained by the capture of neutrons by nuclei followed by the emission of an electron via the reaction n → p + e^- + $\bar{\nu}_e$. This decay (sometimes called a β^- decay) increases the number of protons in the nucleus and consequently produces a new element. It should be noted that these processes are efficient since neutrons can generally interact more easily with nuclei compared to charged particles (i.e. other nuclei) because they do not need to overcome the Coulomb repulsion of the nucleus.

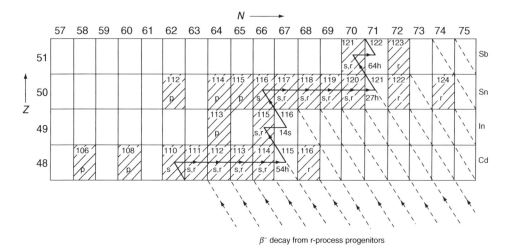

Figure 6.36 Synthesis of the elements Cd through Sb. The stable isotopes are hatched. The solid line shows the path of the *s* process. Figure reproduced and adapted with permission from Pearson, J.M., *Nuclear Physics: Energy and Matter*, Adam Hilger, Bristol (1986).

The formation of the heavier elements via neutron capture can occur by two processes: the s and the r processes. The s (standing for slow) process is present when there is a weak flux of neutrons, while the r (standing for rapid) process is due to a strong flux of neutrons. The s process will be introduced first.

During the various evolutionary or nuclear-burning stages of stars seen in Section 6.7, it was shown that during certain nuclear reactions such as

$$^{12}C + {}^{12}C \rightarrow {}^{23}Mg + n \tag{6.86}$$

$$^{16}O + {}^{16}O \rightarrow {}^{31}S + n \tag{6.87}$$

free neutrons are emitted in the stellar plasma. Several other reactions not seen here can also emit neutrons. The flux of neutrons created by such reactions is relatively weak and can create what are commonly called the s isotopes. Figure 6.36 describes this process for several different elements. To illustrate the s process, the production of the heavy elements starting from the stable ^{114}Cd nucleus will be discussed. The first step in this illustration is the absorption of a neutron by a ^{114}Cd nucleus (see Figure 6.36)

$$^{114}Cd + n \rightarrow {}^{115}Cd + \gamma \tag{6.88}$$

This creates a heavier isotope of cadmium, namely ^{115}Cd. However, ^{115}Cd is an unstable isotope with a half-life of approximately 54 h. Therefore, if there is only a weak neutron flux, this isotope will β⁻ decay before being able to absorb a second neutron. This process will create a ^{115}In nucleus via

$$^{115}\text{Cd} \rightarrow \ ^{115}\text{In} + \text{e}^- + \bar{\nu}_e \tag{6.89}$$

A portion of the ^{115}In nuclei created by this process can also react with the weak neutron flux and create ^{116}In nuclei via the reaction

$$^{115}\text{In} + \text{n} \rightarrow \ ^{116}\text{In} + \gamma \tag{6.90}$$

Again, since ^{116}In is unstable (with a half-life of approximately 14 s) it will decay to give ^{116}Sn nuclei in the following manner

$$^{116}\text{In} \rightarrow \ ^{116}\text{Sn} + \text{e}^- + \bar{\nu}_e \tag{6.91}$$

before it is able to capture an additional neutron. The stable isotopes ^{117}Sn, ^{118}Sn, ^{119}Sn and ^{120}Sn can then be created by successive neutron captures. The s process can then continue to create a large number of heavier elements (and their isotopes) as illustrated in Figure 6.36.

Some neutron-rich isotopes shown in Figure 6.36 such as ^{116}Cd, ^{122}Sn, ^{124}Sn and ^{123}Sb cannot be produced by the s process. For these isotopes to be created, a large flux of neutrons is needed. When a large flux of neutrons is present, the nuclei can absorb several neutrons rapidly (i.e. the r process) before decaying. As more neutrons accumulate in the nucleus, its half-life generally decreases. Eventually, the neutron-rich nucleus will β^- decay when the half-life of the isotope is smaller than the time needed for an additional neutron capture (which itself depends on the intensity of the neutron flux). For example, if ^{115}In nuclei absorb seven neutrons on timescales smaller than the half-lifetime of the various isotopes along the way, and if for the ^{122}In has a half-life relatively small as compared to the average time for another neutron capture to occur, it will then decay to give ^{122}Sn, which is by definition an r isotope.

Some isotopes, such as ^{119}Sn for example, can be created by both the s and r processes. The path by which it is created by the s process was discussed above. It can also be created by the r process when, for example, the flux of neutrons is such that a ^{115}In nucleus absorbs four neutrons before suffering a β^- decay, thus creating a ^{119}Sn nucleus.

Large fluxes of neutrons can occur at the more advanced stages of stellar evolution. They can occur when the central temperatures attain large values (i.e. $T > 10^9 \text{ K}$) so that the high-energy photons present in the radiation field can interact with the nuclei and eject free neutrons via the reaction

$$^{m}\text{El}(Z) + \gamma \rightarrow \ ^{m-1}\text{El}(Z) + \text{n} \tag{6.92}$$

The symbol El represents the element (containing Z protons) with which the high-energy photons interact.

A second way by which large neutron fluxes can be created in stars is during the neutronization phase of evolution of massive stars. As seen earlier, if the star is massive enough it will attain a state in which the protons and electrons in its centre fuse to give a large number of free neutrons

$$\text{p} + \text{e}^- \rightarrow \text{n} + \nu_e \tag{6.93}$$

The large neutron flux created by this process can then cause the r process discussed above.

Finally, it is important to point out that the s process reaches an impasse at ^{209}Bi. The reason being that when ^{209}Bi absorbs a neutron from a weak neutron flux it enters the following cycle of reactions

$$
\begin{aligned}
&^{209}\text{Bi} + \text{n} \rightarrow \, ^{210}\text{Bi} + \gamma \\
&^{210}\text{Bi} \rightarrow \, ^{210}\text{Po} + \text{e}^- + \bar{\nu}_e \\
&^{210}\text{Po} \rightarrow \, ^{206}\text{Pb} + \, ^4\text{He} \\
&^{206}\text{Pb} + \text{n} \rightarrow \, ^{207}\text{Pb} + \gamma \\
&^{207}\text{Pb} + \text{n} \rightarrow \, ^{208}\text{Pb} + \gamma \\
&^{208}\text{Pb} + \text{n} \rightarrow \, ^{209}\text{Pb} + \gamma \\
&^{209}\text{Pb} \rightarrow \, ^{209}\text{Bi} + \text{e}^- + \bar{\nu}_e
\end{aligned}
\tag{6.94}
$$

To overcome this gridlock at ^{209}Bi and to produce heavier isotopes, the r process is required. A large flux of neutrons produced at the most advanced phases of evolution is therefore critical for the presence of the heaviest elements of the periodic table.

6.12.2   The p Process

As seen in Figure 6.36, some proton-rich isotopes such as ^{113}In and ^{112}Sn cannot be explained by the s or r processes. There are 33 such naturally occurring isotopes and they have an abundance about two orders of magnitude lower than the s and r isotopes of the same element. Another mechanism must then be invoked to explain these isotopes. This mechanism is called the p process, p standing for protons because these isotopes are proton rich.

There exists several different reactions that can create p isotopes (i.e. isotopes that are caused by the p process). Therefore, the p process is in reality several different types of interactions that give proton-rich isotopes. First, when the temperature is large (in the central regions of stars at the advanced phases of evolution for instance), energetic photons can eject neutrons from nuclei via the reaction

$$
^m\text{El}(Z) + \gamma \rightarrow \, ^{m-1}\text{El}(Z) + \text{n}
\tag{6.95}
$$

where Z represents the number of protons in the nucleus of some element El. This reaction decreases the relative number of neutrons and as a consequence increases the relative number of protons in the nucleus. It should be noted that the s and r elements serve as seeds for the formation of the p isotopes. Also, since high temperatures are needed for such reactions, it is generally believed that the p process occurs during the explosive stage of massive stars. Since such evolutionary phases are short lived, it assures that photodisintegration of the seeds (i.e. s and r process elements) is not too important.

The reaction above can, for example, transform ^{116}Sn into ^{115}Sn

$$
^{116}\text{Sn} + \gamma \rightarrow \, ^{115}\text{Sn} + \text{n}
\tag{6.96}
$$

Nuclei can also absorb free protons found in the stellar plasma

$$
^m\text{El}(Z) + \text{p} \rightarrow \, ^{m+1}\text{El}(Z+1) + \gamma
\tag{6.97}
$$

This reaction also increases the relative number of protons in the nucleus and produces a new element with an extra proton as compared to the original nucleus $^m\mathrm{El}(Z)$.

Another process that is less probable is the fusion of a proton with a nucleus that then emits a neutron via the reaction

$$^m\mathrm{El}(Z) + \mathrm{p} \rightarrow\ ^m\mathrm{El}(Z+1) + \mathrm{n} \tag{6.98}$$

Finally, the number of protons in nuclei can be increased by positron capture. At the high temperatures reached during the later stages of stellar evolution, electron–positron pairs can be created from energetic photons. The positrons can then be captured by the following reaction

$$^m\mathrm{El}(Z) + \mathrm{e}^+ \rightarrow\ ^m\mathrm{El}(Z+1) + \bar{\nu}_\mathrm{e} \tag{6.99}$$

This reaction transforms a neutron into a proton and can thus create proton-rich nuclei.

6.13   Nuclear Reaction Cross Sections and Rates[††]

In this section, the physics surrounding nuclear reaction rates will be discussed. To better understand the physical dependence of the nuclear reaction rates, an ideal situation illustrated in Figure 6.37 will first be considered. In this figure, the nuclear reaction between two types of nuclei A and B is studied. Nuclei A are travelling with velocity V with respect to nuclei B that are considered stationary. The reaction rate per unit volume r_{AB} is proportional to the flux (F_A) of particles A (i.e. the number of particles A crossing a surface perpendicular to the velocity per unit area, per unit time), to the number density of particles B (n_B) and the cross section $(\sigma(V))$ of the nuclear reaction in question. As discussed in Section 6.4, the cross section depends on the velocity of the incoming particles. The reaction rate per unit volume is therefore

$$r_{AB} = \sigma(V)F_A n_B = \sigma(V)V n_A n_B \tag{6.100}$$

where the flux of particles A is $V n_A$. The numerical value of $n_A n_B$ is equal to the number of interacting pairs (i.e. pairs of A–B particles) per unit volume. Therefore, when the

Figure 6.37   Nuclei A incoming with velocity V upon stationary nuclei B.

incoming particles are identical to the stationary ones, the number of pairs is $n_A^2/2$ and not n_A^2. The nuclear reaction rate may then more generally be written

$$r_{AB} = \frac{\sigma(V)Vn_A n_B}{1+\delta_{AB}} \tag{6.101}$$

where δ_{AB} is Kronecker's delta function (i.e. $\delta_{AB} = 0$ when $A \neq B$ and $\delta_{AB} = 1$ when $A = B$).

This equation gives the rate of reaction for a single value of velocity. In stellar plasma, an integration taking into account the velocity distribution of the nuclei must be undertaken because both particles A and B possess a Maxwellian velocity distribution (assuming that the equation of state is that of an ideal gas). It can be shown that the relative velocity between two types of particles having a Maxwellian distribution, is itself Maxwellian in nature. This Maxwell distribution is represented by the function $\Phi(V)$ where V is the relative velocity of the two interacting particles. The rate of reactions per unit volume may then be written

$$r_{AB} = \frac{\lambda_{AB} n_A n_B}{1+\delta_{AB}} \tag{6.102}$$

where

$$\lambda_{AB} = \int_0^\infty V\sigma(V)\Phi(V)\,dV \tag{6.103}$$

The variable λ_{AB} is used in order to separate the terms related to the nuclear physics of the reaction from those depending on the number densities of the reactants in r_{AB}. This physical quantity, which is called the reaction rate per pair of particles, will be useful when writing nuclear reaction rate equations (see below). The reduced mass of the reacting nuclei intervenes in the distribution $\Phi(V)$. This distribution also depends on the total kinetic energy (E) of these nuclei in the centre-of-mass frame

$$E = \frac{1}{2}\frac{m_A m_B}{m_A + m_B}V^2 = \frac{1}{2}\mu V^2 \tag{6.104}$$

where μ is the reduced mass. The Maxwellian distribution $\Phi(V)$ is given by the following expression

$$\Phi(V) = \sqrt{\frac{2}{\pi}}\left(\frac{\mu}{kT}\right)^{3/2} V^2 e^{-\frac{E}{kT}} \tag{6.105}$$

and it can be shown that λ_{AB} may be written in the form

$$\lambda_{AB} = \left(\frac{8}{\pi\mu(kT)^3}\right)^{1/2} \int_0^\infty E\sigma(E)e^{-\frac{E}{kT}}\,dE \tag{6.106}$$

Nuclear physics shows that nuclear cross sections have the following energy dependence

$$\sigma(E) = \frac{S(E)}{E} e^{-\frac{b}{\sqrt{E}}} \tag{6.107}$$

where

$$b = 2^{3/2}\pi^2 \frac{\sqrt{\mu}Z_A Z_B}{h} \tag{6.108}$$

Here, Z_A and Z_B are the charges of the nuclei under consideration. Their appearance in the cross section formula given above is related to the presence of the barrier potential penetration of the Coulomb potential. Therefore, for a given reaction, b is a constant that depends only on the physical properties of the reacting nuclei. For nonresonant reactions, $S(E)$ is a slowly varying function of energy that will be considered constant here. This function is normally obtained experimentally. As mentioned previously, since the nuclear cross sections at the energies found in stellar cores are very small, the value of $S(E)$ at these low energies cannot be obtained in particle accelerator experiments. However, the value of $S(E)$ may be approximated by extrapolating the experimental data to lower energies.

Figure 6.38 shows the dependence of the cross section for the proton–proton reaction in the centre of the Sun. The term $e^{-\frac{E}{kT}}$ represents the high-energy tail of the Maxwell distribution. Since particles with higher energies more easily penetrate the barrier, the number of particles at high energies is critical for fusion. The term $e^{-\frac{b}{\sqrt{E}}}$ represents the

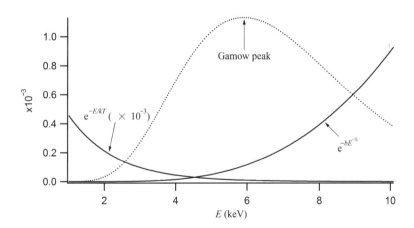

Figure 6.38 The dependence of the nuclear cross section of the proton–proton reaction in the centre of the Sun. The curve $e^{-\frac{E}{kT}}$ represents the high energy tail of the Maxwell distribution while the curve $e^{-\frac{b}{\sqrt{E}}}$ measures the probability of quantum tunnelling. The dotted curve (commonly called the Gamow peak) is the product of these two terms (it is multiplied by 1000 in this figure for visual effect).

probability of barrier penetration as a function of energy. The product of these two functions gives a function commonly called the Gamow peak, named after the Ukrainian-born physicist George Gamow (1904–1968). This scientist was the first to propose that nuclear fusion occurs due to quantum tunnelling (a concept he had introduced in his theory of alpha-decay). The energy related to this peak is commonly called the Gamow energy. In the Sun's centre, the peak for the proton–proton reaction rate is at approximately 6 keV. The reaction rate is therefore determined by a combination of the need for a sufficient number of high-energy nuclei and the need for a relatively large probability of barrier penetration.

Assuming the knowledge of the reactions rates, rate equations (see Eq. 6.111 below) may be written that can give the abundances of the isotopes that intervene in fusion reactions. To illustrate this, deuterium (^{2}H) will be discussed. On the main sequence, deuterium is created via the reaction

$$^1H + {}^1H \rightarrow {}^2H + e^+ + \nu_e \qquad (6.109)$$

while it is destroyed when it is fused with hydrogen

$$^1H + {}^2H \rightarrow {}^3He + \gamma \qquad (6.110)$$

By defining r_{pp} and r_{pd} (where the indices p and d, respectively, represent protons and deuterium) as the respective reaction rates per unit volume of these two nuclear reactions and assuming that deuterium does not react by means other than the two reactions above, the rate of change of the deuterium number density is

$$\frac{dn_d}{dt} = r_{pp} - r_{pd} = \frac{n_p}{2}\lambda_{pp} - n_p n_d \lambda_{pd} \qquad (6.111)$$

When the number density of deuterium nuclei (n_d) is large, r_{pd} is the dominating term and n_d decreases with time. On the other hand, when n_d is small the proton–proton reaction rate becomes larger than the proton–deuterium rate and n_d therefore increases. This equation is thus self-regulating and eventually attains a steady state where

$$\frac{dn_d}{dt} = 0 \qquad (6.112)$$

which leads to

$$\frac{n_d}{n_p} = \frac{\lambda_{pp}}{2\lambda_{pd}} \qquad (6.113)$$

This gives the number of deuterium atoms relative to protons (or hydrogen). Nuclear physics can show that in the centre of stars $\lambda_{pd} \gg \lambda_{pp}$ and consequently deuterium is destroyed there. For example, for the conditions found in the Sun's centre the relative steady-state abundance of deuterium to that of hydrogen is

$$\frac{n_d}{n_p} \approx 10^{-18} \qquad (6.114)$$

From measurements, this ratio is on the order of 10^{-4} in the water found on Earth. Therefore, the abundance of deuterium created at the beginning of the Universe is much larger than the quantity found in stars because of the efficiency of its destruction during the proton–proton chains.

The nuclear energy production rate per unit mass ε that intervenes in the calculation of stellar structure (as seen in Chapter 5) depends on the rate of the nuclear reactions that take place in stars. For example, for the PPI chain, the nuclear energy production rate per unit mass (in units of MeV per gram per second) is given by the following expression

$$\varepsilon_{\text{PPI}} = r_{\text{pp}} \times 1.179\,\text{MeV} + r_{\text{pd}} \times 5.493\,\text{MeV} + r_{33} \times 12.860\,\text{MeV} \tag{6.115}$$

where r_{33} is the rate of the reaction $^3\text{He} + {}^3\text{He} \rightarrow {}^4\text{He} + 2{}^1\text{H}$ and the various numerical values in this equation are equal to the energies emitted by the three corresponding reactions making up the PPI chain. For the first of these reactions ($^1\text{H} + {}^1\text{H} \rightarrow {}^2\text{H} + e^+ + \nu_e$), the average energy taken away by the neutrino was subtracted and the energy due to the positron annihilation was included. Therefore, by calculating the cross sections of the three reactions involved, one may obtain the energy production rate due to the PPI chain. An approximate equation for this quantity was given in Eq. (6.25).

6.14   Summary

Einstein's mass–energy equation: $E = \Delta m c^2 = (m_{\text{initial}} - m_{\text{final}})c^2$ (6.116)

Hydrogen burning: $4{}^1\text{H} \rightarrow {}^4\text{He} + 2e^+ + 2\nu_e + n\gamma$ (where n depends on the chain reaction)

(6.117)

Helium burning (triple-α): $3{}^4\text{He} \rightarrow {}^{12}\text{C} + 3\gamma$ (6.118)

Lifetime of stars on main sequence: $t_{\text{ms}} \approx 10^{10}(M_*/M_\odot)(L_*/L_\odot)^{-1}\,\text{yr}$ (6.119)

Three end states for stars: white dwarf, neutron star and black hole

$$M_{\text{white dwarf}} \leq 1.43\,M_\odot \text{ (Chandrasekhar's limit)} \tag{6.120}$$

$$1.43\,M_\odot < M_{\text{neutron star}} \leq 3\,M_\odot$$
$$\text{(Tolman–Oppenheimer–Volkoff limit)} \tag{6.121}$$

$$M_{\text{black hole}} > 3\,M_\odot \tag{6.122}$$

Distance to near stars (parallax method): $d = \dfrac{1}{\alpha}$ (6.123)

Schwarzschild radius: $R_{\text{Sch}} = \dfrac{2GM}{c^2}$ (6.124)

6.15   Exercises

6.1   Assuming that 10 eV of energy per atom found in the Sun is emitted during some chemical reaction taking place there, calculate the time the Sun could shine at its present intensity if the only energy source was this chemical process. Assume that the Sun is composed of pure hydrogen. Is it then possible that the energy source of the Sun is chemical in nature? Why or why not?

6.2   Complete the following nuclear reactions

$$^{23}\mathrm{Na} + {}^1\mathrm{H} \to {}^{20}\mathrm{Ne} + \underline{\quad} \tag{6.125}$$

$$^{17}\mathrm{F} \to {}^{17}\mathrm{O} + e^+ + \underline{\quad} \tag{6.126}$$

$$^{22}\mathrm{Ne} + {}^4\mathrm{He} \to {}^{26}\mathrm{Mg} + \underline{\quad} \tag{6.127}$$

$$^{21}\mathrm{Ne} + {}^4\mathrm{He} \to {}^{24}\mathrm{Mg} + \underline{\quad} \tag{6.128}$$

6.3   Verify that the global reaction for the PPII and PPIII chain is similar than for the PPI chain.

6.4   Calculate the energy emitted by the triple-α reaction.

6.5   Calculate the energy emitted by the reaction $^1\mathrm{H} + {}^2\mathrm{H} \to {}^3\mathrm{He} + \gamma$.

6.6   Calculate the energy emitted by the reaction $^{12}\mathrm{C} + {}^4\mathrm{He} \to {}^{16}\mathrm{O} + \gamma$.

6.7   Estimate the time the Sun will spend on the horizontal branch supposing that helium burns via the triple-α reaction (note that at this phase of evolution the luminosity of the Sun will be approximately equal to $100\,L_\odot$).

6.8   By making the appropriate unit transformations, show that when the parallax angle of a star is given in units of arcsec, its distance in units of pc is given by the following equation

$$d = \frac{1}{\alpha} \tag{6.129}$$

6.9   You find yourself on a distant planet circularly orbiting a star. You observe a given star and determine its distance through spectroscopic means to be 4 pc. You measure the position of the star in question twice, the second measurement delayed by an interval of time equal to half of the orbital period of the planet. For your second measurement, you measure that you must move the telescope by 6 arcsec to keep the star in the centre of the field of view of the telescope. Calculate the radius of the orbit of the planet in AU.

6.10   A planetary nebula is found at a distance of 140 pc from Earth has an angular diameter of 20 arcmin (see Figure 6.39). What is the geometrical diameter of this nebula? If the measured expansion velocity of the nebula is 22 km/s and assuming that it is spherical, by making proper approximations, estimate its age (i.e. the time since it started expanding).

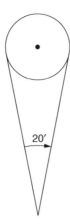

Figure 6.39 Illustration showing the angular dimension of the planetary nebula discussed in Exercise 6.10. The white dwarf is located at the centre of the expanding shell of gas.

6.11   Which type of stellar cluster is bluer? Explain why?

6.12   Find the equation showing the dependence of the average density of black holes with respect to their mass.

6.13   Assuming that Hawking emission obeys the Stefan–Boltzmann law for blackbodies, find an expression for the lifespan of black holes as a function of their mass. Calculate the lifespan for the black holes with masses of m_p, 1000 kg, $M_\oplus$ and $5 M_\odot$.

6.14   A spectral line coming from a quasar is observed to be redshifted due to the super-massive black hole found at its centre. The redshift is such that the observed wavelength is 0.6 % longer than the theoretical value. Assuming that the mass of the supermassive black hole is $10^9 M_\odot$, at what distance (in ly) from the centre of this galaxy is the source of the photons forming the line?

6.15[†]   An astronomer claims to have discovered a Type-II supernova in a globular cluster. Is this possible? Why or why not?

6.16[††]   Assuming that ^{3}He is solely created and destroyed by the following reactions of the PPI chain

$$^1H + {}^2H \rightarrow {}^3He + \gamma \tag{6.130}$$

$$^3He + {}^3He \rightarrow {}^4He + 2{}^1H \tag{6.131}$$

calculate the ratio of the ^{3}He and ^{4}He abundances in terms of the reaction rate per pair of particles (or λ) of these two reactions.

7

Chemically Peculiar Stars and Diffusion[†]

7.1 Introduction and Historical Background

At the end of the ninetieth and at the beginning of the twentieth centuries, astronomers, such as the American researchers Annie Jump Cannon (1863–1941) and Antonia Maury (1866–1952), noticed that the atomic lines in the spectra of certain stars were abnormally strong or weak as compared to typical (or normal) stars. It is now known that these spectral peculiarities are due to the abundances of the elements that can sometimes be very large or very small at the surface of certain stars. These stars are commonly called chemically peculiar stars. Since the radiation field received from stars is formed in the outer layers of stars, the overabundances or underabundances observed are only symptomatic of the surface composition and are not indicative of the average abundances for the star as a whole.

Several theories have been put forth to explain chemically peculiar stars. One of the possibilities to explain the chemically peculiar star phenomenon was the existence of an abnormal atmospheric structure instead of invoking abundance anomalies. It was thought that a modification (of unknown origin) of the physical structure of the atmosphere could cause the presence of abnormal line strengths in the spectra of such stars. However, this potential solution was discarded since it could not adequately explain the spectra of chemically peculiar stars.

A more logical solution to explain chemically peculiar stars is that they possess abnormal elemental abundances. Several processes have been proposed to explain the presence of peculiar abundances required to explain the spectra of these stars. It was first thought that nuclear physics was the source of these abundance anomalies. A theory by which the abundance anomalies were formed by the capture of free neutrons by nuclei was proposed. The nuclei would then decay to give other elements via the decay reaction

An Introduction to Stellar Astrophysics Francis LeBlanc
© 2010 John Wiley & Sons, Ltd

$$n \rightarrow p + e^- + \bar{v}_e \tag{7.1}$$

Such decay reactions were discussed in (optional) Section 6.12.

Another nuclear theory was also proposed to explain chemically peculiar stars. In this theory, α particles would react with nuclei at the surface of stars, thereby changing the composition of the elements there. In this model, the high velocities of the α particles needed to interact with nuclei (i.e. to be able to penetrate the repulsive Coulomb potential) are obtained by the acceleration of the α particles by very large magnetic fields. However, the magnetic fields needed for such nuclear reactions are much larger than those observed at the surface of magnetic stars. Also, since not all chemically peculiar stars possess magnetic fields, this process cannot be the source of the abundance anomalies observed for nonmagnetic stars.

Unfortunately, the nuclear theories proposed above are unable to explain the variety of chemically peculiar stars and the dependence of the observed surface abundances on their effective temperature. Such nuclear theories therefore fail to properly explain the well-established observational anomalies.

Another theory attempting to explain the abundance anomalies by an accretion of enriched matter at the surface of the stars under consideration was also not successful in explaining chemically peculiar stars.

In 1970, the Canadian astrophysicist Georges Michaud (b. 1940) proposed that the abundance anomalies observed in chemically peculiar stars are the result of atomic processes rather than nuclear ones (see Michaud, G. *The Astrophysical Journal*, 160, 641 (1970) for more details). In this theory, the elements are segregated at various depths in the star according to their capability of absorbing photons from the radiation field. When an atom absorbs a photon (following a photoexcitation or photoionisation) the momentum of the photon is transferred to the atom and leads to what is called a radiative force on the atom. Since there exists a positive radiative flux in the outgoing direction in stars, this momentum transfer pushes the atoms toward the surface of the star, while gravity acts in the opposite direction. In this simple description of this theory, when the radiative force is larger than gravity the atoms diffuse towards the surface of the star, while they sink towards the centre when gravity dominates the radiative force for the species under consideration. This can cause elements to accumulate (or depreciate) at certain depths. This process is commonly called atomic diffusion.

Since different species have different absorption capabilities, the species diffuse relative to one another and some elements can become overabundant at the surface while others can become underabundant. A large number of studies have shown that diffusion theory can explain most abundance anomalies observed at the surface of chemically peculiar stars. The diffusion of the elements in stars is not only of importance for the abundances observed at their surface but it can also modify the structure and the evolution of stars. The accumulation (or depreciation) of the elements at various depths in a given star modifies the opacity spectrum, which in turn alters the transfer of radiation. This can have a non-negligible effect on the structure of the star and consequently on the position of its evolutionary track in the H–R diagram.

Atomic diffusion can also play a role in asteroseismology. For example (see Section 5.7.1), the accumulation of iron-peak elements is believed to be responsible for the pulsations of β Cephei stars. These pulsations are driven by radiation that is trapped due to the increase of the opacity caused by this accumulation.

In this chapter, a section describing the different types of chemically peculiar stars will first be presented where the main properties of these stars are reviewed. This will be followed by an advanced section where a simplified version of the diffusion theory is given. Since a major ingredient in the diffusion theory is the radiative acceleration on the atomic species, this physical entity will be discussed in (the advanced) Section 7.4. Finally, other physical processes that can also transport chemical elements in stars will be described.

This chapter has a goal of not only initiating the reader to the field of chemically peculiar stars, but also serves as an excellent application of the various concepts seen earlier in this book. Concepts such as opacity, ionisation, radiative transfer, convection, stellar evolution, etc., all come into play when studying the various types of chemically peculiar stars.

7.2   Chemically Peculiar Stars

Chemically peculiar stars come in many varieties. For example, chemically peculiar stars with different T_{eff} can have very different abundance anomalies. Also, some of these stars possess large magnetic fields while others are nonmagnetic. As will be discussed below, the presence of large magnetic fields has an important effect on the diffusion of the elements in stars. Abundances observed for magnetic stars are therefore different from their nonmagnetic counterparts (with the same T_{eff}). A classification of these astronomical objects is therefore in order.

Besides having abundance anomalies, chemically peculiar stars possess the commonality of having low rotational velocities (typically $V\sin i < 100$ km/s, see Figure 4.9 for the definition of this quantity) as compared to normal stars. Low rotational velocities assure the stability of the stellar medium that is needed for atomic diffusion to take place. For example, when large scale mixing processes such as convection are present in stars, atomic diffusion cannot be efficient there.

Historically, the term chemically peculiar star was associated to upper main-sequence stars. The American astrophysicist George W. Preston (b. 1930) divided these stars into four groups (see Preston, G.W., *Annual Review of Astronomy and Astrophysics*, 12, 257 (1974) for more details). These groups are described below. However, some stars not found in these groups and lying outside the main sequence also possess abundance anomalies. For example, some white dwarfs as well as some horizontal-branch stars have abundance anomalies thought to be due to atomic diffusion. Such stars are also sometimes called chemically peculiar. They will not be discussed here.

The four groups of chemically peculiar stars, were originally named CP1, CP2, CP3 and CP4 stars. However, more prevalent nomenclature used in scientific literature is employed here. CP1 stars are Am- (and Fm-) type stars, m standing for metallic because some metals are overabundant. CP2 stars are Ap (and Bp) stars, p standing for peculiar. These stars have large magnetic fields. CP3 stars are HgMn stars that are stars that are characterised by large overabundances of Hg and Mn. Finally, the CP4 group contains what are called the He-weak stars that are stars with a low He abundance at their surface. Here, we took the liberty to add He-strong and ^{3}He stars in a group called He-abnormal stars. Each of these four groups of stars will now be described.

7.2.1   Am Stars

Am (and Fm) stars are nonmagnetic main-sequence stars in the effective temperature range $7000\,\text{K} < T_{\text{eff}} < 10\,000\,\text{K}$. Their main characteristic is the underabundance of the elements Ca and Sc in their atmosphere. They possess low rotational velocity believed to be due to tidal forces from their companion since most of these stars are found in binary groups.

The outer region of these stars is convective and the abundance anomalies observed at the surface emanate from deeper regions. The first scenario that was thought to be responsible for the observed anomalies is that they are formed by atomic diffusion at the bottom of a surface convective zone due to hydrogen ionisation. Contrarily to normal A-type stars, no helium convection zone exists in Am stars because helium settles to deeper regions of these stars. Helium settling in Am stars is due to the fact that it is not sufficiently supported by radiation as compared to gravity (see Sections 7.3 and 7.4). The elements that accumulate at the bottom of the hydrogen convection zone due to atomic diffusion are then dredged-up and brought to the surface by the convective process. At the bottom of this mixing zone, Ca is not strongly supported by radiation since it is in the Ar-like configuration (i.e. it is twice ionised). As will be seen in Section 7.4, the radiative force of species in noble-gas configurations is relatively weak and they diffuse toward the centre of the star due to gravity. This explains the low abundance of calcium observed at the surface of Am stars.

Recent modelling of the evolution of stars while including atomic diffusion shows that iron accumulation at certain depths of AmFm stars can lead to a convection zone. Convection is triggered there due to the increase of the opacity caused by the overabundance of iron. The depth of the surface convective zone is therefore larger than in the scenario initially proposed to explain the AmFm stars. Because of the high temperature at that depth, Ca is more highly ionised and found in the Ne-like configuration (as opposed to the Ar-like configuration for the scenario described in the previous paragraph). This fact can explain the underabundance of this element since the Ne-like (noble gas) configuration has a relatively small radiative force (see Figure 7.3 for an example of such an occurrence).

7.2.2   Ap Stars

Ap (and Bp) stars are strongly magnetic main-sequence stars in the effective temperature range $7000\,\text{K} < T_{\text{eff}} < 15\,000\,\text{K}$. They possess magnetic fields reaching values up to approximately 10^4 G at their surface. The iron-peak elements are strongly overabundant in the atmosphere of these stars (i.e. up to 100 times the solar abundance) while the rare-earth[1] elements have overabundances that reach 10^5 times their solar values. Helium is underabundant in their atmosphere. The magnetic fields at the surface of these stars have complex configuration but are often approximated by a dipolar field that is inclined with respect to the axis of rotation. This is commonly called the oblique rotator model.

The presence of a strong magnetic field possibly inhibits the convective movements in the atmosphere of these stars. Assuming this is true, as opposed to Am stars where the

---

[1] Rare-earth elements are the elements from lanthanum ($Z = 57$) to lutetium ($Z = 71$).

abundance anomalies emanate from deep inside the star, the atomic diffusion process can be at work in atmospheres of Ap stars. The presence of a magnetic field also modifies the diffusion of the species since the velocity of charged ions is modified via the Lorentz force. The Zeeman effect can also come into play (see Section 7.4). Complex magnetic-field configurations detected at the surface of Ap stars can also lead to patches at their surface where certain elements can be over- or underabundant.

The efficiency of the diffusion process in the atmosphere is also thought to lead to vertical abundance stratification there. A growing amount of observational evidence confirms this tendency. Such abundance stratifications (both horizontal and vertical) modify the atmospheric structure and must be taken into account in detailed spectroscopic studies of these stars.

To conclude this section, a discussion surrounding magnetic fields in stars is in order. Stellar magnetic fields are thought to be of two origins depending on the type of star under consideration. The magnetic field of some stars, such as the Sun, seems to be generated by the currents of plasma inside them. This is commonly called the dynamo theory. Another possibility for the presence of magnetic field in stars is simply a manifestation of a galactic magnetic field found in the original nebulae where the star formed. The contraction of the protostar condenses the magnetic field lines of this galactic field that is therefore intensified by the time the star is born. This theory, which explains the presence of magnetic fields in some stars such as ApBp stars, is called the fossil field theory.

7.2.3 HgMn Stars

HgMn stars are nonmagnetic main-sequence stars in the effective temperature range 10 000 K $< T_{eff} <$ 15 000 K. These stars exhibit very large Hg and Mn overabundances reaching up to respectively, 10^6 and 10^4 their solar value. The rare-earth elements are also overabundant, while helium is generally underabundant.

This type of chemically peculiar star also shows isotopic anomalies. For example, in some HgMn stars the relative abundances of Hg and Pt isotopes are very different from their solar values. These isotopic anomalies are possibly due to a physical process called light-induced drift that will be discussed in Section 7.5.

Since HgMn are relatively hot stars, no hydrogen convection zone should exist in these stars. In this context, their atmosphere should be stable enough for atomic diffusion to take place there.

7.2.4 He-Abnormal Stars

The hottest of the four groups of chemically peculiar stars is the one that contains main-sequence stars with abnormal helium abundance. These helium abnormal stars come in several varieties.

The He-weak stars are found in the effective temperature range 14 000 K $< T_{eff} <$ 20 000 K. They show helium underabundances by factors of 2 to 15 in their atmosphere. They come in both magnetic and nonmagnetic form. The nonmagnetic versions are called P-Ga types (or sometimes phosphorous stars). They possess large overabundances of P (up to a factor of 100) and Ga (up to a factor of 10^5) as compared to their solar values.

Another subclass of the He-weak group is called ^{3}He stars. These stars possess an abnormally high ^{3}He to ^{4}He ratio.

Finally, there also exist He-rich stars where helium is largely overabundant. The He to H ratio can reach up to a factor 10 in the atmosphere of these stars. Radiative acceleration (see advanced Sections 7.3 and 7.4) is not sufficiently large to sustain such large over-abundances of helium. Other mechanisms, such as mass loss are thought to play a role there.

7.3   Atomic Diffusion Theory[††]

Atomic diffusion is the relative movement of atoms in a gas containing at least two species. The diffusion mechanism can be caused by several physical processes, some of which will be discussed here. The diffusion due to gravity, the electric field and radiative acceleration will be presented. Diffusion may occur in stars where the medium is hydrodymanically stable (i.e. in the absence of convection or turbulence for instance). It is therefore not present in all type of stars.

To more readily understand the major factors intervening in atomic diffusion, a simple depiction of the diffusion theory is presented. The first step to understanding diffusion is to study what can be thought of as natural diffusion. Natural diffusion occurs when, in the absence of other forces, a gradient of a given species exists. An example in everyday life is what occurs when a bottle of perfume is opened in the middle of a room. The molecules of the perfume that have evaporated in the atmosphere then naturally diffuse throughout the room because of the gradient of perfume molecules that initially exists surrounding the bottle.[2] The diffusion process ends if a state of equilibrium where a uniform density of perfume molecules is achieved in the room.

In the aim of quantifying the diffusion process caused by a gradient of the number density of a given species, a simple scenario will first be studied here. A gas of constant pressure and temperature composed of two species (named 1 and 2) will be considered. In the scenario studied here, the existence of number density gradients for each species along an axis defined as r is assumed (see Figure 7.1). To conserve a constant pressure in the gas, the gradients of the two species are the opposite of each other

$$\frac{dn_1}{dr} = -\frac{dn_2}{dr} \tag{7.2}$$

These gradients are shown in Figure 7.1 and induce the particles 1 to diffuse in the direction of increasing r (r increases from left to right in this figure) while the particles 2 diffuse in the opposite direction. Species 1 diffuses outward because at a given depth, more particles arrive from deeper regions (or smaller values of r) as compared to those coming from shallower regions because the number density of this species increases with depth.

---

[2]Here, it is assumed that the atmosphere in this room is completely stable since air currents or convective flows could dominate the diffusion phenomenon.

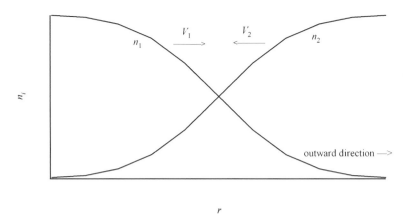

Figure 7.1   Illustration of a gas (at constant pressure and temperature) composed of two types of particles (1 and 2) with gradients along the direction r of opposite value. The variable r increases from left to right and the unitary vector $\hat{r}$ points in the outward direction. The directions of the diffusion velocities are shown in the figure.

The flux of particles 1 (F_1) in the direction $\hat{r}$ at a given depth is equal to

$$F_1 = n_1 V_1 \tag{7.3}$$

where n_1 and V_1 are, respectively, the number density and diffusion velocity of particles 1 at the depth under consideration. An expression for V_1 will now be found for a simplified model of a gas containing two species. In this model, it is assumed that the particles in the gas travel along one of the three perpendicular axis (including an axis parallel to $\hat{r}$) with a speed equal to the average thermal speed $\overline{V_1}$ in the gas. In this scenario, there are three streams each containing one third of the particles. In the stream parallel to $\hat{r}$, 1/6 of the total number of particles travels in the direction $\hat{r}$ and an equal amount travel in the opposite direction.

If the mean free path of these particles is equal to l_1, the number density of particles arriving at the depth under consideration from below and therefore going in the outgoing direction is approximately equal to[3]

$$n_1^+ = \frac{1}{6}\left(n_1 - l_1 \frac{dn_1}{dr} \right) \tag{7.4}$$

while the number density of those coming from above and travelling inward is

$$n_1^- = \frac{1}{6}\left(n_1 + l_1 \frac{dn_1}{dr} \right) \tag{7.5}$$

---

[3]Here the linear approximation $n(r + \Delta r) \approx n(r) + \dfrac{dn}{dr}\Delta r$ was used to find the number density of particles at a distance $\Delta r = \pm l_1$ from the depth under consideration. The fraction 1/6 found in Eq. (7.4) and (7.5) is related to the three stream approximation used.

The flux of particles 1 at the depth under consideration is therefore equal to sum of the fluxes of the two opposing streams of particles travelling along the $\pm\hat{r}$ axis, which leads to the following expression

$$F_1 = n_1 V_1 = \left(n_1^+ - n_1^-\right)\overline{V_1} \approx \frac{1}{6}\overline{V_1}\left(n_1 - l_1\frac{dn_1}{dr}\right) - \frac{1}{6}\overline{V_1}\left(n_1 + l_1\frac{dn_1}{dr}\right) = -\frac{1}{3}\overline{V_1}l_1\frac{dn_1}{dr} \tag{7.6}$$

The two streams perpendicular to $\hat{r}$ do not contribute to the flux in direction $\hat{r}$. These streams are of no interest since for the scenario considered no gradients, and therefore no fluxes, exist along their direction.

Similarly, the flux of the other type of particle (particles 2) in the gas is

$$F_2 = n_2 V_2 \approx -\frac{1}{3}\overline{V_2}l_2\frac{dn_2}{dr} \tag{7.7}$$

and since the gradients of the two types of particles are opposite of one another (Eq. 7.2)

$$F_2 \approx \frac{1}{3}\overline{V_2}l_2\frac{dn_1}{dr} \tag{7.8}$$

The average diffusion velocity $(\overline{V})$ of the particles in the gas can also be estimated by the following equation

$$\overline{V} = \frac{1}{n_1 + n_2}\left(n_1 V_1 + n_2 V_2\right) = -\frac{1}{n_1 + n_2}\left(\frac{1}{3}\overline{V_1}l_1 - \frac{1}{3}\overline{V_2}l_2\right)\frac{dn_1}{dr} \tag{7.9}$$

The velocity of physical interest is the diffusion velocity of the species relative to the bulk of the medium (or relative to the average velocity of the particles in the gas). The flux of particles 1 $(n_1 V_{1,\text{diff}})$ relative to the average velocity, which defines the diffusion velocity $V_{1,\text{diff}}$, is given by the expression

$$n_1 V_{1,\text{diff}} = n_1 V_1 - n_1\overline{V} \approx -\frac{1}{3}\overline{V_1}l_1\frac{dn_1}{dr} - n_1\overline{V} \tag{7.10}$$

It can be shown (see Exercise 7.1) that by using the two equations above, $V_{1,\text{diff}}$ can be written

$$V_{1,\text{diff}} \approx -\frac{1}{3(n_1 + n_2)}\left(n_2\overline{V_1}l_1 + n_1\overline{V_2}l_2\right)\frac{1}{n_1}\frac{dn_1}{dr} \tag{7.11}$$

By defining a physical quantity called the diffusion coefficient D_{12}

$$D_{12} = \frac{1}{3(n_1 + n_2)}\left(n_2\overline{V_1}l_1 + n_1\overline{V_2}l_2\right) \tag{7.12}$$

the diffusion velocity of species 1 becomes

$$V_{1,\text{diff}} \approx -\frac{D_{12}}{n_1}\frac{dn_1}{dr} \tag{7.13}$$

A similar equation is found for the species 2, namely

$$V_{2,\text{diff}} \approx -\frac{D_{12}}{n_2} \frac{dn_2}{dr} \tag{7.14}$$

Since the gradients of the two types of particles are opposite to one another (Eq. 7.2), the diffusion velocity of one species is in the opposite direction to the other one. These last two equations therefore give the diffusion velocities of the two species relative to the bulk of the medium.

A case of great interest in stars is when one species is much less abundant than the other. For instance, if species 2 is much less abundant than species 1 (i.e. $n_2 \ll n_1$), which is commonly called the trace-element approximation, the diffusion coefficient then becomes

$$D_{12} \approx \frac{\overline{V_2} l_2}{3} \tag{7.15}$$

The diffusion velocity of species 2 given above is the one relative to species 1. In stars, the diffusion of the metals can have an important impact on their structure and evolution. In this case, the metals are considered trace elements (species 2) as compared to the dominating species, namely hydrogen (or species 1), and the diffusion velocity of these metals are calculated relative to hydrogen. The hydrogen gas is sometimes called the buffer gas within which the metals diffuse.

In the simplified model presented here, the diffusion coefficient depends on the mean free path, in the buffer gas, for the species under consideration. The diffusion coefficient therefore depends on the interaction of the species within the gas. A species that does not strongly interact with the buffer gas and that therefore has a large mean free path will possess a large diffusion coefficient. All else being equal, this species will diffuse faster in this buffer gas than other species that interact more strongly with it. For example, neutral ions have larger mean free paths than charged ones. Neutral ions are therefore more mobile and generally have larger diffusion velocities than charged ions. This can have an important impact in stellar atmospheres where neutral ions can exist.

As seen above, the presence of an abundance gradient causes diffusion. Diffusion can also be caused by the presence of an external force on the atoms. A trace element (species 2) found in a buffer gas is considered here and where a force that only applies to the trace element is present. When this force (F_2), which is assumed to be directed inward (see Figure 7.2), is applied to species 2 in a gas initially of constant density, pressure, temperature and homogeneous composition, it creates a gradient of its number density leading to a growing number density of particles 2 in the direction of the force. This is similar to the density gradient created by gravity via the hydrostatic equilibrium equation (see Section 2.2). In such a scenario, a gradient of n_2 will develop until (and if) a state of equilibrium is achieved.

Figure 7.2 shows an element of matter containing N_2 atoms of species 2. Since the buffer gas is not affected by the force under consideration, the partial pressure of the dominant species will be assumed to remain constant. At equilibrium the following force equation ensues for the trace element

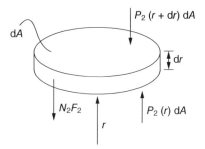

Figure 7.2 Illustration of the forces on the particles of species 2 found in a mass element with volume dAdr found at a distance r from the centre of the star. Since the pressure due to species 1 is assumed constant, only the partial pressure P_2 of species 2 is shown here. The force on the particle 2 inside the mass element is N_2F_2, where N_2 is the total number of particles 2 in the volume dAdr. It should be noted that the force F_2 is arbitrarily chosen to be directed inwards.

$$[P_2(r+dr) - P_2(r)]dA + N_2F_2 = 0 \tag{7.16}$$

where P_2 is the partial pressure due to particles of species 2 and dA is the area of the mass element shown in Figure 7.2. Here, the forces due to the pressure of the buffer gas on the top and bottom of the cylinder cancel out and need not be present in the equation given above. In the ideal-gas approximation, the partial pressure of species 2 is equal to

$$P_2 = n_2kT \tag{7.17}$$

and since

$$n_2 = \frac{N_2}{dAdr} \tag{7.18}$$

the force equation at equilibrium may then be written

$$\frac{1}{n_2}\frac{dn_2}{dr} + \frac{F_2}{kT} = 0 \tag{7.19}$$

Therefore, this equation permits the calculation of the number density of the trace element at equilibrium for a given force, assuming this force does not affect the buffer gas. Of course, this state of equilibrium cannot be attained instantaneously since the migration of particles 2 takes a certain amount of time. By analogy to the equation for the diffusion velocity found in Eq. (7.14), the diffusion velocity during the resettlement of the atoms of the trace element towards the equilibrium configuration can be approximated by the following equation

$$V_{2,\text{diff}} \approx -D_{12}\left(\frac{1}{n_2}\frac{dn_2}{dr} + \frac{F_2}{kT}\right) = -D_{12}\left(\frac{d\ln n_2}{dr} + \frac{F_2}{kT}\right) \tag{7.20}$$

If a state of equilibrium is achieved (i.e. when Eq. 7.19 is valid), the diffusion velocity becomes nil and species 2 stops migrating.

In stars, one of the forces present is evidently gravity. Since hydrogen is the most abundant element in stars, it will compose the buffer gas. Since it is the diffusion of a trace element relative to this buffer gas that is of interest, the force on species i is written relative to the gravity felt by hydrogen atoms

$$F_i = m_i g - m_p g \qquad (7.21)$$

where m_p and m_i are the respective mass of the hydrogen atom and of the atoms of the metal under consideration. This force causes the settling of the elements heavier than hydrogen in stars towards its central region. However, other physical processes can hinder this process. First, when mixing (convection for example) is present, the metals will not settle gravitationally. Also, when atoms acquire momentum following absorption of photons contained in the radiative flux of the star, this leads to a force that can selectively push some elements outward in the star. This process may be expressed as a radiative accelera-tion on the species under consideration (g_{rad}^i, see Section 7.4 for more details). Radiative acceleration depends on the capability of the species under consideration to absorb photons (i.e. on their radiative opacity). Also, since a net flux of radiation flowing in the outgoing direction exists in stars, radiative acceleration is directed outward and induces a positive diffusion velocity. The force felt by the metal under consideration relative to the protons while taking into account gravity and the radiative force therefore becomes

$$F_i = m_i \left(g - g_{rad}^i\right) - m_p g \qquad (7.22)$$

Here it is assumed that the radiative force on the buffer gas (i.e. made up mostly of protons) is negligible. The diffusion velocity of species i relative to the buffer gas may then be written

$$V_{i,\text{diff}} \approx -D_{ip}\left(\frac{d\ln n_i}{dr} + \frac{F_i}{kT}\right) = -D_{ip}\left(\frac{d\ln n_i}{dr} + \frac{m_i\left(g - g_{rad}^i\right) - m_p g}{kT}\right) \qquad (7.23)$$

where D_{ip} is the diffusion coefficient of the trace element in a buffer gas composed of protons. Since $m_i \approx A m_p$, where A is the total number of protons and neutrons found in the nucleus of the atoms of the trace element, the diffusion velocity may be expressed as

$$V_{i,\text{diff}} \approx -D_{ip}\left(\frac{d\ln n_i}{dr} + \left[(A-1)g - A g_{rad}^i\right]\frac{m_p}{kT}\right) \qquad (7.24)$$

Another force that can come into play and affect diffusion in stars is the electric force. In stellar plasma, the free electrons have a tendency to float with respect to protons due to their relatively small mass, thereby causing a slight separation of the electric charge in the plasma. This process creates an electric field E. Since the electrons have a tendency to float toward the surface of the star, the vector of this electric field points in the outgoing direc-tion. At equilibrium, hydrostatic equilibrium equations for both electrons and protons in completely ionised hydrogen stellar plasma may be written as follows (see Exercise 7.2)

$$\frac{d\ln P_e}{dr} = -\frac{m_e g}{kT} - \frac{eE}{kT} \qquad (7.25)$$

$$\frac{d \ln P_p}{dr} = -\frac{m_p g}{kT} + \frac{eE}{kT} \tag{7.26}$$

where P_e and P_p are, respectively, the partial pressure due to electrons and protons. To conserve the electric neutrality of the plasma, the gradients of both the electron and proton number density must be equal. Therefore, the gradients of the partial pressure due to the free electrons and protons must also be equal

$$\frac{d \ln P_e}{dr} = \frac{d \ln P_p}{dr} \tag{7.27}$$

With the three equations above, it is trivial to show that the electric field in completely ionised stellar plasma composed of pure hydrogen is

$$E = \left(\frac{m_p - m_e}{2e}\right) g \approx \frac{m_p g}{2e} \tag{7.28}$$

This electric field points outward and the electric force on the atoms of species i, again written relative to the protons (i.e. the buffer gas), assuming that they have a charge $q = Ze$ is the following

$$F_i = -(q - e)E = -(Z - 1)\frac{m_p g}{2} \tag{7.29}$$

A negative sign is found in the equation above because in the development leading to Eq. (7.24) the force F_2 is directed inward.

This force, when included in Eq. (7.24) along with gravity and radiative acceleration gives the diffusion velocity of a trace element found in a buffer gas composed of protons due to gravity, radiative force and the electric field

$$V_{i,\text{diff}} \approx -D_{ip}\left(\frac{d \ln n_i}{dr} + [(A-1)g - Ag_{\text{rad}}^i]\frac{m_p}{kT} - (Z-1)\frac{m_p g}{2kT}\right) \tag{7.30}$$

This expression may be simplified to the following equation

$$V_{i,\text{diff}} \approx -D_{ip}\left(\frac{d \ln n_i}{dr} + (g - g_{\text{rad}}^i)\frac{Am_p}{kT} - (Z+1)\frac{m_p g}{2kT}\right) \tag{7.31}$$

In the theoretical development elaborated above, a two-component gas composed of a trace element and another species, namely protons, was considered. In this case, the diffusion coefficient depends on the interaction between the trace element and the protons. In the atmosphere of stars, hydrogen is not completely ionised and the interaction between the trace element and neutral hydrogen must also be considered. Moreover, since a large number of species are present in stars, they can interact with one another. However, the effect of this interaction on the diffusion of a given trace element is generally small and can be neglected.

Other physical processes not yet discussed here also come into play. For example, the presence of a stellar magnetic field can affect the diffusion of the elements in stars (see discussion in Section 7.4). Another factor that is relatively rarely important in stars is thermal diffusion. Thermal diffusion is due to the presence of a temperature gradient. The atoms coming up from deeper (and hotter) regions have on average higher thermal velocities than those coming down from cooler regions. This can lead to what is called thermal diffusion. Thermal diffusion depends on the importance of the temperature gradient. These mechanisms and others not mentioned here are outside the scope of this book. For a more comprehensive description of the theory of atomic diffusion, the reader is referred to Alecian, G. and Vauclair, S., *Fundamentals of Cosmic Physics*, 8, 369 (1983) or Vauclair, S., *SAAS-FEE Advanced Course 13, Astrophysical Processes in the Upper Main Sequence*, p. 167 (1983).

Detailed calculations show that diffusion timescales are much shorter than evolutionary ones. Therefore, the elements found in stars have sufficient time to migrate within them during their evolution (providing that the stellar plasma is hydrodynamically stable). As mentioned previously, the accumulation or depreciation of certain elements at certain depths in stars during evolution has an effect on the structure of the star and therefore modifies its evolution. An example of the effect of diffusion on the evolutionary track of the Sun will be shown in the next section.

7.4   Radiative Accelerations[††]

Radiative acceleration is an important factor for the diffusion velocity (see Eq. 7.31). It therefore merits close attention since its value will greatly affect the diffusion velocity of a given species. The value of the radiative acceleration may also determine if that species is over- or underabundant at a particular depth in a star.

The radiative acceleration on a mass element in a star was evaluated in (optional) Section 3.12. This acceleration is the one felt due to the force exerted by the photons on the whole plasma. However, in reality, each species absorbs photons individually and they then each acquire momentum from the radiation field. The value of their radiative acceleration depends on their capacity to absorb photons or in other words on their radiative opacity. The radiative acceleration of a given species i (i.e. a given ion of a certain element for example) at a given depth can be approximated by the following expression

$$g_{rad}^i = \frac{4\pi}{c} \frac{1}{X_i} \int_0^\infty k_\nu^i H_\nu \mathrm{d}\nu \tag{7.32}$$

where X_i, k_ν^i and H_ν are, respectively, the mass fraction and opacity of species i and the Eddington flux at the depth under consideration.

The radiative acceleration of species i not only depends on its opacity, but also on the opacity of all other species present in the stellar plasma via the radiative flux. Consequently, to evaluate the radiative acceleration for a given species, the proper atomic data for all species present must be known (i.e. gf values, ionisation cross sections, etc.). Generally, the radiative acceleration due to bound–bound transitions is much larger than the acceleration due to bound–free transitions. Since both the opacity spectrum and the flux are dominated by relatively narrow atomic lines, the frequency grid used to numerically integrate the equation above must be relatively fine. Such calculations can require considerable computing time. Other less numerically onerous methods also exist to evaluate radiative accelerations (see the special topic below).

The flux depends on the total opacity (which of course includes the opacity of species i). Therefore, since the three quantities X_i, k_ν^i and H_ν found in Eq. (7.32) depend on the abundance of the species under consideration, this leads to a complex relation between g_{rad}^i and the abundance of species i. To understand this dependence, it is useful to examine Eq. (7.32) for stellar layers at large optical depths. As seen in Section 3.8, the flux at large optical depths may be approximated by the following expression

$$H_\nu \approx -\frac{1}{3k_\nu\rho}\frac{dB_\nu}{dT}\frac{dT}{dz} \tag{7.33}$$

where k_ν and ρ are, respectively, the local total monochromatic opacity and mass density. The total opacity may be divided into two parts, the opacity due to the species under consideration (k_ν^i) and the opacity due to all other species that is often called the background opacity (k_ν^{back})

$$k_\nu = k_\nu^i + k_\nu^{back} \tag{7.34}$$

Therefore, at large optical depths the radiative acceleration on species i, while only keeping the terms explicitly dependent on its abundance, is proportional to

$$g_{rad}^i \propto \frac{1}{X_i}\int_0^\infty \frac{k_\nu^i}{k_\nu^i + k_\nu^{back}}d\nu \tag{7.35}$$

For a very weak value for the abundance of species i, the flux may be considered independent of this abundance and since k_ν^i is proportional to X_i, the radiative acceleration becomes constant and attains its maximal value. If the abundance is increased, g_{rad}^i decreases since the photons available must be shared among a larger number of atoms of the species under consideration. At very large abundances of the species, the total opacity can become strongly dependant on this species. If the abundance of the species increases, so does the total opacity within its atomic lines. The monochromatic flux becomes smaller there and the radiative acceleration of the species decreases.

The dependence of the radiative acceleration on the abundance is nonlinear. An example is shown in Figure 7.3 for iron. Rather than the acceleration of a single ion, this figure shows the average radiative acceleration of the element (see Eq. 7.36 below). Two general tendencies can be seen in this figure. First, the radiative acceleration decreases as the abundance increases. Secondly, the radiative acceleration varies with depth mainly because different ionisation stages appear (and disappear) as temperature increases. Different ionisation stages have different opacity spectrum, thereby leading to variations in the radiative acceleration of the element as a function of depth in stars. The radiative acceleration shows a dip at depths where the element is found in a noble gas configuration because such ions have a weak opacity spectrum and therefore absorb less photons. Another factor that affects the variation of the radiative accelerations with respect to depth is that the radiation field changes.

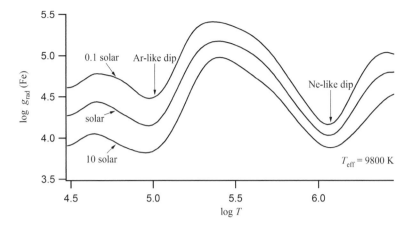

Figure 7.3 Radiative acceleration of iron (g_{rad}(Fe)) as a function of depth (shown in log T that increases with depth) inside a star with $T_{eff} = 9800$ K. The three curves shown here are for abundances for iron of 0.1, 1 and 10 times its solar value. The two dips in the radiative acceleration, due to the growing proportion of atoms found in the noble-gas configuration, are indicated in this figure.

The average radiative acceleration of an element A is a weighted sum of the accelerations of its ions. The accelerations of the ions are weighted by the product of their ionisation fraction (f_i) and their diffusion coefficient (D_i)

$$g_{rad}(A) = \frac{\sum_i f_i D_i g^i_{rad}}{\sum_i f_i D_i} = \frac{\sum_i f_i D_i g^i_{rad}}{\overline{D}} \tag{7.36}$$

where $\overline{D}$ is the average diffusion coefficient for element A and $g_{rad}(A)$ is its radiative acceleration. The radiative accelerations of the ions are weighted by $f_i D_i$ because the momentum transferred to ions due to radiation is proportional to the D_i (see Eq. (7.31) for the diffusion velocity).

A rough estimation of the quantity of a given species that can be supported by radiation can be obtained by using Eq. (7.24). By neglecting the term of the derivative of n_i, the diffusion velocity (Eq. 7.24) tends toward nil when $g^i_{rad} \approx g$. Under this approximation and assuming that the value of the abundance at which the radiative acceleration equals gravity is reached, that species will stop migrating at that point. The value for the abundance at which this happens (at a given depth) gives an estimation of the maximum amount of matter of the species under consideration that can be supported there by radiative acceleration.

The evaluation of precise radiative accelerations is more complex than that described up to now. The momentum acquired by a given ion following the absorption of a photon can be spent (the ions lose their excess of momentum following collisions with the other particles in the stellar plasma) while this ion stays in its ionisation state. But since ions can be ionised or can recombine, the momentum can in such cases be spent in a state of ionisation that has a different diffusion coefficient from that of the ion in which the photon was absorbed. This process is called the redistribution of momentum among the ions and it can strongly affect the value of the radiative acceleration.

The diffusion coefficient of charged ions is approximately proportional to Z_i^{-2}, where Z_i is the charge of the ion. The diffusion coefficient of neutral ions is approximately two orders of magnitude larger than those of once-ionised ions. Therefore, if an ion that has just absorbed a photon is quickly thereafter ionised, the radiative acceleration is then in reality spent when the atom is in a higher charged state of ionisation that is less mobile. This effectively diminishes the radiative acceleration as compared to the case where the absorbing ion would have spent the acquired momentum in its initial state of ionisation. On the other hand, a recombination leads to an amplification of the radiative acceleration. The precise determination of redistribution of the momentum among the various ions for a given element is very complex because it depends on the ionisation, recombination and collisional rates for each energy level of each ion at the stellar depth under consideration. Redistribution is most important in the outer regions of stars where neutral ions are present. This is due to the fact that neutral ions have a much higher mobility than charged ions.

Another example of the importance of the relative value of the diffusion coefficients of the various ions is the acceleration due to bound–free transitions. When an ion absorbs a photon that causes its ionisation, the momentum is spent in the newly formed ion. The radiative acceleration from photoionisation associated to the bound-free opacity of an ion with charge Z is therefore redistributed to the ion with charge $Z + 1$.

Another correction must also be brought to the acceleration due to photoionisation. The electron that is ejected during the photoionisation process possesses momentum. The effective momentum given to the newly formed ion must therefore be corrected to take into account the momentum of the ejected electron. Such corrections require complex quantum-mechanical calculations. To avoid such complications, this correction is some-times estimated by using the correction term for a simple case such as for the fundamental energy level of the hydrogen atom (which has an analytical form) for all of the species considered.

Special Topic – Various Methods for Calculating Radiative Accelerations

There are several methods that are commonly used for evaluating radiative accel-erations. The most direct, but the most numerically onerous, is the numerical integration of Eq. (7.32) on a sufficiently fine frequency grid. This method requires the calculation of the opacity spectrum for each ion along with the radiative field on the chosen frequency grid. This grid must be fine enough to properly sample the opacity spectrum for each species. This method is sometimes called the opacity sampling method. The number of frequencies of grids commonly used typically contain up to the approximately 10^5 points. Since the radiative field must be evalu-ated at each of these frequencies through the resolution of the radiative-transfer equation, this procedure requires a large amount of computing time.

In order to diminish the amount of computing time, radiative accelerations may be pretabulated on a given temperature-density grid. These quantities can then be interpolated at a given depth in a stellar model. This method is employed by The Opacity Project that furnishes the radiative accelerations of the most abundant metals for use by the scientific community (see Seaton, M.J., *Monthly Notices of the Royal Astronomical Society*, 289, 700 (1997) for more details).

A third method, called the parametric method, can also be used to quickly and easily calculate radiative accelerations. This method is based on approximate parametric formulae for radiative accelerations for both bound–bound and bound–free transitions. For example, it can be shown that the acceleration of an ion i due to bound–bound transitions, assuming that their lines have a Lorentz profile (see Section 4.3.1), may be approximated by the following equation

$$g_{rad}^i \approx g_{rad,0}^i \left(1+\frac{C_i}{C_{sat}^i}\right)^{-\frac{1}{2}} \tag{7.37}$$

where C_i is the number of ions of the species per hydrogen atom in the plasma. The parameter $g_{rad,0}^i$ represents the radiative acceleration for a very small abundance (i.e. the maximum radiative acceleration for the species). It depends on the gf values of the transitions of the ion under consideration. The parameter C_{sat}^i controls the saturation of the lines. Saturation effects depend on the width of the lines. All else being equal, wider lines are less saturated than narrower ones since they can absorb photons from a larger portion of the electromagnetic spectrum. These two parameters are calculated with appropriate weighted sums of the atomic data for the lines of the ion.

The parametric method is very numerically efficient since no numerical integration is necessary. The parameters $g_{rad,0}^i$ and C_{sat}^i found in Eq. (7.37) may be pretabulated for a number of species by using the appropriate atomic data. These parameters can then be used in astrophysical applications such as stellar-evolution codes wishing to study atomic diffusion. The radiative accelerations can then be obtained without having to access the full array of atomic data normally necessary for radiative acceleration calculations. For more information about this method, the reader is referred to Alecian, G. and LeBlanc, F, *Monthly Notices of the Royal Astronomical Society*, 332, 891 (2002).

Another important factor that influences radiative accelerations and atomic diffusion in stars is the presence of a magnetic field. Stellar magnetic fields affect atomic diffusion in two ways. First, and as discussed in Section 4.3.3, Zeeman splitting caused by the presence of a magnetic field widens the atomic lines. Wider lines can absorb photons from a larger portion of the spectrum and this diminishes the saturation of these lines. This amplifies the radiative acceleration of the species under consideration. Secondly, the diffusing ions are evidently affected by the presence of a magnetic field. Their trajectory is altered by the Lorentz force. Therefore, atomic diffusion in regions of a star where the magnetic field is horizontal to the surface differs from the diffusion where the magnetic field lines are perpendicular to it. This is believed to lead to the abundance patches observed on the surface of Ap stars.

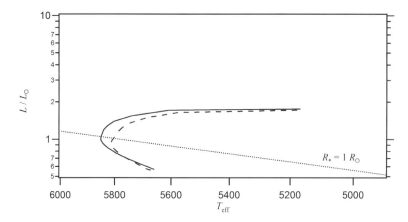

Figure 7.4 Evolutionary tracks for the Sun while including atomic diffusion (dashed line) and without it (solid line). The model with diffusion goes up to an age of 9.4 Gyr and the one without diffusion up to 11.4 Gyr. Also shown in the figure is the position of stars on the H–R diagram with a solar radius (dotted line). Data courtesy of Mathieu Vick, Jacques Richer and Georges Michaud obtained with the Montréal evolution code.

As mentioned above, the diffusion of the elements in stars affects their structure and evolution. Since atomic diffusion is a time-dependent process, during the construction of an evolutionary model of a star, a large number of stellar models must be calculated. A stellar model must be calculated at each chosen time step because the abundances change due to diffusion. These time steps must be chosen in such a way as to be able to appropriately follow the diffusion of the elements. In other words, the time steps must be smaller that the typical diffusion timescales inside stars. Also, at each time step, the radiative accelerations must be evaluated in the new structural model using the values of the new abundances (of course this is done at each geometrical layer of the model because the abundances vary with depth). Such complex numerical modelling requires considerable computing time.

Figure 7.4 shows the effect of atomic diffusion on the evolution of the Sun. For the simulation shown with atomic diffusion, the abundances of 28 metals (this number includes some isotopes of certain elements) are modified at each time step. These calculations were done with the Montréal evolution code that can include atomic diffusion. For more information about such calculations, the reader is referred to Turcotte, S., Richer, J., Michaud, G., Iglesias, C.A. and Rogers, F.J., *The Astrophysical Journal*, **504**, 539 (1998).

7.5   Other Transport Mechanisms[††]

Several other physical mechanisms that are sometimes present in stars compete with atomic diffusion. As mentioned previously, in convection zones atomic diffusion can not take place. Anomalies formed at the bottom of convection zones by diffusion can, however, manifest themselves in these zones and above them because elements can be dredged-up by the mixing process. Other phenomena such as mass loss can also affect the migration of the elements in stars. Meridional circulation and turbulence are other physical processes that can also play a role. Meridional circulation is a large-scale movement of matter that is defined by currents that are parallel to the meridians of stars. The currents of matter penetrate deeply within stars. Meanwhile, the presence of turbulence can also affect the abundances observed at the surface of certain stars.

The discussion of all of these competing physical processes is outside the scope of this book. However, in order to give the reader some insight into the complexities that intervene when trying to model chemically peculiar stars, two transport mechanisms that can be present inside such stars are described below. These two physical processes are light-induced drift and ambipolar diffusion of hydrogen.

7.5.1   Light-Induced Drift

In addition to abundance anomalies of the elements, certain stars show isotopic anomalies. In other words, the isotopic mix at the surface of these stars is different from, for instance, the isotopic ratios observed in the Sun. Some HgMn stars show isotopic anomalies of Hg or Pt. For instance, at the surface of the HgMn star χ Lupi approximately 99 % of mercury is in the form of ^{204}Hg, while this isotope accounts for only 7 % of the mercury observed in the Sun. A physical process possibly responsible for such isotopic anomalies is light-induced drift. This section is dedicated to describing this transport mechanism.

Light-induced drift occurs when there is a difference in the average diffusion coefficient of a given species between the atoms travelling in the outgoing and ingoing directions in a star. If it is assumed that atomic line broadening is dominated by the Doppler effect (see Section 4.3.2), atoms travelling in the ingoing and outgoing direction absorb photons at wavelengths on opposite sides of the natural wavelength of a given atomic line. In other words, the atoms 'see' a wavelength in the radiation field that depends on their thermal velocity. If a gradient (with respect to wavelength) exists in the radiative flux inside the Doppler width of the line, there are more atoms exited for one of these directions (see Figure 7.5). If the atoms are approximated by hard spheres, the exited atoms, having a larger radius, are less mobile (or have a shorter mean free path and thus a smaller diffusion coefficient) than atoms found in the initial energy level of the line under consideration. A drift of the species then ensues in the direction that has a lesser number of exited atoms. This phenomenon is called light-induced drift. As mentioned, it is caused by a directional asymmetry in the diffusion coefficient due to a flux gradient inside the atomic line widths.

In the scenario described above, for light-induced drift to be important, a significant gradient of the flux within the atomic line widths must be present. The corresponding energy levels (and therefore the natural wavelengths of the bound–bound transitions) of isotopes are slightly shifted relative to one another. Therefore, for a given isotope of an element with at least two isotopes, the flux asymmetry needed for light-induced drift can be caused by the corresponding atomic line of its isotope or isotopes. In the simplest case where only two isotopes exist, each of their atomic lines may blend with the corresponding lines of the other isotope, thereby causing an asymmetry of the flux within their line widths. If this asymmetry is large enough, it leads to a non-negligible drift of the two species (in opposite directions). One of these isotopes can then accumulate at the surface of the star

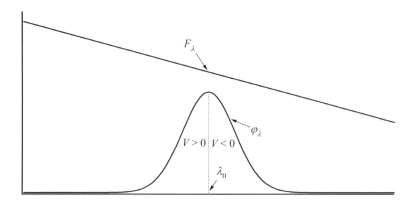

Figure 7.5   Illustration of the light-induced drift process. Shown in this figure is the line absorption profile (φ_λ) and the monochromatic flux (F_λ). More photoexcitations will occur for atoms having a positive velocity (i.e. going outward in the star) than for those going inward ($V < 0$) because of the slope of the flux within the line width. For this case, it leads to a negative light-induced drift velocity because of the asymmetry caused in the diffusion coefficient. Also shown in this figure is the natural wavelength (λ_0) of the atomic line.

while the other one can drift inward. This physical process can therefore cause isotopic anomalies at the stellar surface that can be detected by spectroscopy. Light-induced drift can, for example, accelerate the separation of ^{3}He and ^{4}He (which leads to the so-called ^{3}He stars) as compared to the separation caused by gravitational settling due to their different masses.

7.5.2   Ambipolar Diffusion of Hydrogen

In stars, hydrogen is progressively ionised as a function of geometrical depth due to the increasing temperature. A negative gradient in the number density of protons (as a function of r) therefore exists and this leads to a diffusion of the protons in the outward direction, while neutral hydrogen diffuses inward (see Figure 7.6). The diffusion velocities illustrated in this figure are important in the ionisation zone of hydrogen where the gradients of the respective number densities are large. This phenomenon is called the ambipolar diffusion of hydrogen.

Ambipolar diffusion of hydrogen can have two important effects in stars. First, since the protons are charged particles they can effectively 'drag' other ions as they diffuse toward the surface. This can therefore alter the abundances of the metals found at the stellar surface. Secondly, in the presence of a magnetic field, the flow of the protons is affected by the Lorentz force. It can be shown that this can cause a downward force on the medium, which amplifies the effective gravity felt by the stellar medium. This process therefore modifies the pressure stratification and the physical structure of the star in the vicinity of the hydrogen-ionisation zone.

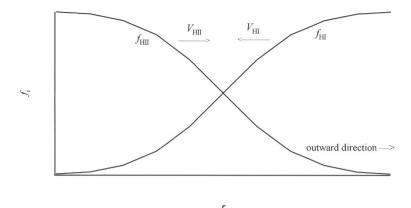

Figure 7.6 Illustration the ionisation fractions of HI and HII inside a star. The variable r increases from left to right, therefore the surface of the star is found on the right end of this figure. The directions of the diffusion velocities are shown in the figure. The protons diffuse toward the surface while HI diffuses toward its centre. This physical process is called ambipolar diffusion of hydrogen.

7.6 Summary

Types of CP stars: AmFm, HgMn, ApBp and He-abnormal stars.

Diffusion velocity: $V_{i,\text{diff}} \approx -D_{ip}\left(\dfrac{\mathrm{d}\ln n_i}{\mathrm{d}r} + \left(g - g_{\text{rad}}^i\right)\dfrac{Am_\mathrm{p}}{kT} - (Z+1)\dfrac{m_\mathrm{p}g}{2kT}\right)$ (7.38)

Radiative acceleration on ion i: $g_{\text{rad}}^i = \dfrac{4\pi}{c}\dfrac{1}{X_i}\int_0^\infty k_\nu^i H_\nu \mathrm{d}\nu$ (7.39)

Average radiative acceleration on element A: $g_{\text{rad}}(A) = \dfrac{\displaystyle\sum_i f_i D_i g_{\text{rad}}^i}{\displaystyle\sum_i f_i D_i} = \dfrac{\displaystyle\sum_i f_i D_i g_{\text{rad}}^i}{\overline{D}}$ (7.40)

7.7 Exercises

7.1 Show that Eq. (7.9) and (7.10) lead to

$$V_{1,\text{diff}} = -\frac{1}{3(n_1+n_2)}\left(n_2\overline{V_1}l_1 + n_1\overline{V_2}l_2\right)\frac{1}{n_1}\frac{\mathrm{d}n_1}{\mathrm{d}r} = -\frac{D_{12}}{n_1}\frac{\mathrm{d}n_1}{\mathrm{d}r}$$ (7.41)

7.2 Show that when an electric field E is present (pointing in the outgoing direction of the star), the equations relating the pressure of free electrons and protons in a completely ionised hydrogen plasma are

$$\frac{\mathrm{d}\ln P_{\mathrm{e}}}{\mathrm{d}r} = -\frac{m_{\mathrm{e}}g}{kT} - \frac{eE}{kT} \tag{7.42}$$

$$\frac{\mathrm{d}\ln P_{\mathrm{p}}}{\mathrm{d}r} = -\frac{m_{\mathrm{p}}g}{kT} + \frac{eE}{kT} \tag{7.43}$$

7.3   Explain why light-induced drift is less important for charged ions than for neutral ones.

Answers to Selected Exercises

1.3 $M = 0.5$, $d = 205\,\text{ly}$

1.7 $32\,000\,\text{K}$

1.8 $12.34\,\text{eV}$

1.9 0.446

1.10 $n_{\text{tot}} = 9.77 \times 10^{14}\,\text{cm}^{-3}$ and $\rho = 1.21 \times 10^{-9}\,\text{gcm}^{-3}$

1.12 $n_{\text{e}} = 1.46 \times 10^{19}\,\text{cm}^{-3}$ and $T = 51200\,\text{K}$

2.1 $P(r) = \rho_0 g \left(\dfrac{r^2}{2R_f} - r + \dfrac{R_f}{2} \right)$

2.3 $\Omega = -\dfrac{26}{35} \dfrac{GM_*^2}{R_*}$

2.5 $T \approx 30\,\text{K}$

3.1 $J_v(\tau) = \dfrac{b_v(\tau)}{3}$, $H_v(\tau) = \dfrac{a_v(\tau)}{3}$

3.4 $10^8\,\text{cm}$ or 0.14% of $R_\odot$

3.6 $I_v^0 e^{-(k_{v,1}\rho_1 d_1 + k_{v,2}\rho_2 d_2)}$

An Introduction to Stellar Astrophysics Francis LeBlanc
© 2010 John Wiley & Sons, Ltd

3.7 $z_0 = 12\zeta^{2/3}$

4.5   The star is receding from the observer at 24.6 km/s.

4.8   4.8%

4.9   2.25 Å

4.13   Three ionisation stages

5.1 $\rho_c = \dfrac{15 M_*}{8 \pi R_*^3}$

5.10 $M_* = \dfrac{4}{\pi} R_*^3 \rho_c$

6.6   7.16 MeV

6.7 10^7 yr

6.10   18 000 yr

6.12 $\rho = \dfrac{3 c^6}{32 \pi G^3} \dfrac{1}{M^2}$

Appendix A
Physical Constants

Speed of light	c	$2.99793458 \times 10^{10}$ cm/s
Gravitational constant	G	6.67259×10^{-8} cm^3/g/ s^2
Planck constant	h	$6.6260755 \times 10^{-27}$ erg s
Boltzmann constant	k	1.380658×10^{-16} erg/K
		8.617385×10^{-5} eV/K
Elementary charge	e	$4.8032068 \times 10^{-10}$ esu
Atomic mass unit	u	1.660540×10^{-24} g
		931.5 MeV/c^2
Mass of electron	m_e	$9.1093897 \times 10^{-28}$ g
		0.511 MeV/c^2
Mass of proton	m_p	$1.6726231 \times 10^{-24}$ g
		938.3 MeV/c^2
Mass of neutron	m_n	$1.6749286 \times 10^{-24}$ g
		939.6 MeV/c^2
Mass of ^{1}H atom	m_H	$1.6735344 \times 10^{-24}$ g
Stefan–Boltzmann constant	σ	5.67051×10^{-5} erg/cm^2/K/s
Thomson scattering constant	σ_T	6.6524×10^{-25} cm^2

An Introduction to Stellar Astrophysics Francis LeBlanc
© 2010 John Wiley & Sons, Ltd

Appendix B
Units in the cgs and SI Systems

Dimension	cgs unit	Conversion factor (cgs to SI)	SI unit
Time	s	1	s
Length	cm	10^{-2}	m
Mass	g	10^{-3}	kg
Energy	erg	10^{-7}	J
Power	erg/s	10^{-7}	W
Force	dyn	10^{-2}	N
Pressure	dyn/cm^2	10^{-1}	N/m^2
Magnetic flux density	G	10^{-4}	T

An Introduction to Stellar Astrophysics Francis LeBlanc
© 2010 John Wiley & Sons, Ltd

Appendix C
Astronomical Constants

Solar mass	$M_\odot$	$1.9891 \times 10^{33}\,\mathrm{g}$
Solar radius	$R_\odot$	$6.95508 \times 10^{10}\,\mathrm{cm}$
Solar luminosity	$L_\odot$	$3.8458 \times 10^{33}\,\mathrm{erg/s}$
Solar effective temperature	$T_{\mathrm{eff}\,\odot}$	$5777\,\mathrm{K}$
Earth's mass	$M_\oplus$	$5.9742 \times 10^{27}\,\mathrm{g}$
Earth's radius	$R_\oplus$	$6.378136 \times 10^8\,\mathrm{cm}$
Astronomical unit	AU	$1.4959787066 \times 10^{13}\,\mathrm{cm}$
Light-year	ly	$9.460730472 \times 10^{17}\,\mathrm{cm}$
Parsec	pc	$3.0856776 \times 10^{18}\,\mathrm{cm}$
		$3.26167\,\mathrm{ly}$

An Introduction to Stellar Astrophysics Francis LeBlanc
© 2010 John Wiley & Sons, Ltd

Appendix D

Ionisation Energies (in eV) for the First Five Stages of Ionisation for the Most Important Elements

	I	II	III	IV	V
H	13.598				
He	24.587	54.418			
Li	5.392	75.640	122.454		
Be	9.323	18.211	153.897	217.713	
B	8.298	25.155	37.931	259.366	340.22
C	11.260	24.383	47.888	64.492	392.08
N	14.534	29.601	47.449	77.472	97.89
O	13.618	35.117	54.936	77.413	113.90
F	17.423	34.971	62.708	87.140	114.24
Ne	21.565	40.963	63.45	97.12	126.21
Na	5.139	47.286	71.620	98.91	138.40
Mg	7.646	15.035	80.144	109.265	141.27
Al	5.986	18.829	28.448	119.99	153.83
Si	8.152	16.346	33.493	45.142	166.77
P	10.487	19.769	30.203	51.444	65.03
S	10.360	23.338	34.79	47.222	72.59
Cl	12.968	23.814	39.61	53.465	67.8
Ar	15.760	27.630	40.74	59.81	75.02
K	4.341	31.63	45.806	60.91	82.66
Ca	6.113	11.871	50.913	67.27	84.50
Sc	6.561	12.800	24.757	73.489	91.65
Ti	6.828	13.576	27.492	43.267	99.3
V	6.746	14.66	29.311	46.71	65.28
Cr	6.767	16.486	30.96	49.16	69.46
Mn	7.434	15.640	33.668	51.2	72.4
Fe	7.902	16.188	30.652	54.8	75.0
Co	7.881	17.083	33.50	51.3	79.5
Ni	7.640	18.169	35.19	54.9	75.5

An Introduction to Stellar Astrophysics Francis LeBlanc
© 2010 John Wiley & Sons, Ltd

Appendix E

Solar Abundances for the Most Important Elements

Element	Abundance[#]	N_{elem}/N_{tot}
H	12.00	9.097×10^{-1}
He	10.99	8.890×10^{-2}
Li	1.16	1.315×10^{-11}
Be	1.15	1.285×10^{-11}
B	2.60	3.621×10^{-10}
C	8.56	3.303×10^{-4}
N	8.05	1.021×10^{-4}
O	8.93	7.742×10^{-4}
F	4.56	3.303×10^{-8}
Ne	8.09	1.119×10^{-4}
Na	6.33	1.945×10^{-6}
Mg	7.58	3.458×10^{-5}
Al	6.47	2.684×10^{-6}
Si	7.55	3.228×10^{-5}
P	5.45	2.564×10^{-7}
S	7.21	1.475×10^{-5}

(continued overleaf)

An Introduction to Stellar Astrophysics Francis LeBlanc
© 2010 John Wiley & Sons, Ltd

Table   (*continued*)

Element	Abundance[#]	N_{elem}/N_{tot}
Cl	5.50	2.877×10^{-7}
Ar	6.56	3.303×10^{-6}
K	5.12	1.199×10^{-7}
Ca	6.36	2.084×10^{-6}
Sc	3.10	1.145×10^{-9}
Ti	4.99	8.890×10^{-8}
V	4.00	9.097×10^{-9}
Cr	5.67	4.255×10^{-7}
Mn	5.39	2.233×10^{-7}
Fe	7.54	3.154×10^{-5}
Co	4.92	7.566×10^{-8}
Ni	6.25	1.618×10^{-6}
Cu	4.21	1.475×10^{-8}
Zn	4.60	3.621×10^{-8}

[#]The abundances in this column are written as a function of the number of hydrogen atoms N_H atoms, which is set at log $(N_H) = 12.00$. This way of expressing the chemical abundances is very common in astrophysics.

Appendix F
Atomic Masses

Species	Atomic mass (in units of u[#])
e^-	5.4857990×10^{-4}
n	1.0086649
p	1.0072765
^{1}H	1.0078250
^{2}H	2.0141018
^{3}H	3.0160493
^{3}He	3.0160293
^{4}He	4.0026032
^{6}Li	6.0151223
^{7}Li	7.0160040
^{8}Be	8.0053051
^{9}Be	9.0121821
^{12}C	12.000000
^{13}C	13.003355
^{14}N	14.003074
^{15}N	15.000109
^{16}O	15.994915
^{17}O	16.999132
^{18}O	17.999161
^{20}Ne	19.992440
^{24}Mg	23.985042
^{28}Si	27.976927
^{32}S	31.972071
^{56}Fe	55.934942

[#] u = atomic mass unit = 1.661×10^{-24} g = $931.5 \, \text{MeV}/c^2$.

An Introduction to Stellar Astrophysics Francis LeBlanc
© 2010 John Wiley & Sons, Ltd

Appendix G

Physical Parameters for Main-Sequence Stars

Spectral type	T_{eff} (K)	$M/M_\odot$	$L/L_\odot$	$R/R_\odot$
O5	42 000	60	400 000	12
B0	30 000	17.5	40 000	7.4
B5	15 200	5.9	730	3.9
B8	11 400	3.8	140	3.0
A0	9790	2.9	48	2.4
A5	8180	2.0	12	1.7
F0	7300	1.6	5.7	1.5
F5	6650	1.4	3.0	1.3
G0	5940	1.05	1.4	1.1
G5	5560	0.92	0.73	0.92
K0	5150	0.79	0.46	0.85
K5	4410	0.67	0.18	0.72
M0	3840	0.51	0.070	0.60
M2	3520	0.40	0.034	0.50
M5	3170	0.21	0.0066	0.27

An Introduction to Stellar Astrophysics Francis LeBlanc
© 2010 John Wiley & Sons, Ltd

Appendix H

Periodic Table of the Elements

1 H $1s^1$																	2 He $1s^2$
3 Li $[He]2s^1$	4 Be $[He]2s^2$											5 B $[He]2s^2p^1$	6 C $[He]2s^2p^2$	7 N $[He]2s^2p^3$	8 O $[He]2s^2p^4$	9 F $[He]2s^2p^5$	10 Ne $[He]2s^2p^6$
11 Na $[Ne]3s^1$	12 Mg $[Ne]3s^2$											13 Al $[Ne]3s^23p^1$	14 Si $[Ne]3s^23p^2$	15 P $[Ne]3s^23p^3$	16 S $[Ne]3s^23p^4$	17 Cl $[Ne]3s^23p^5$	18 Ar $[Ne]3s^23p^6$
19 K $[Ar]4s^1$	20 Ca $[Ar]4s^2$	21 Sc $[Ar]3d^14s^2$	22 Ti $[Ar]3d^24s^2$	23 V $[Ar]3d^34s^2$	24 Cr $[Ar]3d^54s^1$	25 Mn $[Ar]3d^54s^2$	26 Fe $[Ar]3d^64s^2$	27 Co $[Ar]3d^74s^2$	28 Ni $[Ar]3d^84s^2$	29 Cu $[Ar]3d^{10}4s^1$	30 Zn $[Ar]3d^{10}4s^2$	31 Ga $[Ar]3d^{10}4s^24p^1$	32 Ge $[Ar]3d^{10}4s^24p^2$	33 As $[Ar]3d^{10}4s^24p^3$	34 Se $[Ar]3d^{10}4s^24p^4$	35 Br $[Ar]3d^{10}4s^24p^5$	36 Kr $[Ar]3d^{10}4s^24p^6$
37 Rb $[Kr]5s^1$	38 Sr $[Kr]5s^2$	39 Y $[Kr]4d^15s^2$	40 Zr $[Kr]4d^25s^2$	41 Nb $[Kr]4d^45s^1$	42 Mo $[Kr]4d^55s^1$	43 Tc $[Kr]4d^55s^2$	44 Ru $[Kr]4d^75s^1$	45 Rh $[Kr]4d^85s^1$	46 Pd $[Kr]4d^{10}5s^0$	47 Ag $[Kr]4d^{10}5s^1$	48 Cd $[Kr]4d^{10}5s^2$	49 In $[Kr]4d^{10}5s^25p^1$	50 Sn $[Kr]4d^{10}5s^25p^2$	51 Sb $[Kr]4d^{10}5s^25p^3$	52 Te $[Kr]4d^{10}5s^25p^4$	53 I $[Kr]4d^{10}5s^25p^5$	54 Xe $[Kr]4d^{10}5s^25p^6$
55 Cs $[Xe]6s^1$	56 Ba $[Xe]6s^2$	57 La $[Xe]5d^16s^2$	72 Hf $[Xe]4f^{14}5d^26s^2$	73 Ta $[Xe]4f^{14}5d^36s^2$	74 W $[Xe]4f^{14}5d^46s^2$	75 Re $[Xe]4f^{14}5d^56s^2$	76 Os $[Xe]4f^{14}5d^66s^2$	77 Ir $[Xe]4f^{14}5d^76s^2$	78 Pt $[Xe]4f^{14}5d^96s^1$	79 Au $[Xe]4f^{14}5d^{10}6s^1$	80 Hg $[Xe]4f^{14}5d^{10}6s^2$	81 Tl $[Xe]4f^{14}5d^{10}6s^26p^1$	82 Pb $[Xe]4f^{14}5d^{10}6s^26p^2$	83 Bi $[Xe]4f^{14}5d^{10}6s^26p^3$	84 Po $[Xe]4f^{14}5d^{10}6s^26p^4$	85 At $[Xe]4f^{14}5d^{10}6s^26p^5$	86 Rn $[Xe]4f^{14}5d^{10}6s^26p^6$
87 Fr $[Rn]7s^1$	88 Ra $[Rn]7s^2$	89 Ac $[Rn]6d^17s^2$															

58 Ce $[Xe]4f^15d^16s^2$	59 Pr $[Xe]4f^35d^06s^2$	60 Nd $[Xe]4f^45d^06s^2$	61 Pm $[Xe]4f^55d^06s^2$	62 Sm $[Xe]4f^65d^06s^2$	63 Eu $[Xe]4f^75d^06s^2$	64 Gd $[Xe]4f^75d^16s^2$	65 Tb $[Xe]4f^95d^06s^2$	66 Dy $[Xe]4f^{10}5d^06s^2$	67 Ho $[Xe]4f^{11}5d^06s^2$	68 Er $[Xe]4f^{12}5d^06s^2$	69 Tm $[Xe]4f^{13}5d^06s^2$	70 Yb $[Xe]4f^{14}5d^06s^2$	71 Lu $[Xe]4f^{14}5d^16s^2$
90 Th $[Rn]6d^27s^2$	91 Pa $[Rn]5f^26d^17s^2$	92 U $[Rn]5f^46d^17s^2$	93 Np $[Rn]5f^46d^17s^2$	94 Pu $[Rn]5f^66d^07s^2$	95 Am $[Rn]5f^76d^07s^2$	96 Cm $[Rn]5f^76d^17s^2$	97 Bk $[Rn]5f^96d^07s^2$	98 Cf $[Rn]5f^{10}6d^07s^2$	99 Es $[Rn]5f^{11}6d^07s^2$	100 Fm $[Rn]5f^{12}6d^07s^2$	101 Md $[Rn]5f^{13}6d^07s^2$	102 No $[Rn]5f^{14}6d^07s^2$	103 Lr $[Rn]5f^{14}6d^17s^1$

References

Alecian, G. and Vauclair, S., *Fundamentals of Cosmic Physics*, 8, 369 (1983)

Alecian, G. and LeBlanc, F., *Monthly Notices of the Royal Astronomical Society*, 332, 891 (2002)

Arnould, M. and Samyn, M., La physique nucléaire en astrophysique, *EDP Sciences*, Les Ulis (2002)

Arp, H.C., *The Astrophysical Journal*, 133, 874 (1961)

Cameron, C., Ph.D. Thesis, University of British Columbia (2009)

Cameron, C., Matthews, J., et al. (in preparation)

Chandrasekhar, S., *The Astrophysical Journal*, 101, 328 (1945)

Chandrasekhar, S., *Radiative Transfer*, Dover, New York (1960)

Cox, A.N., *Allen's Astrophysical Quantities*, Springer, New York (2004)

Cunto, W., Mendoza, C., Ochsenbein, F., and Zeippen, C.J., *Astronomy & Astrophysics*, 275, L5 (1993)

Doggett, J.B. and Branch, D., *The Astronomical Journal*, 90, 2303 (1985)

Eisberg, R. and Resnick, R., *Quantum Physics of Atoms, Molecules, Solids, Nuclei and Particles*, John Wiley & Sons, Ltd, New York (1985)

Fontaine, G., Brassard, P., Charpinet, S., Quirion, P.-O., and Randall, S.K., *ASP Conference Series* (in press)

Hauschildt, P.H., Allard, F., and Baron, E., *The Astrophysical Journal*, 512, 377 (1999)

Hawking, S.W., *Nature*, 248, 30 (1974)

Hessman, F.V. and Guenther, E.W., *Astronomy & Astrophysics*, 321, 497 (1997)

Iben, I., *Annual Review of Astronomy and Astrophysics*, 5, 571 (1967)

Iliadis, C., *Nuclear Physics of Stars*, Wiley-VCH, Weinheim (2007)

Kippenhahn, R. and Weigert, A., *Stellar Structure and Evolution*, Springer-Verlag, Berlin (1990)

Laughlin, G., Bodenheimer, P., and Adams, F.C., *The Astrophysical Journal*, 482, 420 (1997)

Martin, B.R. and Shaw, G., *Particle Physics*, John Wiley & Sons, Ltd, Chichester (2008)

Michaud, G., *The Astrophysical Journal*, 160, 641 (1970)

Mihalas, D., *Stellar Atmospheres*, W.H. Freeman and Company, San Francisco (1970 and 1978)

Pearson, J.M., *Nuclear Physics: Energy and Matter*, Adam Hilger, Bristol (1986)

Percy, J.R., *Understanding Variable Stars*, Cambridge University Press, Cambridge (2007)

Preston, G.W., *Annual Review of Astronomy and Astrophysics*, 12, 257 (1974)

Reid, M.J. and Goldston, J.E., *The Astrophysical Journal*, 568, 931 (2002)

Renzini, A. and Fusi Pecci, F., *Annual Review of Astronomy and Astrophysics*, vol. 26, 199 (1988)

Salaris, M. and Cassisi, S., *Evolution of Stars and Stellar Populations*, John Wiley & Sons, Ltd, Chichester (2005)

Seaton, M.J., *Monthly Notices of the Royal Astronomical Society*, 289, 700 (1997)

Seaton, M.J., Yan, Y., Mihalas, D. and Pradhan, A.K., *Monthly Notices of the Royal Astronomical Society*, 266, 805 (1994)

Stahler, S.W. and Palla, F., *The Formation of Stars*, Wiley-VCH, Weinheim (2004)

Tassoul, J.-L. and Tassoul, M., *A Concise History of Solar and Stellar Physics*, Princeton University Press, Princeton (2004)

Turcotte, S., Richer, J., Michaud, G., Iglesias, C.A., and Rogers, F.J., *The Astrophysical Journal*, 504, 539 (1998)

Vauclair, S., *SAAS-FEE Advanced Course 13, Astrophysical Processes in the Upper Main Sequence*, p. 167 (1983)

An Introduction to Stellar Astrophysics Francis LeBlanc
© 2010 John Wiley & Sons, Ltd

Bibliography

Aller, L.H., *Atoms, Stars and Nebulae*, Cambridge University Press, Cambridge (1991)

Abell, G.O., Morrison, D., and Wolff, S.C., *Exploration of the Universe*, Saunders College Publishing, Philadelphia (1991)

Audouze, J. and Vauclair, S., *L'astrophysique nucléaire*, Presses Universitaires de France. Paris (2003)

Arnett, D., *Supernovae and Nucleosynthesis*, Princeton University Press, Princeton (1996)

Bahcall, J.N., *Neutrino Astrophysics*, Cambridge University Press, Cambridge (1989)

Böhm-Vitense, E., *Introduction to Stellar Astrophysics*, 3 volumes, Cambridge University Press, Cambridge (1989)

Bowers, R. and Deeming, T., *Astrophysics I: Stars*, Jones and Bartlett Publishers, Boston (1984)

Boyd, R., *An Introduction to Nuclear Astrophysics*, The University of Chicago Press, Chicago (2008)

Carroll, B.W. and Ostlie, D.A., *An Introduction to Modern Astrophysics*, Pearson – Addison Wesley, Reading (2007)

Chaisson, E. and McMillan, S., *Astronomy Today*, Prentice Hall, Englewood Cliffs (1993)

Chandrasekhar, S., *An Introduction to the Study of Stellar Structure*, Dover, New York (1958)

Clayton, D.D., *Principles of Stellar Evolution and Nucleosynthesis*, The University of Chicago Press, Chicago (1983)

Cox, J.P., *Theory of Stellar Pulsations*, Princeton University Press, Princeton (1980)

de la Cotardière, P., *Dictionnaire de l'astronomie*, Larousse. Paris (1988)

Duric, N., *Advanced Astrophysics*, Cambridge University Press, Cambridge (2004)

Fermi, E., *Thermodynamics*, Dover, New York (1956)

Forestini, M., *Principes fondamentaux de structure stellaire*, Gordon and Breach Science Publishers, Paris (1999)

Frolov, V.P. and Nonikov, I.D., *Black Hole Physics*, Kluwer Academic Publishers, Dordrecht (1998)

Gray, D.F., *The Observation and Analysis of Stellar Photospheres*, Cambridge University Press, Cambridge (2005)

Hansen, C.J., Kawaler, S.D., and Trimble, V., *Stellar Interiors Physical Principle, Structure, and Evolution*, Springer, New York (2004)

Harpaz, A., *Stellar Evolution*, A.K. Peters, Ltd., Wellesley (1994)

Harwit, M., *Astrophysical Concepts*, Springer, New York (2006)

Irwin, J., *Astrophysics Decoding the Cosmos*, John Wiley & Sons, Ltd, Chichester (2007)

Jaschek, C. and Jaschek, M., *The Classification of Stars*, Cambridge University Press, Cambridge (1987)

Kaler, J.B., *Stars and Their Spectra an Introduction to the Spectral Sequence*, Cambridge University Press, Cambridge (1997)

Kaler, J.B., *The Cambridge Encyclopedia of Stars*, Cambridge University Press, Cambridge (2006)

Kaufmann, W.J., *Black Holes and Warped Spacetime*, W.H. Freeman and Company, New York (1979)

Kitchin, C.R., *Optical Astronomical Spectroscopy*, Taylor & Francis, New York (1995)

Krane, K.S., *Introductory Nuclear Physics*, John Wiley & Sons, Ltd, New York (1988)

Lang, K.R., *Astrophysical Formulae: A Compendium for the Physicist and Astrophysicist*, Springer-Verlag, Berlin (1974)

Luminet, J.-P., *Les trous noirs*, Éditions Balfond, Paris (1987)

Maoz, D., *Astrophysics in a Nutshell*, Princeton University Press, Princeton (2007)

Monier, R., *Les étoiles et le milieu interstellaire*, Ellipses, Paris (2006)

Motz, L., *Astrophysics and Stellar Structure*, Ginn and Company, Waltham (1970)

Novotny, E., *Introduction to Stellar Atmospheres and Interiors*, Oxford University Press, New York (1973)

Padmanabhan, T., *An Invitation to Astrophysics*, World Scientific, Hackensack (2006)

Palen, S., *Schaum's Outline of Theory and Problems of Astronomy*, McGraw-Hill, New York (2002)

Pecker, J.C. and Schatzman, E., *Astrophysique générale*, Masson, Paris (1959)

Phillips, A.C., *The Physics of Stars*, John Wiley & Sons, Ltd, Chichester (1999)

Prialnik, D., *An Introduction to the Theory of Stellar Structure and Evolution*, Cambridge University Press, Cambridge (2000)

Proust, D. and Breysacher, J., *Les étoiles*, Éditions du Seuil, Paris (1996)

Raine, D. and Thomas, E., *Black Holes*, Imperial College Press, London (2005)

Reddish, V.C., *The Physics of Stellar Interiors. An Introduction*, Edinburgh University Press, Edinburgh (1974)

Reeves, H., *Évolution stellaire et nucléosynthèse*, Gordon & Breach, Paris (1968)

Reif, F., *Fundamentals of Statistical and Thermal Physics*, McGraw-Hill, New York (1965)

Robinson, K., *Spectroscopy: The Key to the Stars*, Springer, London (2007)

Rose, W.K., *Advanced Stellar Astrophysics*, Cambridge University Press, Cambridge (1998)

Rybicki, G.B. and Lightman, A.P., *Radiative Processes in Astrophysics*, John Wiley & Sons, Ltd, New York (1979)

Ryden, B. and Peterson, B.M., *Foundations of Astrophysics*, Pearson Addison-Wesley, San Francisco (2010)

Schwarzschild, M., *Structure and Evolution of the Stars*, Princeton University Press, Princeton (1958)

Shapiro, S.L. and Teukolsky, S.A., *Black Holes, White Dwarfs and Neutron Stars, The Physics of Compact Objects*, Wiley-Interscience, New York (1983)

Shu, F.H., *The Physical Universe: An Introduction to Astronomy*, University Science Books, Mill Valley (1982)

Sobolev, V., *Cours d'astrophysique théorique*, Éditions Mir, Moscow (1990)

Taylor, R.J., *The Stars: Their Structure and Evolution*, Cambridge University Press, Cambridge (1994)

Tennyson, J., *Astronomical Spectroscopy*, Imperial College Press, London (2005)

Turck-Chièze, S., Däppen, W., Fossat, E., Provost, J., Schatzman, E., and Vignaud, D., The Solar Interior, Physics Reports, 230, 57 (1993)

Weiss, A., Hillebrandt, W., Thomas, H.-C., and Ritter, H., *Cox & Giuli's Principles of Stellar Structure*, Cambridge Scientific Publishers, Cambridge (2004)

Weymann, R.J., Swihart, T.L., Williams, R.E., Cocke, W.J., Pacholczyk, A.G. and Felton, J.E., *Lecture Notes on Introductory Theoretical Astrophysics*, Pachart Publishing House, Tucson (1976)

Index

Printed and bound by CPI Group (UK) Ltd, Croydon, CR0 4YY